STRUCTURAL DRAFTING

DAVID L. GOETSCH

STRUCTURAL DRAFTING

Dedicated to:

Dr. Carl Proehl – University of West Florida
Dr. Polly Einbecker – Nova University
Dr. Earl N. Gulledge – Okaloosa Walton Junior College
Dr. J. E. McCracken – Okaloosa Walton Junior College

DELMAR PUBLISHERS INC. • ALBANY, NEW YORK 12205

STRUCTURAL DRAFTING

DAVID L. GOETSCH

COPYRIGHT © 1982
BY DELMAR PUBLISHERS INC.

All rights reserved. No part of this work covered by the copyright hereon may be reproduced or used in any form or by any means — graphic, electronic, or mechanical, including photocopying, recording, taping, or information storage and retrieval systems — without written permission of the publisher.

10 9 8 7 6 5 4 3 2 1

LIBRARY OF CONGRESS CATALOG CARD NUMBER: 80-70232
ISBN: 0-8273-1930-4

Printed in the United States of America
Published simultaneously in Canada by
Delmar Publishers, a division of
Van Nostrand Reinhold, Ltd.

Preface

Structural drafting is a specialized drafting field that is part of the heavy construction industry. The old adage "it must be drawn before it can be built" is particularly true in structural drafting. STRUCTURAL DRAFTING is designed to assist the drafting student in developing the knowledge and skills necessary to begin a career in structural drafting at a productive level.

Students completing all of the reinforcement activities contained in this textbook will have developed skills equivalent to those possessed by a structural drafter with at least one full year of work experience. The text is divided into six sections with a total of twenty-five units of instruction. The first five units provide background information pertinent to a study of structural drafting. Units six through twenty-three are knowledge and skills development lessons. Units twenty-four and twenty-five teach the student how to find, get, and keep a job in structural drafting after acquiring the necessary skills.

The book is designed for self-directed, individualized instruction so that any course relying on it as the required textbook can be offered on a multilevel basis with other drafting courses. Each unit contains performance objectives, a presentation of the material, original drawings prepared by the author, a summary, review questions, and reinforcement activities.

The author, Dr. David L. Goetsch, has worked in structural drafting for fourteen years. He has owned his own drafting and design consulting service and taught structural drafting at the high school, vocational school, and community college level for five years. He is currently Chairman of Industrial Education and Head of Drafting and Design at Okaloosa-Walton Junior College in Niceville, Florida. His drafting program was recently selected for inclusion in the Florida Department of Education's Catalogue of Innovative Programs and maintains a 95 percent successful placement record of its graduates.

Acknowledgments

Contributors of Photographs and Technical Material

The author wishes to thank those people, organizations, and companies that provided materials for use in this book. Sources of such materials are listed below.

American Concrete Institute
American Institute of Steel Construction: Appendix A
Bethlehem Steel Corporation
Goodheart-Wilcox Company, Inc.
National Forest Products Association: Appendix B
Southern Prestressed Concrete Company, Inc.
Weyerhaeuser Company

Charles D. Willis

Sherrell Helm, Tom Benton, Clark Williams, Dieter Maucher, Roger Tonnessen, and Doug Smith of Southern Prestressed Concrete, Inc.

Linden Lyons of Mallet Engineering Company

The author also wishes to thank his wife, Deb, for providing valuable photographic and illustration assistance.

Classroom Testing

The material in this text was classroom tested at Okaloosa-Walton Junior College, Niceville, Florida.

Reviewer

A technical review of the text was completed by:

    Charles D. Willis
    Head, Department of Construction
    John Brown University
    Siloam Springs, Arizona 72761

Delmar Staff

Project Editor — Marjorie A. Bruce
Editor — Barbara A. Christie

Contents

PREFACE iv
ACKNOWLEDGMENTS v

SECTION 1 OVERVIEW OF STRUCTURAL DRAFTING

UNIT 1   INTRODUCTION TO STRUCTURAL DRAFTING 1
Structural Drafting Defined — Types of Structural Drawings — Employers of Structural Drafters — Structural Drafting Techniques — Linework — Line Types — Lettering — Scale Use — Paper Sizes — Title Blocks and Borders — Summary — Review Questions — Reinforcement Activities

UNIT 2   TYPICAL STRUCTURAL DRAFTING DEPARTMENT 20
Drafting Department Organization — Drafting Clerk — Junior Drafter — Drafter — Senior Drafter — Checker — Chief Drafter — Summary — Review Questions — Reinforcement Activities

UNIT 3   DRAWING, CHECKING, CORRECTING, AND REVISING PROCESSES 28
Original Drawing — Checking — Correcting — Revising — Summary — Review Questions — Reinforcement Activities

UNIT 4   PRODUCT FABRICATION AND SHIPPING 42
Structural Steel Product Fabrication — Precast Concrete Fabrication — Poured-in-Place Concrete Fabrication — Glued-Laminated Wood Fabrication — Shipping Products to the Jobsite — Summary — Review Questions — Reinforcement Activities

UNIT 5   STRUCTURAL CONNECTORS 60
Connecting Structural Members — Bolted Connections — Split Ring and Shear Plate Connections — Welded Connections — Weld Types and Symbols — Riveted Connections — Summary — Review Questions — Reinforcement Activities

SECTION 2 STRUCTURAL STEEL DRAFTING

UNIT 6   STRUCTURAL STEEL FRAMING PLANS 73
Structural Steel Drawings — Structural Steel Framing Products — Heavy-Load/Long-Span Framing Products — Structural Steel Framing Plans — Drawing Structural Steel Framing Plans — Summary — Review Questions — Reinforcement Activities

UNIT 7   STRUCTURAL STEEL SECTIONS 91
Structural Steel Sections Defined — Full Sections — Partial Sections — Offset Sections — Section Conventions — Drawing Structural Steel Sections — Summary — Review Questions — Reinforcement Activities

UNIT 8   STRUCTURAL STEEL CONNECTION DETAILS 103
Structural Steel Connection Details — Summary — Review Questions — Reinforcement Activities

UNIT 9	STRUCTURAL STEEL FABRICATION DETAILS	115

Structural Steel Shop Drawings Defined — Structural Steel Fabrication Details Defined — Constructing Structural Steel Fabrication Details — Summary — Review Questions — Reinforcement Activities

UNIT 10	STRUCTURAL STEEL BILLS OF MATERIALS	128

Bills of Materials — Preparing Bills of Materials — Summary — Review Questions — Reinforcement Activities

SECTION 3 STRUCTURAL PRECAST CONCRETE DRAFTING

UNIT 11	PRECAST CONCRETE FRAMING PLANS	141

Precast Concrete Framing Plans — Drawing Framing Plans — Summary — Review Questions — Reinforcement Activities

UNIT 12	PRECAST CONCRETE SECTIONS	167

Precast Concrete Sections Defined — Full Sections — Partial Sections — Offset Sections — Section Conventions — Drawing Precast Concrete Sections — Summary — Review Questions — Reinforcement Activities

UNIT 13	PRECAST CONCRETE CONNECTION DETAILS	181

Precast Concrete Connection Details Defined — Column Baseplate Connections — Drawing Baseplate Connection Details — Bolted Beam-to-Column Connection Details — Drawing Bolted Beam-to-Column Connection Details — Welded Connections — Drawing Welded Connections — Precast Concrete Haunch Connections — Summary — Review Questions — Reinforcement Activities

UNIT 14	PRECAST CONCRETE FABRICATION DETAILS	195

Shop Drawings Defined — Fabrication Details Defined — Constructing Fabrication Details — Summary — Review Questions — Reinforcement Activities

UNIT 15	PRECAST CONCRETE BILLS OF MATERIALS	219

Definition and Purpose — Constructing Bills of Materials — Counting Material Quantities — Summary — Review Questions — Reinforcement Activities

SECTION 4 STRUCTURAL POURED-IN-PLACE CONCRETE

UNIT 16	POURED-IN-PLACE CONCRETE FOUNDATIONS	237

Poured-in-Place Concrete Construction — Poured-in-Place Concrete Drawings — Sheet Layout and Scales — Symbols and Abbreviations — Mark Numbering Systems — Schedules — Poured-in-Place Foundation Drawings — Summary — Review Questions — Reinforcement Activities

UNIT 17	POURED-IN-PLACE CONCRETE WALLS AND COLUMNS	249

Poured-in-Place Concrete Walls and Columns — Wall and Column Engineering Drawings — Wall and Column Placing Drawings — Summary — Review Questions — Reinforcement Activities

UNIT 18	POURED-IN-PLACE CONCRETE FLOOR SYSTEMS	262

Poured-in-Place Concrete Floor Systems — One-Way Solid Slab and Beam — One-Way Ribbed or Joist Slab — Two-Way Solid Slab and Beam — Two-Way Flat-Plate Floor Systems — Waffle Slab — Poured-in-Place Concrete Floor System Drawings — Summary — Review Questions — Reinforcement Activities

UNIT 19	POURED-IN-PLACE STAIRS AND RAMPS	270
	Types of Stairs – Stair Design – Stair Design Computations – Poured-in-Place Concrete Ramps – Summary – Review Questions – Reinforcement Activities	

SECTION 5 STRUCTURAL WOOD DRAFTING

UNIT 20	STRUCTURAL WOOD FLOOR SYSTEMS	277
	Joist and Girder Floor Systems – Plywood Floor Systems – Bridging – Floor Truss Systems – Floor Joist Selection – Summary – Review Questions – Reinforcement Activities	
UNIT 21	STRUCTURAL WOOD WALLS	285
	Structural Wood Walls – Platform Framing – Balloon Framing – MOD 24 Framing – Bracing – Summary – Review Questions – Reinforcement Activities	
UNIT 22	STRUCTURAL WOOD ROOFS	293
	Common Roof Classifications – Roof Slope and Pitch – Eave and Ridge Details – Wooden Roof Framing – Roof Configuration Diagrams – Summary – Review Questions – Reinforcement Activities	
UNIT 23	STRUCTURAL WOOD POSTS, BEAMS, GIRDERS, AND ARCHES	303
	Post-and-Beam Construction – Laminated Arches – Laminated Beams and Girders – Post, Beam, Girder, and Arch Drawings – Summary – Review Questions – Reinforcement Activities	

SECTION 6 EMPLOYMENT IN STRUCTURAL DRAFTING

UNIT 24	FINDING A JOB IN STRUCTURAL DRAFTING	313
	Employment in Structural Drafting – Primary Employers of Structural Drafters – Structural Consulting Engineering Firms – Structural Steel Fabrication Companies – Precast Concrete Companies – Secondary Employers of Structural Drafters – Locating Potential Employers – The Yellow Pages – The Sunday Want-Ads – State Employment Office – Local Chamber of Commerce – Summary – Review Questions – Reinforcement Activities	
UNIT 25	GETTING AND KEEPING A JOB IN STRUCTURAL DRAFTING	318
	Letter of Introduction – Resume – Portfolio of Drawings – The Interview – Interview Follow-Up – The Wise Job Seeker – Summary – Review Questions – Reinforcement Activities	

GLOSSARY	325
APPENDIX A	327
APPENDIX B	343
INDEX	353

Section 1
Overview of Structural Drafting

Unit 1
Introduction to Structural Drafting

OBJECTIVES

Upon completion of this unit, the student will be able to

- define structural drafting.
- identify the different types of structural drawings.
- list the most common employers of structural drafters.
- demonstrate proper structural drafting techniques in the areas of linework, lettering, and scale use.

STRUCTURAL DRAFTING DEFINED

In heavy construction, anything composed of parts is called a *structure*. All products of the heavy construction industry, bridges, buildings, towers, and countless other possibilities, are composed of parts, making them structures. In order for the various parts to be designed, manufactured, and put together to form a completed structure, drawings must be made. This need for drawings forms the foundation for drafting as an important occupation in the heavy construction industry. Figures 1-1 through 1-6 show examples of typical structures that were built according to plans drawn by structural drafters.

Fig. 1-1 Intercity bridge over Columbia River (Prestressed Concrete Institute)

2 Section 1 Overview of Structural Drafting

Fig. 1-2 Private home (Prestressed Concrete Institute)

Fig. 1-3 Cleveland State Office Building (Bethlehem Steel Corporation)

Fig. 1-4 Picnic and concession shelters (Prestressed Concrete Institute)

Fig. 1-5 Pedestrian walkway (Bethlehem Steel Corporation)

Fig. 1-6 Rock Island Municipal Parking structure (Prestressed Concrete Institute)

TYPES OF STRUCTURAL DRAWINGS

Structural drafters are called upon to prepare two separate types of drawings: engineering drawings and shop drawings. *Engineering drawings* are used to provide an overall picture of a job for sales, marketing, estimating, or engineering purposes. *Shop drawings* are much more detailed. They are used for designing, fabricating, manufacturing, and erecting the structural products that go into a job. Engineering and shop drawings may be combined into one set to form *working drawings*.

Engineering drawings for heavy construction jobs are usually prepared by structural drafters employed by architects, engineers, contractors, or sales engineers. Information found on engineering drawings includes the following:

- Locations of major structural components such as columns, beams, and girders
- Basic dimensions
- Typical sections
- Notes to clarify complicated situations

Figures 1-7 and 1-8 show examples of structural engineering drawings.

Section 1 Overview of Structural Drafting

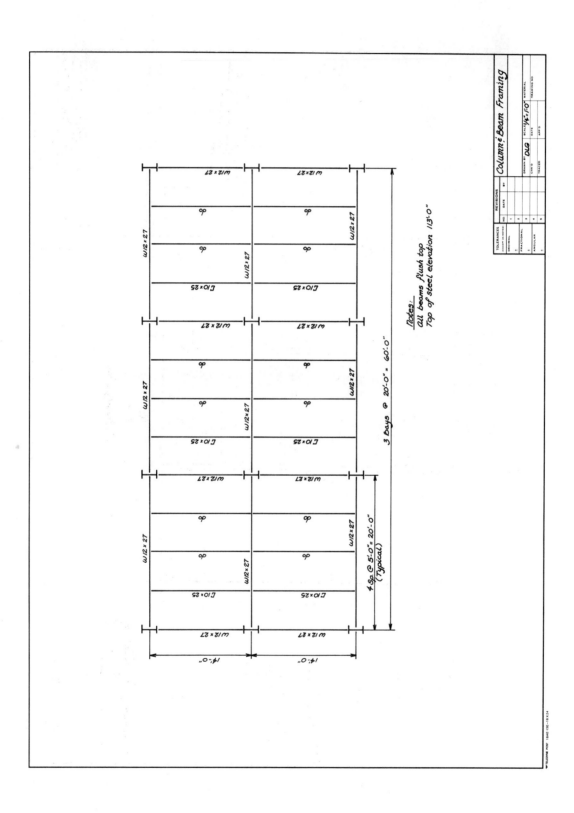

Fig. 1-7 Structural steel engineering drawing

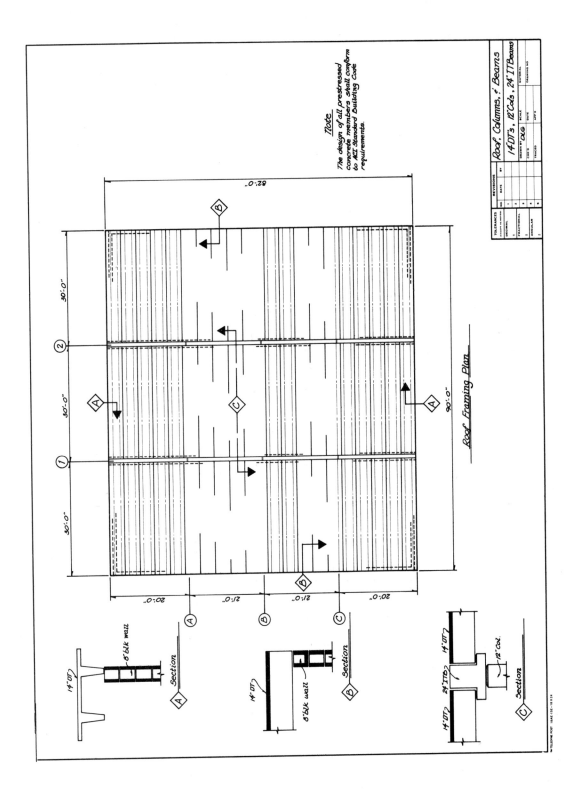

Fig. 1-8 Precast concrete engineering drawing

Shop drawings for heavy construction jobs are usually prepared by structural drafters employed by companies that actually manufacture structural products. For the most part, these are structural steel fabrication companies and precast concrete manufacturers. Shop drawings include the following:

- Precise connection details
- Fabrication details for every structural member in a job
- Details of miscellaneous metal connectors required
- Bills of material listing every item that will be used during fabrication of the products and erection of the final structure.

Figures 1-9 and 1-10 show examples of structural fabrication details which are the primary component of a set of shop drawings.

EMPLOYERS OF STRUCTURAL DRAFTERS

Structural drafters are usually employed in one of two ways. The first is preparing engineering and shop drawings of wood, concrete, or steel structures for structural consulting engineering firms. The second is preparing shop drawings for structural steel or precast concrete manufacturers.

The drafting student preparing to enter the world of structural drafting should develop knowledge and skills in several areas. These areas include structural steel, precast concrete, poured-in-place concrete, and structural wood drafting. In certain employment situations, such as with a consulting structural engineering firm, the structural drafter is often called upon to prepare plans for structures involving all four types of structural products.

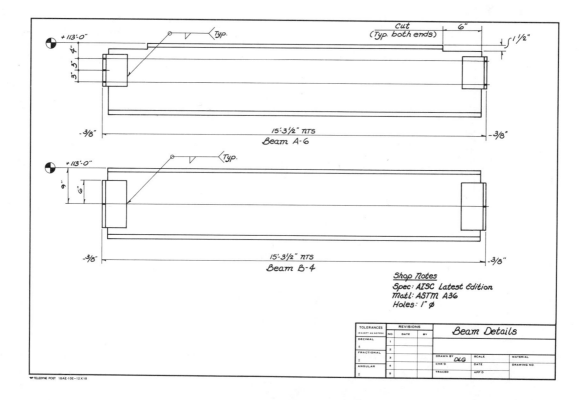

Fig. 1-9 Structural steel fabrication detail

STRUCTURAL DRAFTING TECHNIQUES

Before developing specific knowledge and skills in the areas of structural steel, precast concrete, and poured-in-place concrete, the student must learn several general items that apply to all areas of structural drafting. These items are grouped under the general heading *structural drafting techniques* and include the following:

- Structural drafting linework
- Structural drafting lettering
- Structural drafting scale use
- Structural drafting paper sizes
- Structural drafting title blocks and borders

LINEWORK

As is the case in all types of drafting, structural drafting has a set of line types that are commonly used. This set includes object lines, hidden lines, phantom lines, centerlines, dimension lines, extension lines, cutting plane lines, and break lines. Figure 1-11 illustrates each of these types of lines drawn to an acceptable width. It is evident from Figure 1-11 that object lines may be drawn to one of several widths. Actually, in structural drafting, this is true of all lines. The examples provided in Figure 1-11 are meant to serve as guidelines for reference when preparing structural drawings. However, it should be noted that though the shapes of the lines should agree with the examples provided, the widths may actually vary slightly in use. In application, line widths are sometimes varied. This is done to emphasize one aspect of a drawing or de-emphasize another aspect. The width of a line is determined by how the line is to be used and the individual circumstances of the drawing.

LINE TYPES

The most commonly used line is the object line. An object line is a continuous solid line used to show the outline of the object being drawn. Object lines are drawn in slightly varied widths depending upon the amount of emphasis desired by the drafter, Figure 1-12.

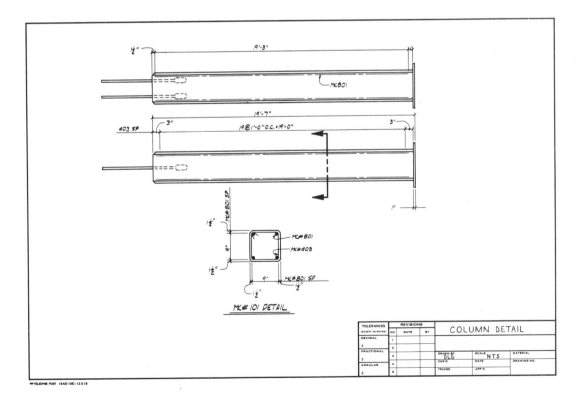

Fig. 1-10 Precast concrete fabrication detail

8 Section 1 Overview of Structural Drafting

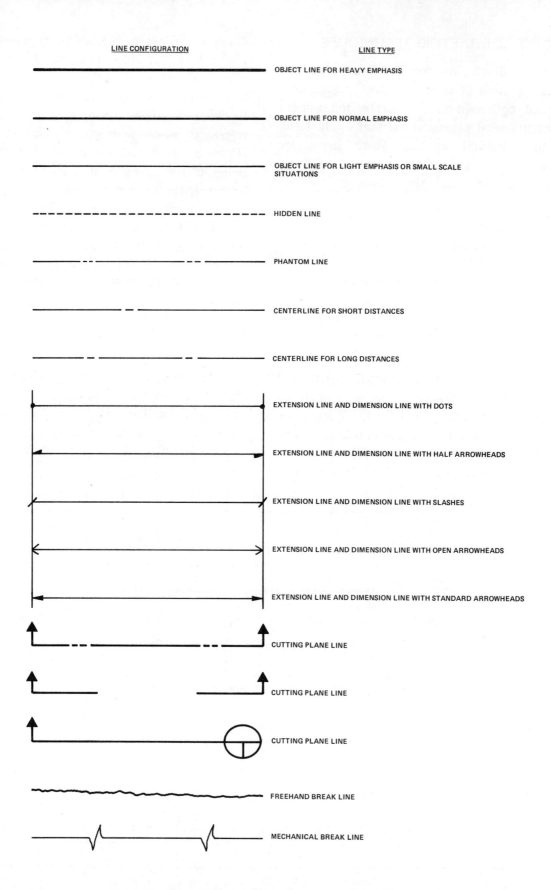

Fig. 1-11 Commonly used line types in structural drafting

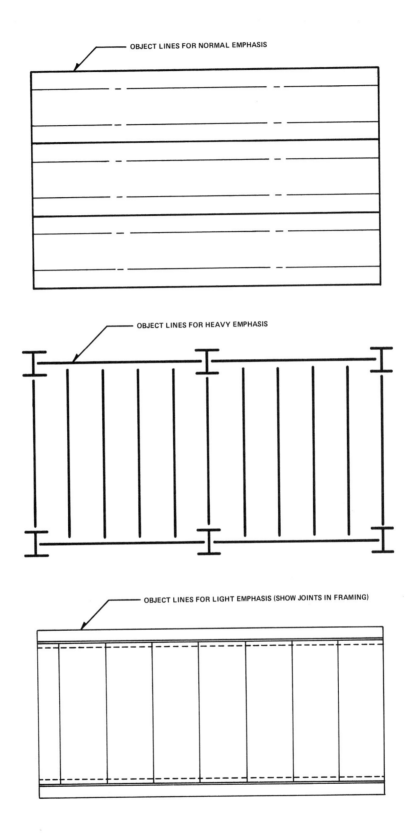

Fig. 1-12 Object line samples

Hidden lines are used almost as frequently as object lines. They show the edges or outlines of structural components that are important but could not be seen by an actual viewer of a given part of a structure. Hidden lines are composed of a number of very short line segments. These line segments should be drawn with sharp, crisp pencil strokes to achieve consistency and uniformity, Figure 1-13.

Phantom lines are used to show parts of a structure that do not actually appear in a given view for the purpose of clarity. For example, phantom lines may be used to show the outline of a roof that is actually above a framing plan or the outline of a wall that is not included on a framing plan, Figure 1-13.

Centerlines are used frequently in structural drafting to locate centers of columns on framing plans or holes on fabrication details. Centerlines are thin lines broken by a short dash. Centerlines are often extended beyond the object being drawn and used as extension lines to enclose dimension lines. Figure 1-13 illustrates how centerlines are used on structural framing plans.

Dimension lines are enclosed by extension lines and are frequently used in structural drafting to indicate size and distance. Dimension lines are thin, continuous lines that end in any one of a variety of arrowheads or arrowhead substitutes, Figure 1-14.

Cutting plane lines are thick lines that are used to, in a sense, cut through an object for the purpose of clarification. A cutting plane line indicates the area that has been sliced through and a direction for the viewers sight. A sectional view is then drawn showing the viewer what would actually be seen if the object were sliced as indicated. Figure 1-14 shows two examples of how cutting plane lines are commonly used in structural drafting.

Break lines may be constructed mechanically or freehand. In either case, they are used to cut out unnecessary or lengthy portions of a drawing. This allows the drafter to show only those portions of a detail that are needed to convey an idea, thereby economizing on space and drafting time. Figure 1-13 shows an example of a freehand break line.

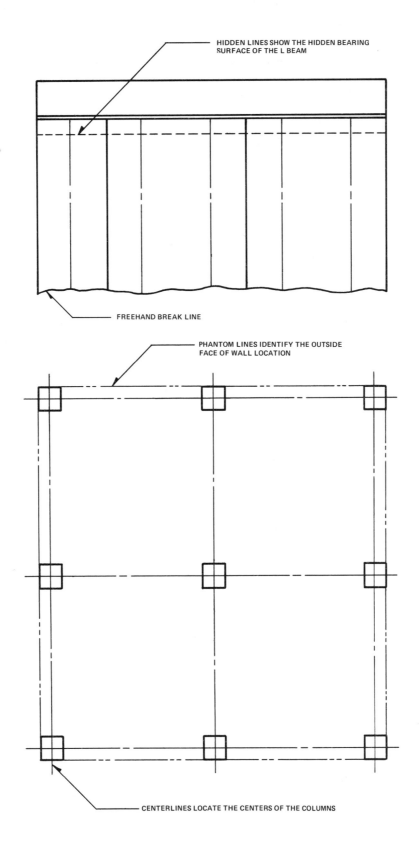

Fig. 1-13 Hidden, phantom, freehand break, and centerline samples

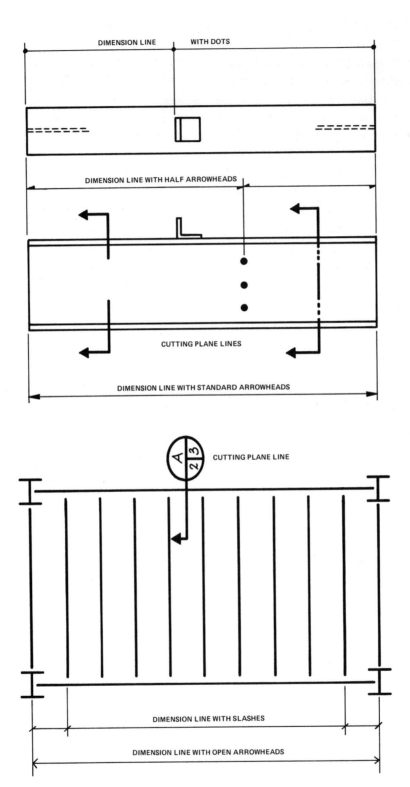

Fig. 1-14 Dimension and cutting plane line

LETTERING

Clear, legible drafting lettering is one of the structural drafter's most important skills. Every structural drawing requires a great deal of lettering. A well-developed, legible style of lettering adds a professional quality to drawings. In addition, it makes drawings easier to work with and read. As in architecture, structural drafting lettering styles vary from person to person, and each structural drafter develops his or her own distinctive style.

Unlike mechanical drafters, structural drafters are not confined to the traditional block style of lettering. However, any tendency toward overly fancy lettering should be avoided. The best structural drafting lettering is simple, easy to read, and can be done rapidly.

Versatility in lettering is important to the structural drafter. Although no particular style or standards of lettering are strictly set forth and enforced in structural drafting, individual companies may set their own standards in their company's drafting practices manual. The structural drafting student should practice vertical lettering, lettering that inclines to the right, uppercase lettering, and lowercase lettering. When the student is called upon to work on a drawing prepared by another drafter, he or she will then be able to match the lettering style. This is a very common occurrence on the job.

In learning to letter, the drafting student should remember three words: Practice, Practice, AND Practice. Plenty of practice and an effort to improve on every individual letter in the alphabet will insure that the drafting student develops an acceptable lettering style. Figure 1-15 shows several different styles of structural drafting lettering.

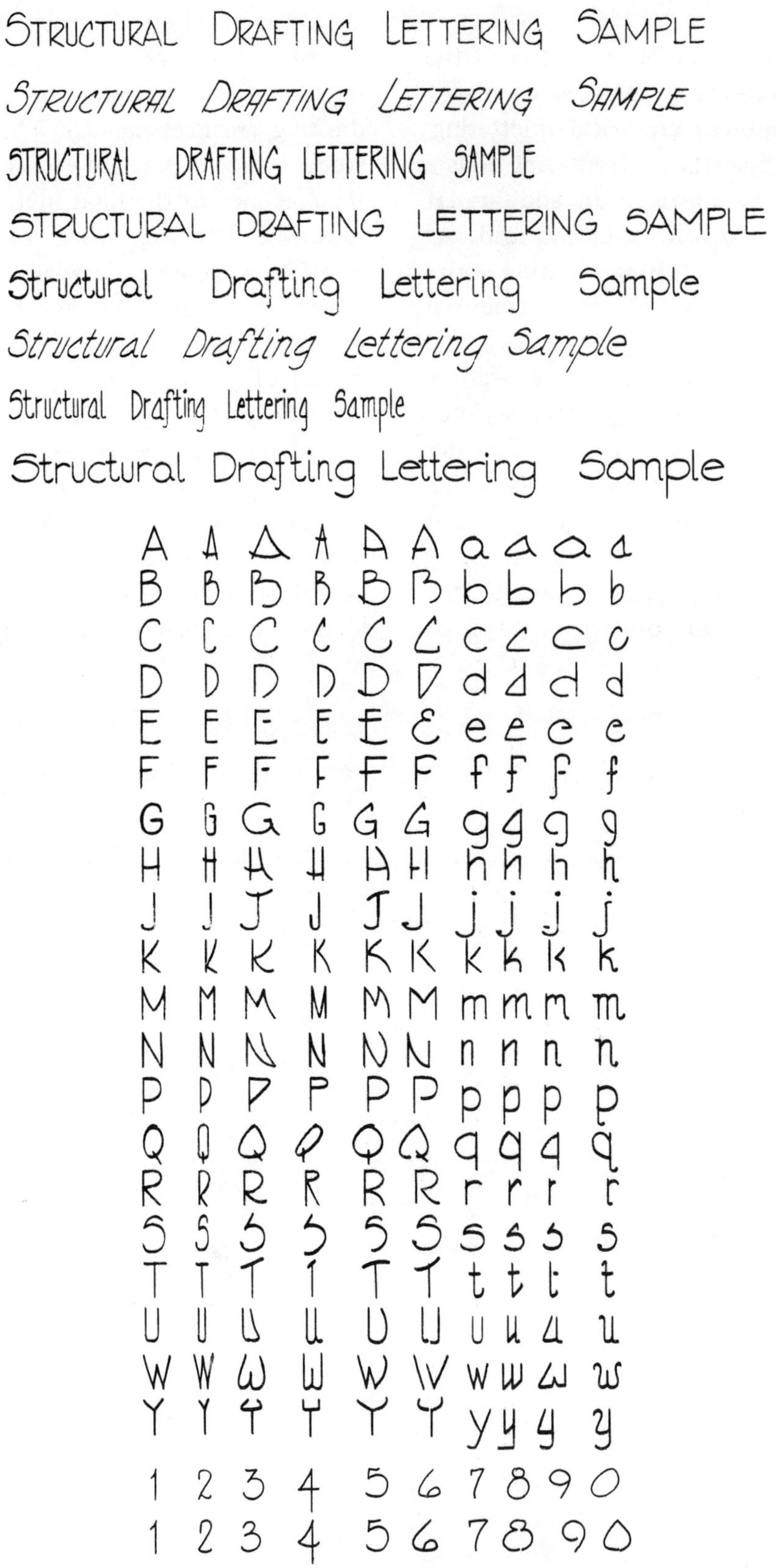
STRUCTURAL DRAFTING LETTERING SAMPLE
STRUCTURAL DRAFTING LETTERING SAMPLE
STRUCTURAL DRAFTING LETTERING SAMPLE
STRUCTURAL DRAFTING LETTERING SAMPLE
Structural Drafting Lettering Sample
Structural Drafting Lettering Sample
Structural Drafting Lettering Sample
Structural Drafting Lettering Sample
A A A A A A a a a a
B B B B B B b b b b
C C C C C C c c c c
D D D D D D d d d d
E E E E E E e e e e
F F F F F F f f f f
G G G G G G g g g g
H H H H H H h h h h
J J J J J J j j j j
K K K K K K k k k k
M M M M M M m m m m
N N N N N N n n n n
P P P P P P p p p p
Q Q Q Q Q Q q q q q
R R R R R R r r r r
S S S S S S s s s s
T T T T T T t t t t
U U U U U U u u u u
W W W W W W w w w w
Y Y Y Y Y Y y y y y
1 2 3 4 5 6 7 8 9 0
1 2 3 4 5 6 7 8 9 0

Fig. 1-15 Structural drafting lettering samples

SCALE USE

Structural drafting involves drawing the plans for very large structures on relatively small sheets of paper. To make a large building fit on a small sheet of paper, the structural drafter uses a scale. The two scales commonly used in structural drafting are the architect's scale, which is the most often used, and the engineer's scale.

The *architectural scale* is divided into ten separate scales that set off multiples of an inch and fractions of an inch equal to one foot. An architect's scale contains various sized scales ranging from the smallest 3/32" = 1'-0" to 3" = 1'-0", Figure 1-16. Two additional scales may be obtained from the architect's scale, full scale and 1/16" = 1'-0". The full scale is a normal ruler-type scale and is not often used. The scale, 1/16" = 1'-0", may be obtained by using the full scale and setting every 1/16" off as being one foot. This would be done in the event that an extremely large structure was required to fit on one small sheet of paper. The architect's scale is read by placing it along the line to be measured, aligning the nearest full foot mark with the end of the line, and reading any leftover inches and fractions on the additional foot that is provided just beyond the zero point on each scale, Figure 1-17.

3/32" = 1'-0"	....	3/16" = 1'-0"
1/8" = 1'-0"	....	1/4" = 1'-0"
3/8" = 1'-0"	....	3/4" = 1'-0"
1/2" = 1'-0"	....	1" = 1'-0"
1 1/2" = 1'-0"	....	3" = 1'-0"
	FULL SIZE	

Fig. 1-16 Architectural scale

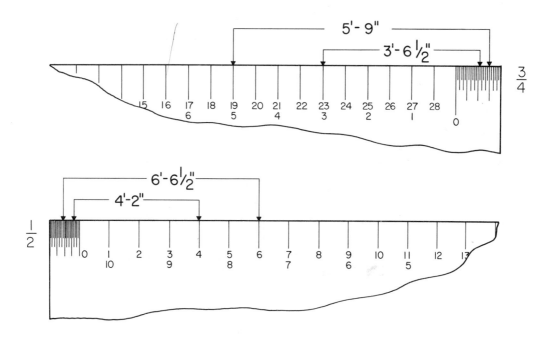

Fig. 1-17 Reading the architect's scale

The *engineer's scale* is used when a site plan, plot plan, or property boundaries must be drawn. It is a decimal scale that ranges from 1″ = 10′ to 1″ = 60′, Figure 1-18. The scales listed in Figure 1-18 may also be increased by tens since the engineer's scale is a decimal scale. For example, the scale 1″ = 10′ may also be used as 1″ = 100′, 1000′, etc. The engineer's scale is read by placing it along the line that is to be measured, reading the number of full feet off of the scale, and estimating the decimals of a foot when the end of the line falls between full foot marks, Figure 1-19.

PAPER SIZES

Most structural drafting firms use paper that has been precut to standard sizes. Sizes B, C, and D are the most commonly used sizes of paper in structural drafting. B-size paper is 12″ wide x 18″ long. C-size paper is 18″ wide x 24″ long, and D-size paper is 24″ wide x 36″ long. These figures include the area outside of the border. The most commonly used sizes of paper in the structural drafting classroom are B and C.

Figure 1-20 contains examples of B-size and C-size paper with 1/2″ borders. These formats should be used along with the title block shown in Figure 1-21 for all drafting activities requiring borders and a title block.

ENGINEER'S SCALE

1″ = 10′	1″ = 20′
1″ = 30′	1″ = 40′
1″ = 50′	1″ = 60′
1″ = 100′	1″ = 200′
1″ = 300′	1″ = 400′
1″ = 500′	1″ = 600′

Fig. 1-18 Engineer's scale

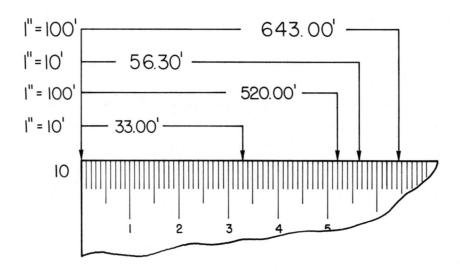

Fig. 1-19 Reading the engineer's scale

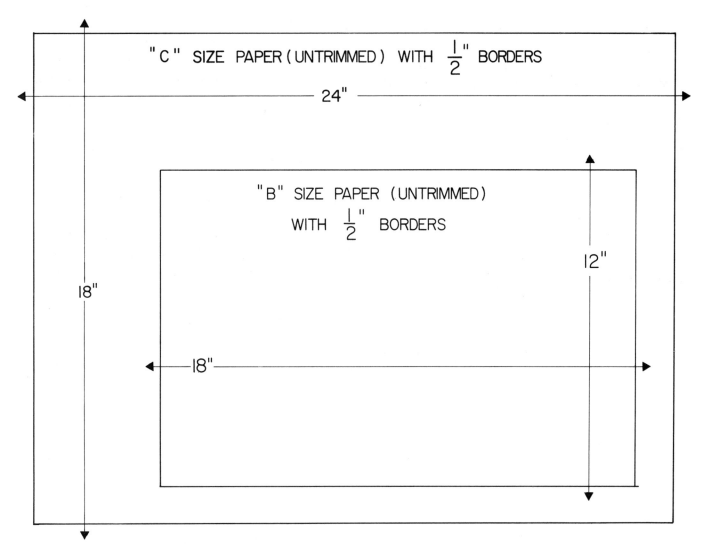

Fig. 1-20 Sample paper sizes with border formats

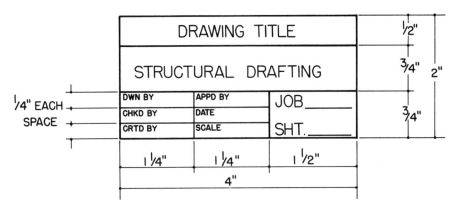

Fig. 1-21 Sample title block

TITLE BLOCKS AND BORDERS

Most structural drafting firms purchase precut paper with proper borders and a company title block printed on the paper. However, in the structural drafting classroom, the student may be required to apply his or her own borders and title blocks. The border lines should be drawn to approximate the width of an object line for either normal or heavy emphasis. Figure 1-21 contains a sample of a title block that may be used by students on B, C, or D paper.

SUMMARY

... The need for structural drafting is based on the requirement to have drawings in order to properly design, manufacture, and erect structures.
... There are two types of structural drawings: engineering drawings and shop drawings.
... Engineering drawings and shop drawings may be combined into one set to form structural working drawings.
... Engineering drawings are prepared by structural drafters employed by architects, engineers, contractors, and sales personnel and contain a minimum of detail.
... Shop drawings are prepared by structural drafters employed by companies that actually manufacture steel and concrete products.
... Structural drafters are most commonly employed by consulting engineering firms and manufacturers of structural products.
... Linework in structural drafting may vary in width according to the needs of the job, but must be sharp, clear, and rapidly done.
... Line types most commonly used in structural drafting are object, hidden, phantom, center, dimension, and cutting plane lines.
... Structural drafting lettering is not confined to the traditional block style, but should be simple, clear, and easy to read.
... Learning to letter requires much practice combined with an effort to improve.
... Structural drafters must be able to use the architect's and the engineer's scales.
... Structural drafters use B, C, and D sizes of paper.
... Title blocks for drawings may be prepared according to specifications set forth by the instructor, a structural drafting firm, or samples in this textbook.

REVIEW QUESTIONS

1. Why are structural drawings needed in the heavy construction industry?
2. Name two types of structural drawings.
3. What do engineering drawings and shop drawings form when combined into one set?
4. Name four different situations in which a structural drafter would be called upon to prepare engineering drawings.
5. Shop drawings are prepared by what types of companies?
6. Name the two most common employers of structural drafters.
7. Construct an example of an object line, a hidden line, and a centerline.
8. Letter the alphabet (all caps) in proper structural drafting lettering.
9. What is the most important requirement in learning to letter well?
10. List the three most commonly used sizes of paper in structural drafting.

REINFORCEMENT ACTIVITIES

1. Properly prepare a B-size sheet of drafting paper with 1/2" borders and title block as shown in Figure 1-21.

2. On the B-size sheet of paper prepared in Activity #1, draw a vertical object line dividing the paper in half. On the left half, reconstruct the structural drafting line types from Figure 1-11.

3. On the right half of the B-size sheet of paper prepared in Reinforcement Activity #1, complete Lettering Activity #1. This will be the first of several lettering activities that will be required following completion of learning units in this text. The activities are designed to give the student sufficient practice to develop an acceptable structural drafting lettering style upon completion of his or her course.

LETTERING ACTIVITY #1 SPECIFICATIONS

   a. In all caps, letter the alphabet twice in letters 1/2" tall. Refer to Figure 1-15 for samples.
   b. In lowercase, letter the alphabet twice in letters 1/4" tall. Refer to Figure 1-15 for samples.
   c. Letter the numerals from 0 to 9 four times each, using characters 3/16" tall.
   d. Letter the following sentence twice in all caps using characters 3/16" tall:
   "With enough practice and a continuous effort to improve, anyone can learn to letter well."

4. Prepare another B-size sheet of paper according to the specifications in Reinforcement Activity #1. Draw a vertical object line dividing the paper in half. On the left-hand side of the paper at the top, letter the title, Scale Use Activity—Architect's Scale, in all caps 1/4" tall. On the right side of the paper at the top, letter the title, Scale Use Activity—Engineer's Scale, in all caps 1/4" tall.

5. On the left-hand side of the paper prepared above, complete the Scale Use Activity—Architect's Scale. Directly below the scale use title, draw a line that is exactly 5 inches long, full scale. Using an architect's scale, measure the line to all scales shown in Figure 1-16 and record your answers in neat structural drafting lettering.

6. On the right-hand side of the paper prepared in Activity #4, complete the Scale Use Activity—Engineer's Scale. Directly below the scale use title, draw a line that is exactly 5 inches long, full scale. Using an engineer's scale, measure the line to all scales shown in Figure 1-18 and record your answers in neat structural drafting lettering.

Unit 2
Typical Structural Drafting Department

OBJECTIVES

Upon completion of this unit, the student will be able to

- sketch an organizational chart for a typical structural drafting department.
- write a job description for an entry-level position in structural drafting.
- list the primary duties of a junior drafter, drafter, senior drafter, checker, and chief drafter in a typical structural drafting department.

DRAFTING DEPARTMENT ORGANIZATION

One of the main complaints of students entering their first structural drafting position is that they do not know what to expect on the job. Most have the technical drafting skills required. However, many do not know exactly how or where they fit in. It is important for the structural drafting student to understand the structure of a typical drafting department. The inconsistency on the part of employers in position descriptions and position titles has further complicated the problem. However, there is enough consistency so that similarities can be drawn and generalities stated.

The typical drafting department in a company involved in structural drafting will include a drafting clerk, junior drafters, drafters, senior drafters, checkers, and a chief drafter. Although the position titles may vary from company to company, the positions themselves as well as the responsibilities and tasks required remain acceptably constant, Figure 2-1.

DRAFTING CLERK

Persons employed as drafting clerks are sometimes also referred to as a reproduction clerk,

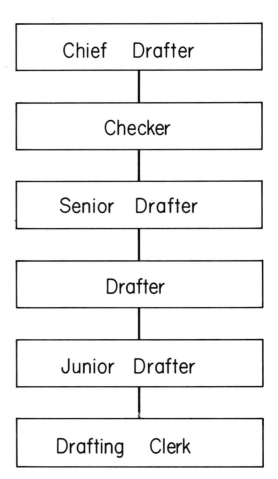

Fig. 2-1 Organizational chart — typical structural drafting department

printing clerk, or drafting/engineering secretary. The following are the primary responsibilities of persons in this position:

- Running prints (see Figure 2-2 for typical print request)
- Typing internal and external correspondence
- Filing
- Running errands
- Maintaining the supply room

The structural drafting student, having completed the instruction in this textbook, will usually start at a higher level than drafting clerk. However, this position should not be overlooked as a possible stepping stone to a career in drafting. Many successful structural drafters have started in the field as a drafting clerk.

REQUEST FOR PRINTS

JOB NUMBER _____ DATE _____

SHEET NUMBER _____

NUMBER OF PRINTS REQUIRED _____

STAMPS REQUIRED

_____ DATE ONLY
_____ FOR YOUR APPROVAL
_____ RESUBMITTED
_____ REVISED
_____ APPROVED FOR PRODUCTION/FABRICATION
_____ FOR FIELD USE
_____ PRELIMINARY
_____ INCOMPLETE

REMARKS

Fig. 2-2 A work order for the drafting clerk

JUNIOR DRAFTER

Persons employed as junior drafters are sometimes referred to as detailers, apprentice drafters, drafting trainees, or beginning drafters. The primary responsibilities of persons in this position are as follows:

- Running prints when the drafting clerk is overloaded (see Figure 2-3)
- Preparing elementary engineering and shop drawings
- Drawing details
- Performing revisions and corrections
- Preparing bills of materials

Most structural drafting students, having completed a course of study in structural drafting but lacking work experience in the field, will begin their career as a junior drafter.

DRAFTER

There is very little variation in position titles for persons employed in structural drafting at this level. The term *drafter* and the previously used term *draftsman* are almost universal. The primary responsibilities of persons employed as drafters are the following:

- Preparing engineering and shop drawings in accordance with architectural plans, contractor's sketches, engineer's sketches, sales personnel sketches, or verbal instructions from any of these sources
- Assisting junior drafters assigned to cooperative projects
- Adhering as closely as possible to projected timetables and work schedules

Fig. 2-3 Running prints is sometimes the job of the junior drafter. (Deborah M. Goetsch)

SENIOR DRAFTER

Senior drafters are sometimes referred to as designers or team leaders. The primary responsibilities of persons in this position include the following:

- Preparing more complicated engineering and shop drawings in accordance with company standard drafting procedures and the raw data available for a job
- Supervising drafters and junior drafters assigned to a job or drafting team (Figure 2-4)
- Insuring that projected timetables and work schedules are met
- Performing minor checking duties
- Acting as liaison between the checker and project engineer assigned to a job (Figure 2-5)

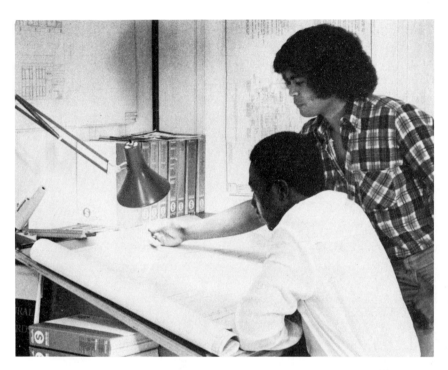

Fig. 2-4 The senior drafter consults with one member of the drafting team. (Deborah M. Goetsch)

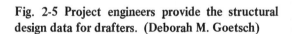

Fig. 2-5 Project engineers provide the structural design data for drafters. (Deborah M. Goetsch)

CHECKER

Checkers are sometimes referred to as senior designers and are persons of much experience and expertise in structural drafting. The following are the primary responsibilities of persons in this position:

- Checking engineering and shop drawings for dimensional accuracy
- Adherence to company drafting procedures
- Adherence to information presented in the raw data for a job
- General drafting technique

In addition to these, the checker is sometimes required to help drafters with less experience to interpret engineering calculations and design.

CHIEF DRAFTER

The chief drafter is sometimes referred to as the head drafter. Persons serving in this position are the administrators or managers of drafting departments. The following are the primary responsibilities of the chief drafter:

- Supervising all drafting department personnel
- Scheduling and assigning work, insuring that all functions of the drafting department are carried out properly and on time
- Reviewing new projects and estimating the amount of time that will be required to complete drawings, Figure 2-6
- Requesting supplies for the drafting department
- Conducting interviews of prospective drafting department employees

Fig. 2-6 New jobs are reviewed thoroughly before work is assigned to drafters. (Deborah M. Goetsch)

SUMMARY

... It is important for the structural drafting student to understand the organizational structure of a typical drafting department.
... Position titles vary somewhat from company to company, but most companies have positions that equate to drafting clerk, junior drafter, drafter, senior drafter, checker, and chief drafter.
... The primary responsibilities of the drafting clerk are running prints, preparing and distributing correspondence, filing, and running errands.
... The primary responsibilities of the junior drafter are performing corrections and revisions, preparing elementary drawings, and assisting the drafting clerk when the work load is heavy.
... The primary responsibilities of the drafter are preparing engineering and shop drawings from architectural, engineering, contractor, or sales sketches.
... The primary responsibilities of the senior drafter are preparing complicated engineering and shop drawings, coordinating the efforts of the drafting team, and assisting less-experienced drafters in meeting timetables and work schedules.
... The primary responsibilities of the checker are checking engineering and shop drawings and helping drafters with less experience to interpret design calculations.
... The primary responsibilities of the chief drafter are the overall supervision and operation of the drafting department, examining new jobs to determine how much time and how many people will be required to complete drawings, and interviewing prospective employees.

REVIEW QUESTIONS

1. What positions are usually found in a typical structural drafting department?
2. List the primary responsibilities of the drafting clerk.
3. List the primary responsibilities of the junior drafter.
4. List the primary responsibilities of the drafter.
5. List the primary responsibilities of the senior drafter.
6. List the primary responsibilities of the checker.
7. List the primary responsibilities of the chief drafter.

REINFORCEMENT ACTIVITIES

Unit 2 Reinforcement Activities are designed to help the student continue developing fundamental skills of proper lettering, linework, and scale use as introduced in Unit 1.

1. Prepare a B-size sheet of paper to match the one contained in Figure 2-7. Begin to develop versatility in lettering by completing Lettering Activity #2. Try to match the lettering styles in Figure 2-7 and repeat each letter or sentence as many times as is indicated. The larger letters are to be 1/4" to 1/2" tall. The smaller letters are to be 1/8" to 3/16" tall.

2. Continue to develop linework, lettering, and scale use skills by transforming the contractor's sketch in Figure 2-8 into a mechanical drawing. Use a B-size or C-size sheet of paper with a title block as in Figure 1-21 and 1/2" borders. Use a scale of 3/8" = 1'-0".

3. Continue to develop linework, lettering, and scale use skills by transforming the engineer's sketch in Figure 2-9 into a mechanical drawing. Use a B-size or C-size sheet of paper with a title block as in Figure 1-21 and 1/2" borders. Use a scale of 3/16" = 1'-0".

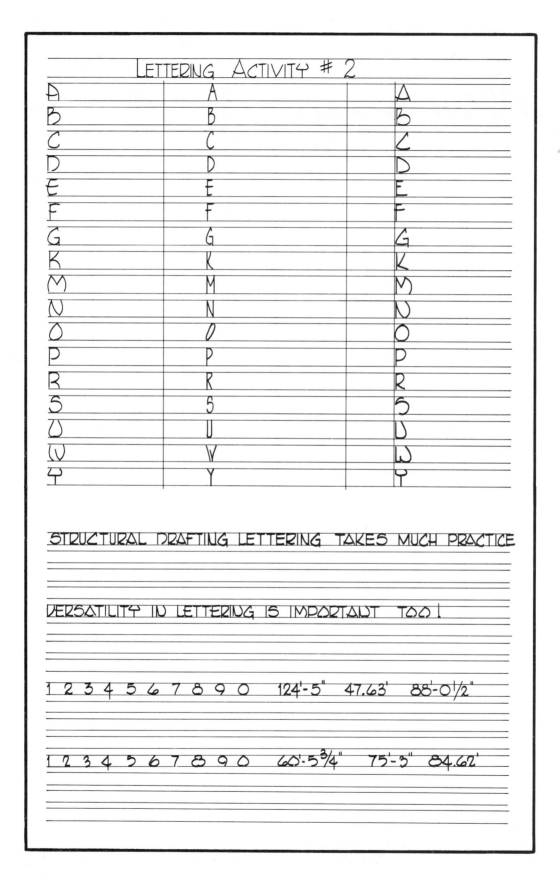

Fig. 2-7 Guide for Reinforcement Activity #1

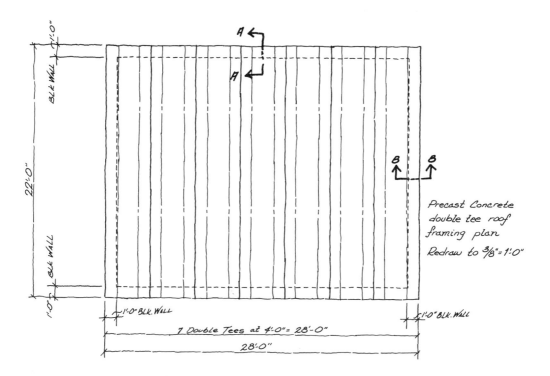

Fig. 2-8 Contractor's sketch for Reinforcement Activity #2

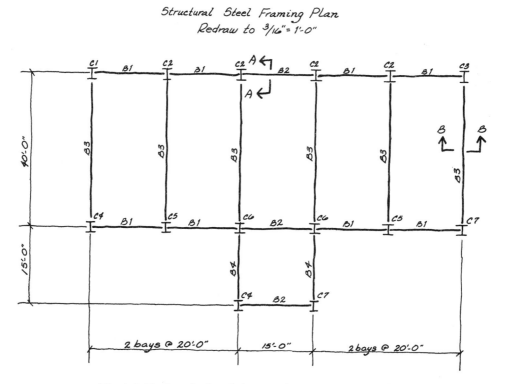

Fig. 2-9 Engineer's sketch for Reinforcement Activity #3

Unit 3
Drawing, Checking, Correcting, and Revising Processes

OBJECTIVE

Upon completion of this unit, the student will be able to

- explain the original drawing process, checking process, correcting process, and revising process in structural drafting.

ORIGINAL DRAWING

Structural drafters prepare original engineering and shop drawings based on raw data that may be supplied by a number of different sources. Some of the more common sources are architects, engineers, contractors, and sales personnel. The raw data itself may consist of a complete set of thoroughly prepared architectural plans with comprehensive specifications. On the other hand, the raw data may consist of as little as a freehand sketch made in the field.

The process that takes place in the structural drafting department in preparing original drawings is the same. The chief drafter collects, examines, and analyzes all of the data that is available for a job. A drafting team is then selected by the chief drafter to prepare the original drawings. The team is given a projected timetable for completion.

The duties of each team member will vary according to their experience and abilities and the job size. In a large job, the bulk of the drawing will be assigned to drafters and senior drafters. The senior drafters perform the most complicated tasks. Figure 3-1 shows an architectural drawing of the floor plan for a small bank. Such drawings are commonly provided as the raw data. From this, a structural drafter must develop the structural drawings that are needed to manufacture products and construct the building.

Figure 3-2 contains a sample prestressed concrete engineering drawing that was prepared based on the architect's drawing in Figure 3-1. It is a roof framing plan of structural prestressed concrete

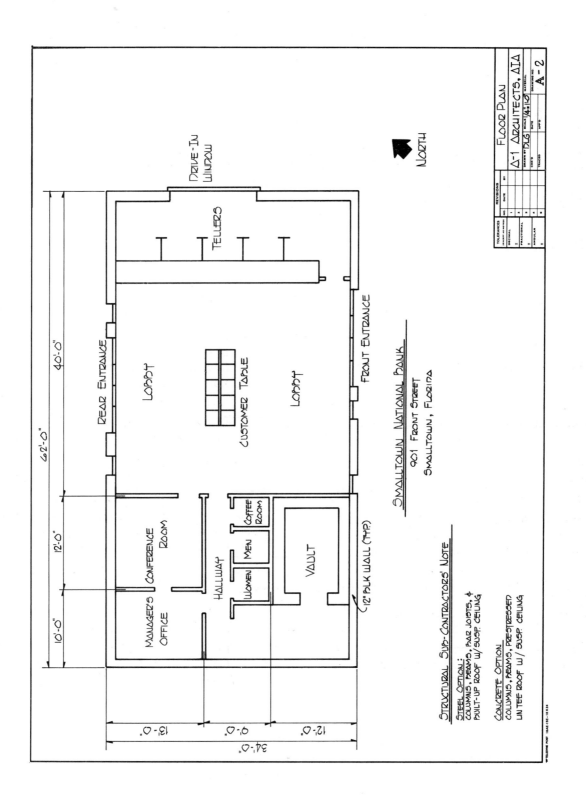

Fig. 3-1 Architect's floor plan for a small bank

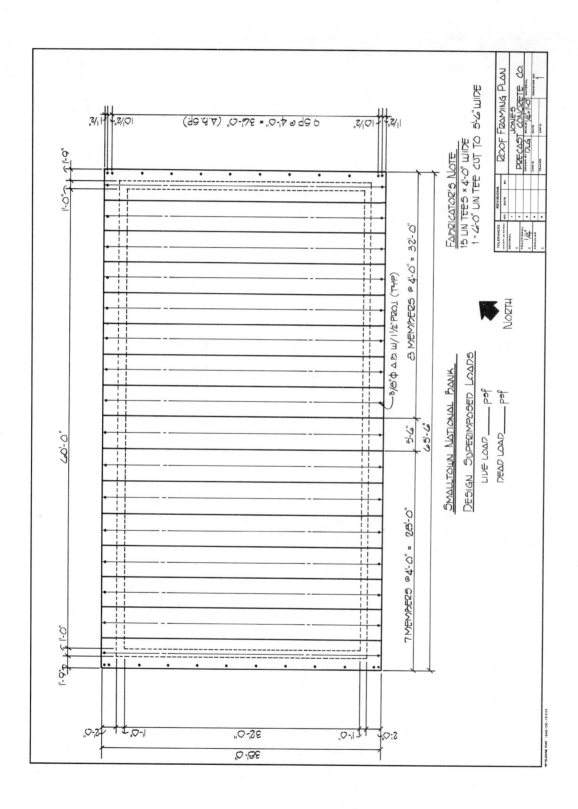

Fig. 3-2 Prestressed concrete engineering drawing based on architectural drawing in Figure 3-1

Lin Tee members. Figure 3-3 is an example of a structural engineering drawing involving a combination of poured-in-place concrete and structural steel. The drawing consists of the layout information for concrete columns and beams and steel bar joists. It was prepared based on the architect's drawing in Figure 3-1.

CHECKING

Once an original drawing has been completed, it must be checked. A checker is assigned to each job that is brought into the drafting department. A checker insures that the drawings have been properly prepared before they are turned over to the originator of the job for approval. The checker checks all drawings against standard company drafting procedures, accepted drafting practices, and the raw data used in preparing the original drawings.

A common checking practice is to mark through incorrect information and write in the correct information. This is done on a print of the original drawing, not on the original. Once a drawing has been completely checked, it is marked *checkprint #1* and given to a junior drafter to be corrected. This process continues with every sheet in a set of drawings until all mistakes have been identified and corrected. This might involve as many as ten checkprints for each sheet in the job or as few as one. When there is more than one checkprint, each successive print is numbered and kept in order and on file until the job is completed and the building has been constructed.

Figures 3-4 and 3-5 are examples of checkprints that have been marked by a checker and are ready to be corrected by a junior drafter. Both samples are part of a complete set of structural steel drawings.

32 Section 1 Overview of Structural Drafting

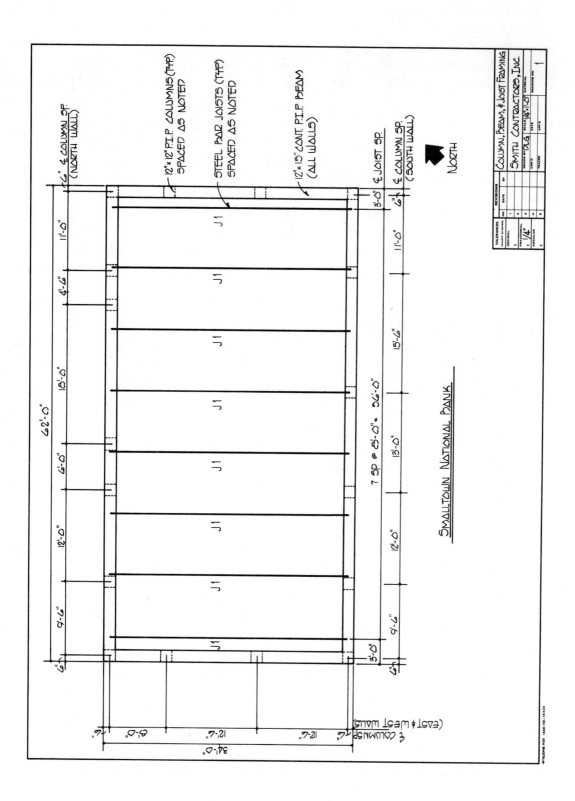

Fig. 3-3 Structural engineering drawing based on architectural drawing in Figure 3-1

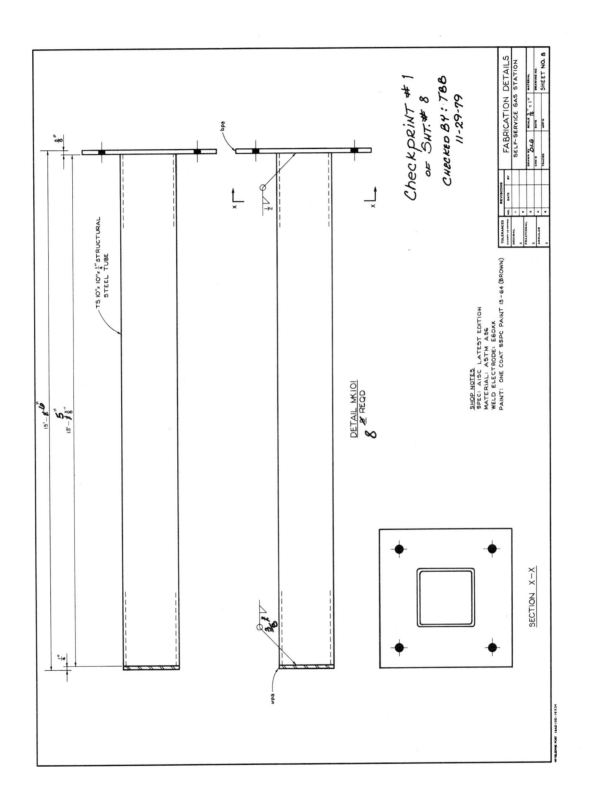

Fig. 3-4 Checkprint of a column detail

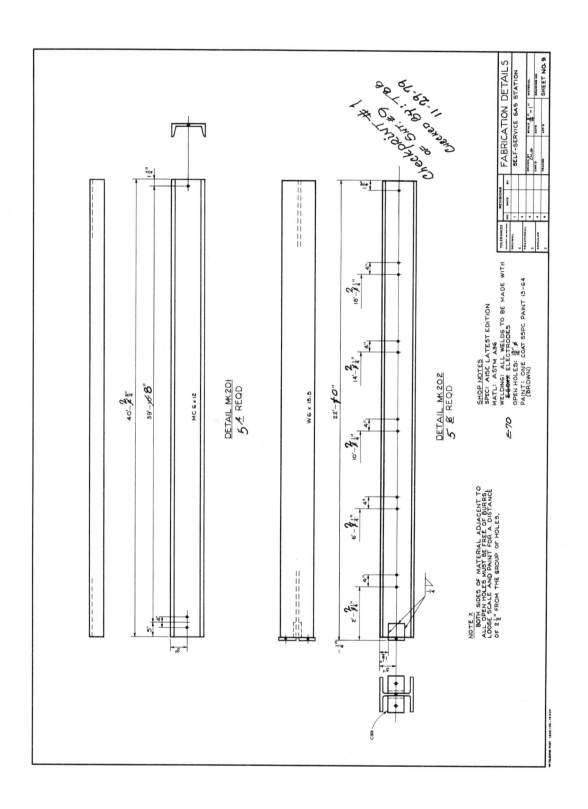

Fig. 3-5 Checkprint of a beam detail

CORRECTING

Correcting drawings is usually the job of the junior drafter. By performing corrections, junior drafters become familiar with structural drawings and drafting practices. This familiarity makes it easier for them to develop more advanced skills and upgrade themselves to a drafter position.

Junior drafters receive checkprints and any necessary instructions from the checker before they begin corrections. Once they have been instructed, they get the appropriate original drawings from the files and make the corrections indicated on the checkprints. The junior drafter crosses off the corrections as they are made, runs a new print of the corrected original, signs the first checkprint, and returns the checkprints to the checker. The checker then insures that all corrections were properly made. When satisfied that a print is completely correct, the checker initials the title block. This signifies that it has been checked, corrected, and is now ready to be approved.

When all sheets of a job are ready for approval, the chief drafter instructs the drafting clerk to run a set of prints and stamp them *for your approval.* These prints are forwarded to the architect, contractor, or engineer who originated the job to be examined and approved. The originator may approve the plans, disapprove the plans, or approve the plans with revisions. Figure 3-6 shows a sample structural drawing that has been checked, corrected, and prepared for approval of the originator. Notice the checker's initials in the title block and the approval stamp.

REVISING

The originator of a job may approve, disapprove, or, as is often the case, approve the structural drawings with revisions. A *revision* is a change in design, configuration, or plan that was made after the original drawings were developed. Revisions are caused by the originator of a job, not the company preparing the structural drawings.

Revisions are marked on the set of approval plans by the originator. The plans are then marked *approved with revisions* and returned to the company that prepared the original drawings to be corrected. It is important to note the distinction between revisions and corrections. Corrections are caused by mistakes made in preparing the original drawings. Revisions are the result of changes by the originator of a job in his or her design, plan, or approach. Time spent in performing revisions to the original drawings is usually charged to the person causing the revisions.

Figure 3-7 contains an example of an original drawing that has been revised by a structural drafter. Note the revision triangles that identify the number of each revision and the revision notes that explain each revision.

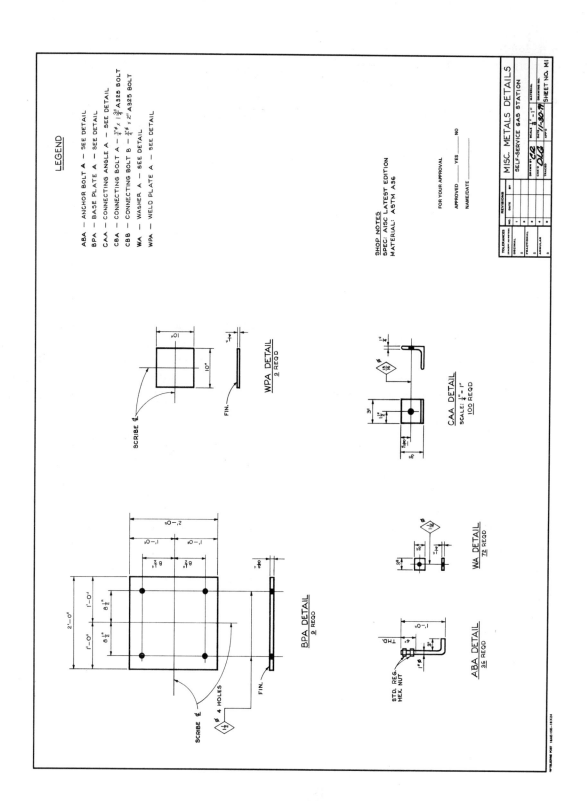

Fig. 3-6 Structural drawing ready to be submitted for approval

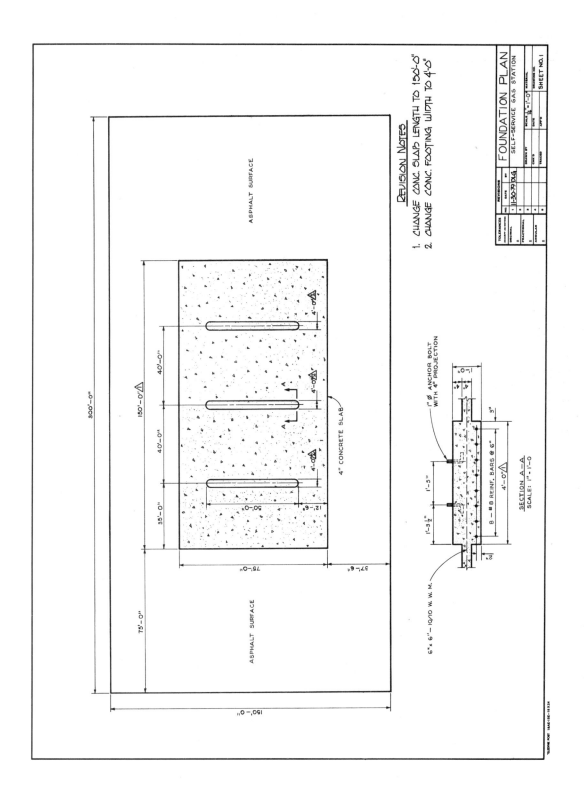

Fig. 3-7 Revised original drawing for poured-in-place concrete job

SUMMARY

... Raw data for preparing structural drawings comes from architects, engineers, contractors, or sales personnel.
... Structural drawings might be produced based on very complete architectural plans or on very quickly prepared freehand sketches.
... Every original structural drawing must be checked for accuracy, conformance to raw data specifications, and proper drafting practices.
... Checkers make their marks on prints of original drawings called checkprints.
... Corrections are usually performed by junior drafters.
... Once drawings are corrected, they must be printed and forwarded to the originator for approval.
... The originator of a structural job must check the drawings and approve them, disapprove them, or approve them with revisions.
... A revision differs from a correction in that it is caused by a change in plans rather than a mistake.

REVIEW QUESTIONS

1. Explain where structural drafters get the raw data they use in preparing the structural drawings for a job.
2. Who must approve structural drawings once they are complete?
3. Explain how a correction and a revision differ.
4. What step immediately follows after preparation of the original drawings?
    a. Checking
    b. Revisions
    c. Approval
    d. Product fabrication
5. Checkers make their marks on:
    a. original drawings.
    b. specifications.
    c. checkprints.
    d. file copies.
6. Corrections to original drawings are usually performed by:
    a. drafters.
    b. junior drafters.
    c. chief drafters.
    d. checkers.

REINFORCEMENT ACTIVITIES

1. Continue to develop good, fast structural drafting lettering by completing Lettering Activity #3. This activity may be done on an A-size sheet of paper with borders 1/2" in from each side and no title block. Referring to lettering samples in the previous two units, letter the following paragraph in your best structural drafting lettering (characters 1/8" to 3/16" tall):

   A STRUCTURAL DRAFTER'S LETTERING IS HIS OR HER TRADEMARK. A GOOD STRUCTURAL DRAFTER HAS GOOD LETTERING. THIS IS IMPORTANT BECAUSE LETTERING IS THE MOST VISUALLY SIGNIFICANT COMPONENT OF A STRUCTURAL DRAWING. CLEAR, CRISP, ATTRACTIVE LETTERING ADDS A TOUCH OF PROFESSIONALISM TO A SET OF STRUCTURAL DRAWINGS. FIRST IMPRESSIONS ARE LASTING, AND THE FIRST IMPRESSION AN ARCHITECT, CONTRACTOR, OR ENGINEERING FIRM WILL HAVE OF A COMPANY INVOLVED IN THE HEAVY CONSTRUCTION INDUSTRY WILL COME FROM THAT COMPANY'S DRAWINGS. NOTHING WILL ENHANCE THE APPEARANCE OF A SET OF STRUCTURAL PLANS MORE THAN ATTRACTIVE LETTERING.

   1 2 3 4 5 6 7 8 9 0

   123'-4"   234'-5"   345'-6"   456'-0 1/2"

2. Begin to understand the original drawing and correcting process by redrawing and correcting the marked-up print in Figure 3-8. This project should be done on a C-size sheet of paper with a title block as in Figure 1-21 and 1/2" borders. While preparing the drawing, concentrate on developing your lettering, linework, and scale use skills. The drawing should be done to a scale of 3/16" = 1'-0".

3. Begin to understand the revising process by redrawing the revised print in Figure 3-9. Identify each revision with a number inside of a revision triangle and explain each with a set of revision notes. This project should be done on a C-size sheet of paper with a title block as in Figure 1-21 and 1/2" borders. While preparing the drawing, concentrate on developing your lettering, linework, and scale use skills. The drawing should be done to a scale of 3/16" = 1'-0".

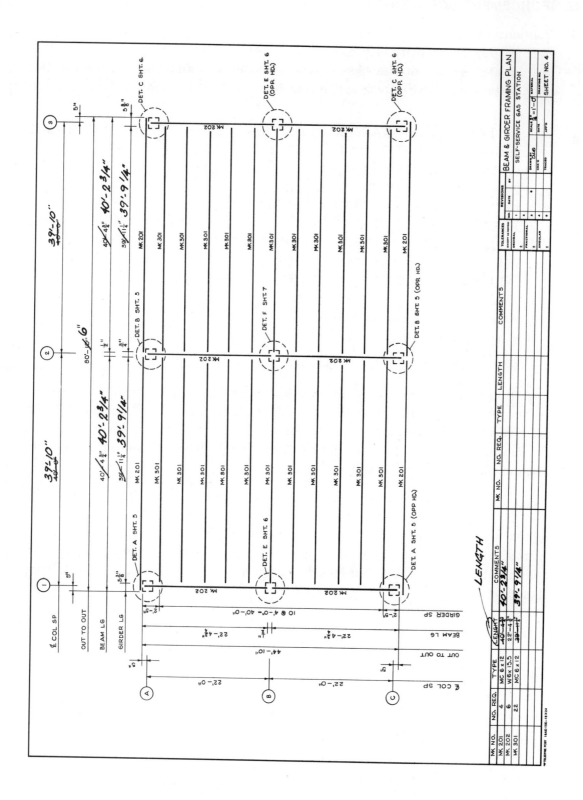

Fig. 3-8 Drafting problem for Reinforcement Activity #2

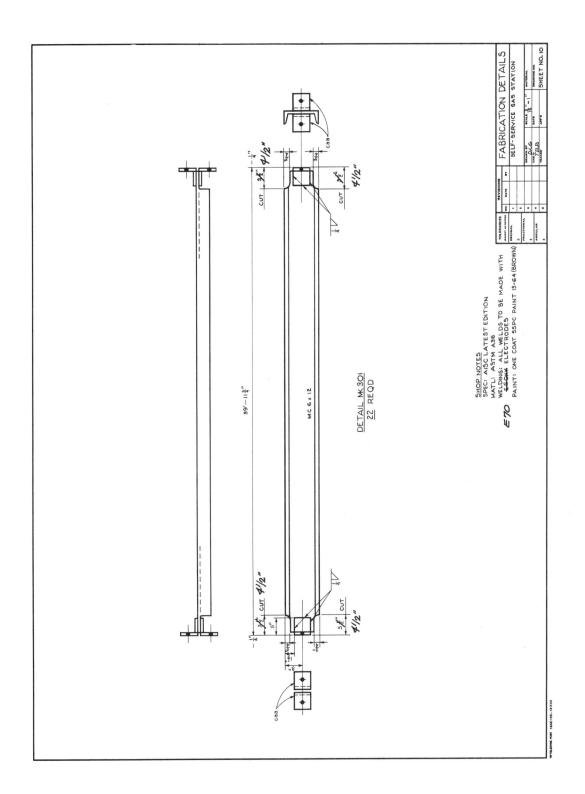

Fig. 3-9 Drafting problem for Reinforcement Activity #3

Unit 4
Product Fabrication and Shipping

OBJECTIVES

Upon completion of this unit, the student will be able to

- explain the product fabrication processes for structural steel, precast concrete, and poured-in-place concrete.
- explain how structural steel and precast concrete products are shipped to the jobsite.

STRUCTURAL STEEL PRODUCT FABRICATION

Structural steel is produced in standard shapes in rolling mills, Figures 4-1, 4-2, and 4-3. From there, it is shipped to steel construction companies to be altered and used according to the needs of each individual job the company contracts. Dimensions for detailing and properties for designing steel components of a structure are listed in the *Manual of Steel Construction*. This manual is put out by the American Institute of Steel Construction (AISC). It is the primary reference source for structural steel drafters and engineers.

Excerpts from the *Manual of Steel Construction* have been included in the Appendix of this textbook. It should be referred to by students when determining the dimensions of steel shapes contained in drawings. For example, if the width and depth of a W14x426 steel beam were required, the information could be found by using the Appendix in this text or the *Manual of Steel Construction* labeled *W Shapes – Dimensions For Detailing*. The notation W14x426 appears under the heading *Designation*. The *Depth Column* lists the desired depth as 18 5/8" and the *Width Column* lists the desired width as 16 3/4".

There are a number of steps that a structural steel shape must go through during fabrication. The following are the most common steps:

- handling and cutting
- punching and drilling
- straightening
- bending
- bolting
- riveting
- welding
- finishing

Not every steel product undergoes all of these processes. Only those processes needed to prepare the product for delivery to and erection at the jobsite are used.

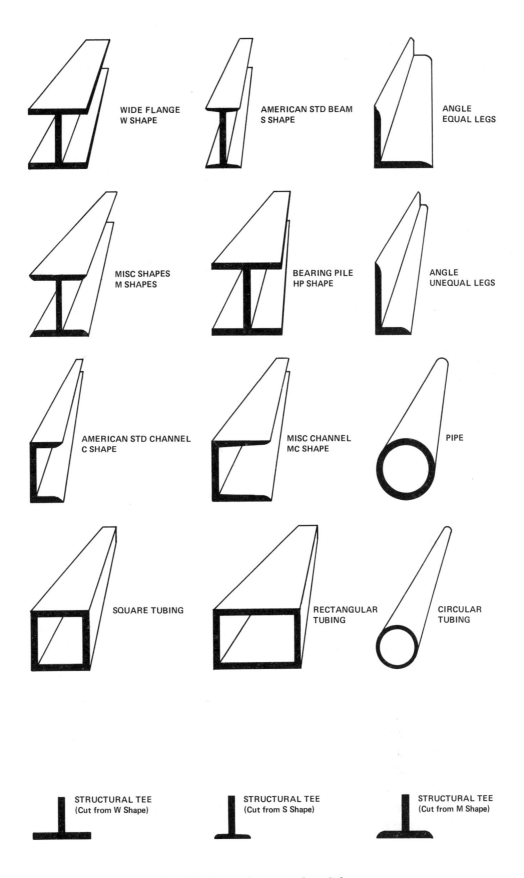

Fig. 4-1 Standard structural steel shapes

Fig. 4-2 Steel ingots are reduced to standard sizes and shapes. (Bethlehem Steel Corporation)

Fig. 4-3 White-hot ingots are moved through the rolling mill on conveyor lines. (Bethlehem Steel Corporation)

Handling and Cutting

Structural steel products are very heavy. Therefore, they require special methods for handling, cutting, and transporting. Special handling is accomplished by forklifts, cranes, overhead hoists, or straddle carriers, Figure 4-4.

Cutting steel products can be done in one of several ways depending on the product shape and size. Thin, flat shapes can be cut on a shear. Medium-sized members such as beams and columns are often cut with a special hot cutting saw, Figure 4-5, while thicker structural shapes are usually cut with a flame cutting torch.

Punching and Drilling

Structural steel shapes are often connected at the jobsite by bolting or less frequently by riveting. The holes required in a structural member may be drilled or punched. Punching is confined to the thinner structural members. When punching is not possible, the desired holes may be made by drilling.

Fig. 4-4 Heavy steel members are stacked with an overhead hoist. (Bethlehem Steel Corporation)

Single holes are punched in structural members by a machine called a *detail punch*. Several holes may be punched in a piece of steel at the same time by a machine called a *multiple punch*. Punch machines are versatile and fast. However, they do occasionally bend the structural member being punched. The thicker the material, the more prone the punch is to disfigure it. This makes drilling the more practical process for putting holes in thick products.

There are several different types of drills used in structural steel fabrication shops. Some of these are the drill press, radial arm drill, multiple spindle drills, drills on jibs, and gantry drills. In modern steel fabrication shops, much of the drilling is automated. This makes the drilling fast, accurate, and thereby cuts down on waste due to human error.

Straightening and Bending

Steel members that have been punched or mishandled often become bent and must be straightened before they can be used. The most common machine used for straightening structural steel shapes is the *bend press* or *gag press*. Structural steel shapes are straightened by resting one edge of the member against a rigid retainer and then applying pressure with a high-strength plunger apparatus. This machine is also used for bending members when long radius curves are required.

Bolting, Riveting, and Welding

There are three basic methods used in fastening structural steel shapes together to form a completed structure: bolting, riveting, and welding (Unit 5). Bolting and welding are frequently used alone or in combination, but riveting is no longer considered a major connecting process.

Bolting of structural steel shapes is a common fastening method, especially for connections that are made in the field. Bolts may be applied by hand or, as is more often the case, with power wrenches. Most structural steel applications in heavy construction require special high-strength bolts that are tightened by a compressed-air device known as an *impact wrench*.

Welding is a common fastening method that is particularly useful in making shop connections and permanent field connections. Structural steel shapes that are to be welded are carefully marked off or layed out to insure that all welds are accurately applied. Welding processes may be performed by hand or they may be automated. Most modern shops have cut down on the time and waste from human error by switching from hand welding to automated welding.

Until the year 1950, riveting was the primary method for connecting structural members. However, due to improvements in welding processes and the arrival of reliable, high-strength bolting processes, riveting has become less and less used. It is no longer considered a major connecting process, but since many riveted structures are still in use, the drafting student should be aware of this process. All three of the above fastening processes are discussed in depth in Unit 5.

Fig. 4-5 Steel shapes are cut to desired lengths by hot sawing. (Bethlehem Steel Corporation)

Finishing

The final process that structural steel members must undergo during fabrication is finishing. The finishing process involves smoothing rough edges or surfaces to insure the proper flatness for bearing purposes or fit. Ends of beams that have been cut, edges and tops of bearing plates, and ends of columns that will rest on a baseplate are commonly finished. Finishing is done by any one of the following methods: sawing, milling, filing, high-pressure blasting, or various other means.

PRECAST CONCRETE FABRICATION

Like structural steel, precast concrete products are manufactured in standard shapes that may be altered to suit the needs of the individual job, Figures 4-6 through 4-13. Precast concrete can be broken into two separate categories: prestressed products and reinforced products.

Double Tees

NOTE
Double-Tee lengths and stranding patterns vary according to the needs of each individual job and design situation.

Typical Uses
Roof and floor systems for commercial and industrial buildings; Wall panels for commercial and industrial buildings; Pier decks, tank covers, catwalks, tunnel covers, and conveyor trestle decks

Standard Sizes
10" deep by 4' wide
14" deep by 4' wide
16" deep by 4' wide
24" deep by 8' wide

Fig. 4-6 Standard precast concrete double-tee members

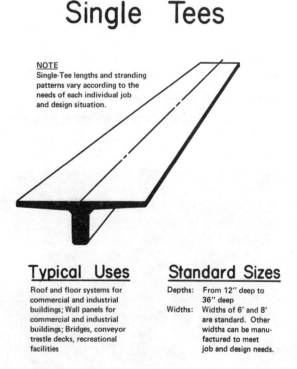

Single Tees

NOTE
Single-Tee lengths and stranding patterns vary according to the needs of each individual job and design situation.

Typical Uses
Roof and floor systems for commercial and industrial buildings; Wall panels for commercial and industrial buildings; Bridges, conveyor trestle decks, recreational facilities

Standard Sizes
Depths: From 12" deep to 36" deep
Widths: Widths of 6' and 8' are standard. Other widths can be manufactured to meet job and design needs.

Fig. 4-7 Standard precast concrete single-tee members

Flat Slabs

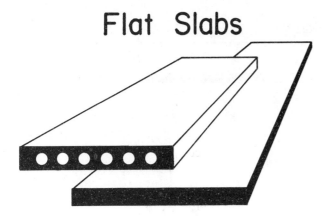

Typical Uses
Roof or floor systems where a flat ceiling or minimum structural depth is particularly desired. Also used for wall panels, tank covers, and tunnel covers.

Standard Sizes
Solid slabs: 3" to 6" deep and widths to 10'-0"
Cored slabs: 6", 8", and 10" by 10'-0" wide

NOTE
Flat-slab lengths and stranding patterns vary according to the needs of each individual job and design situation.

Fig. 4-8 Standard precast concrete flat-slab members

Joists

Typical Uses
Roof and floor systems with prefabricated or poured decks; Also used as fins, frames, posts, and columns.

Standard Sizes
8" and 12" deep in the keystone shapes.
16" deep in the tee shapes.

NOTE
Joist lengths and stranding patterns vary according to the needs of each individual job and design situation.

Fig. 4-9 Standard precast concrete joist members

Building Beams

Typical Uses
Primary structural beams for all types of roof and floor systems and other structural framing systems

Standard Sizes
Sizes and shapes of all beams can be adjusted to meet the design requirements of each individual job.

Fig. 4-10 Standard precast concrete building beams

Piles And Columns

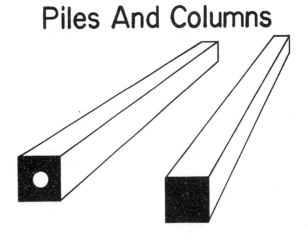

Typical Uses
Bridge and pier bearing piles; Foundation piles and building columns

Standard Sizes
Sizes and shapes of piles and columns may be manufactured to meet the needs of the individual job and design situation.

NOTE
Pile and column lengths and reinforcing or stranding vary according to the needs of each individual job and design situation.

Fig. 4-11 Standard precast concrete piles and columns

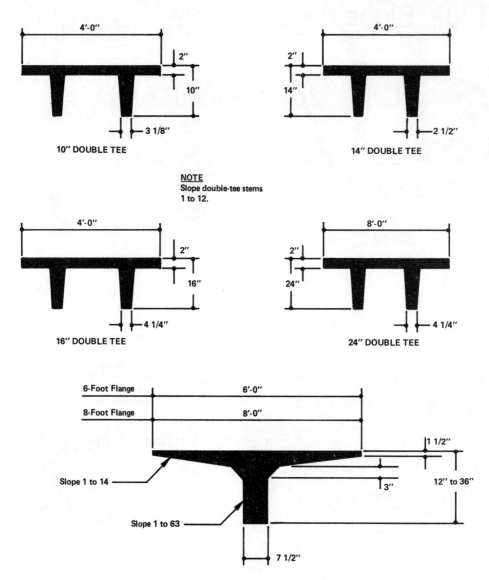

Fig. 4-12 Standard precast concrete products — dimensions for detailing

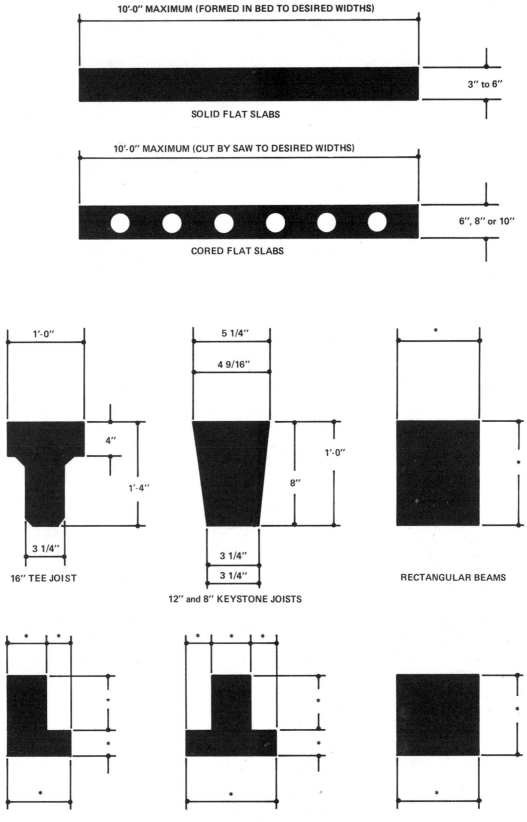

* DIMENSIONS MAY BE VARIED TO MEET THE REQUIREMENTS OF THE JOB.

Fig. 4-13 Standard precast concrete products – dimensions for detailing

Prestressed concrete products are poured in steel beds or forms through which high-strength steel strands have been passed. Before the concrete is poured into the bed, the strands are stretched to a predetermined amount of tension. The concrete is then poured into the bed around the strands and allowed to harden. Once the concrete has sufficiently hardened, the tension on the strands is released and the concrete member is placed in compression or prestressed. The principle is commonly illustrated by attempting to lift a row of books on a shelf, by pressing against each end and lifting, Figure 4-14. If enough pressure is exerted inward from both ends, there is sufficient compression to lift all of the books at once.

Prestressed concrete members are further strengthened by the addition of *nonstressed reinforcing bars* (rebars) or corners, around openings, or in other high-stress situations. Rebars are designated by numbers that correspond with their diameters measured in eighths of an inch. For example, a #3 bar is 3/8" in diameter, a #4 bar is 4/8" or 1/2" in diameter, a #8 bar is 8/8" or 1" in diameter, and so on.

Reinforced concrete products are also poured in beds or forms, but without the prestressing strands. Instead, they are reinforced solely with wire mesh and reinforcing bars. Prestressed and reinforced concrete products are removed from the beds after hardening by cranes or overhead lifts, stacked on wooden blocks for storage, and shipped to the jobsite by truck or train. Figures 4-15 through 4-23 illustrate both categories of precast concrete fabrication.

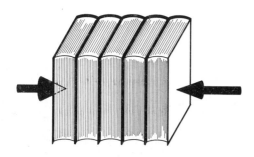

Fig. 4-14 The prestressing principle

Fig. 4-15 Steel abutment for pulling prestressing strands. (Deborah M. Goetsch)

Fig. 4-16 Precast concrete double-tee form (Deborah M. Goetsch)

Unit 4 Product Fabrication and Shipping 51

Fig. 4-17 Precast concrete beam (Deborah M. Goetsch)

Fig. 4-18 Reinforcing bars stacked for use in a precast concrete job (Deborah M. Goetsch)

Fig. 4-19 Precast concrete column (Deborah M. Goetsch)

Fig. 4-20 Precast concrete flat-slab forms (Deborah M. Goetsch)

Fig. 4-21 Precast double-tee members are stacked awaiting shipping. (Deborah M. Goetsch)

Fig. 4-22 Precast concrete flat slabs stacked and labeled for shipping (Deborah M. Goetsch)

Fig. 4-23 High-strength steel prestressing strands (Deborah M. Goetsch)

POURED-IN-PLACE CONCRETE FABRICATION

Poured-in-place concrete is similar to reinforced concrete; in fact, it is reinforced concrete. The difference between precast reinforced concrete and poured-in-place reinforced concrete is that the latter is poured at the jobsite. Precast reinforced concrete is poured at a plant or shop and shipped to the jobsite for erection. Poured-in-place reinforced concrete is usually poured in wooden forms as a part of the structure. These wooden forms are built in place at the jobsite and torn down after the concrete has hardened. Figures 4-24 and 4-25 show examples of wooden forms.

In structural steel and precast concrete construction, the general contractor works with a steel or concrete subcontractor. In poured-in-place concrete construction, there is no concrete or steel subcontractor. Using shop drawings, the contractor's workers build the necessary forms, place the prebent reinforcing bar cages into the forms, and pour the concrete. Figures 4-24 through 4-27 show examples of poured-in-place concrete fabrication.

Fig. 4-24 Wooden form for a poured-in-place concrete beam

Fig. 4-25 Wooden form for a poured-in-place column base (Deborah M. Goetsch)

Fig. 4-26 Reinforcing bars to be used in a poured-in-place concrete job (Deborah M. Goetsch)

Fig. 4-27 Wooden form for poured-in-place concrete bank vault (Deborah M. Goetsch)

GLUED-LAMINATED WOOD FABRICATION

Several separate pieces of wood may be pressed together to form a laminated structural member that is stronger than its individual parts. Laminated wood members such as columns, beams, and arches are usually made of kiln-dried lumber. The moisture content of this lumber is less than 15 percent. Special high-strength, waterproof glues are used under pressure to permanently seal individual pieces of lumber into one solid piece. These single pieces may be bent or shaped to the desired form with special equipment in a fabrication shop, Figure 4-28.

Fig. 4-28 Fabrication of glued-laminated structural wood members (Weyerhaeuser Company)

SHIPPING PRODUCTS TO THE JOBSITE

In steel and precast concrete and heavy timber construction, the structural products are shipped to the jobsite after being manufactured or fabricated in a shop. Most companies have a shipping deck or yard. From this deck, products that are ready for delivery are loaded onto trucks or trains by cranes or overhead hoists. Due to the size of the products, loading equipment, and vehicles used for shipping, this loading area must be very large.

Shipping is an important phase in the heavy construction industry and sometimes involves the drafter. If a large structural member is to be shipped to the jobsite by truck, the state road department or highway patrol will sometimes request sketches. These sketches assist them in recommending the best possible route for the truck. Maximum clearances under bridges or tunnels as well as the truck's minimum turning radius must be considered before shipment. In such cases, the drafter will be required to prepare a basic drawing of the truck showing length, width, and height dimensions. These dimensions must take into consideration the product that is being shipped, Figure 4-29.

SUMMARY

... Structural steel is produced in standard shapes including: W shapes, S shapes, M shapes, HP shapes, angles, C shapes, MC shapes, pipe, square tubing, rectangular tubing, circular tubing, and structural tees cut from W, S, and M shapes.

... The most common structural steel fabrication processes are: handling, cutting, punching, drilling, straightening, bending, bolting, riveting, welding, and finishing.

... Handling of structural steel products is accomplished by forklifts, cranes, overhead hoists, or straddle carriers.

... Cutting of structural steel products is accomplished by shearing, hot sawing, or with a cutting torch.

... Holes are placed in structural steel products by punching (for thin members) or drilling (for thick members).

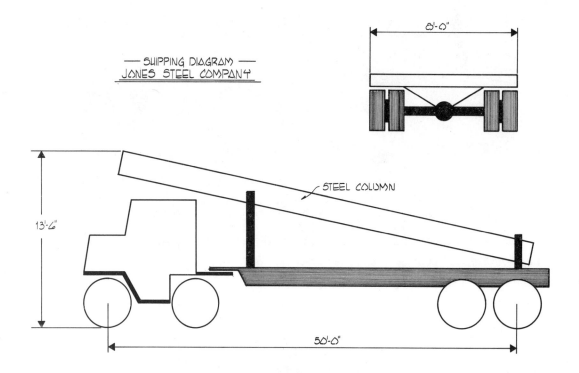

Fig. 4-29 Drafter's shipping diagram

... Structural steel members may be straightened or bent on a machine known as a bend press or gag press.
... There are three basic methods for fastening structural steel members: bolting, riveting, and welding. Bolting and welding are the most frequently used.
... The final phase in fabricating steel products is finishing which is accomplished by: sawing, milling, filing, high-pressure blasting, and various other means.
... Precast concrete products are manufactured in two categories: prestressed concrete and reinforced concrete.
... Prestressed concrete products are manufactured in steel beds and contain high-stressed steel strands that place the concrete member in compression.
... Reinforced concrete products are poured in steel or wooden forms and are reinforced with steel reinforcing bars, but are not prestressed.
... Common prestressed and reinforced concrete products include: columns, beams, double tees, single tees, flat slabs, joists, bridge girders, and stadium seats.
... Poured-in-place concrete members are reinforced concrete products that are poured, usually in temporary wooden forms, at the jobsite.
... Structural wood members are either solid timber or built-up glu-lam members. Common glu-lam products are beams, columns, and arches.
... Drafters are sometimes called upon to make shipping diagrams so that the highway patrol and state road department can map out special routes when large structural products are to be shipped by truck.

REVIEW QUESTIONS

1. Sketch the basic configuration of the following structural steel shapes:
   a. W shape
   b. S shape
   c. M shape
   d. HP shape
   e. C shape
   f. MC shape
   g. pipe
   h. square tubing
2. List the ten most common structural steel fabrication processes.
3. List three methods of cutting structural steel products.
4. List the three basic methods for fastening structural steel members.
5. List four different methods that might be used during the finishing process.
6. List the two categories of precast concrete products.
7. Sketch the basic configuration of the following precast concrete products:
   a. Double tee
   b. Single tee
   c. Square column
   d. Rectangular beam
   e. Flat slab
   f. Cored flat slab
   g. Keystone joist
   h. Bridge girder
8. Explain the difference between precast reinforced concrete fabrication and poured-in-place reinforced concrete fabrication.
9. Explain how drafters sometimes become involved in the shipping process.
10. Handling of heavy structural products is accomplished by:
    a. forklifts.
    b. cranes.
    c. overhead hoists.
    d. straddle carriers.
    e. all of the above.

11. Holes are placed in thin structural steel members by:
    a. drilling.
    b. shearing.
    c. punching.
    d. torching.
12. The machine used for bending or straightening structural steel members is:
    a. a gag press.
    b. a bend press.
    c. an offset press.
    d. a vise press.
    e. a and b.
    f. c and d.

REINFORCEMENT ACTIVITIES

1. Continue to develop your structural drafting lettering style and speed by completing Lettering Activity #4. Prepare a C-size sheet of paper with borders 1/2" in from each side and no title block. On this sheet, copy the entire lettering activity shown in Figure 4-30 in your best structural drafting lettering. Characters should be 1/8" to 3/16" tall.

2. Continue to develop your lettering, linework, and scale use skills by drawing the precast concrete section shown in Figure 4-31. The drawing should be placed on a C-size sheet of paper in vertical format with no title block or border at a scale of 1 1/2" = 1'-0".

3. Continue to develop your lettering, linework, and scale use skills by drawing the structural steel shop drawing shown in Figure 4-32. The drawing should be placed on a C-size sheet of paper in horizontal format with no title block or border at a scale of 1 1/2" = 1'-0".

4. Begin to develop skills in the use of the AISC *Manual of Steel Construction* by completing the following activity. Prepare a B-size sheet of paper with a title block and borders as shown in Figure 1-21. At a scale of 1 1/2" = 1'-0", construct and dimension a detail similar to the example in Figure 4-33 for each of the following steel shapes:

   W 36 x 300
   W 30 x 132
   W 24 x 62
   W 14 x 117
   W 8 x 36

LETTERING ACTIVITY #4

COPY THE FOLLOWING GENERAL NOTES & LEGEND IN YOUR BEST STRUCTURAL DRAFTING LETTERING

GENERAL NOTES

1. GENERAL CONTRACTOR SHALL FIELD-CHECK AND VERIFY ALL DIMENSIONS AND CONDITIONS AT JOBSITE.
2. ERECTION INCLUDES PLACING MEMBERS AND MAKING MEMBER CONNECTIONS ONLY.
3. ERECTION BY OTHERS, PRODUCTS F.O.B. TRUCKS JOBSITE.
4. NO GROUTING, POINTING, OR FIELD-POURED CONCRETE BY JONES PRECAST CONCRETE COMPANY.
5. RELEASE STRENGTH: 3500 PSI (UNLESS OTHERWISE NOTED)
6. CEILING FINISH: SUSPENDED
7. BLOCKOUTS SMALLER THAN 10" x 10" TO BE CUT IN THE FIELD BY PROPER TRADES (EXCEPT AS NOTED ON THE SHOP DRAWINGS).
8. WALL PANEL FINISHES: INTERIOR PANELS — EXPOSED
   EXTERIOR PANELS — CRUSHED GRAVEL
9. THE FOLLOWING ITEMS ARE TO BE SHIPPED LOOSE TO THE JOBSITE:
   24 WAA, 15 WAB, 7 WAC

LEGEND

ba - BLOCKOUT - a - 24" x 2" x 5'-0 3/4"
301 - NUMBER 3 BAR BY 16'-6" long (STRAIGHT)
302 - NUMBER 3 BAR BY 5'-7" long (BENT) SEE DETAIL ON SHEET M-1
WAA - WELD ANGLE - A - SEE DETAIL ON SHEET M-1
WAB - WELD ANGLE - B - SEE DETAIL ON SHEET M-1
WAC - WELD ANGLE - C - SEE DETAIL ON SHEET M-1

DIRECTIONS FOR THE REMAINDER OF THE ACTIVITY

USE THE SPACE REMAINING ON THE SHEET TO IMPROVE YOUR LETTERING SPEED. TAKE THE ENTRIES IN THE GENERAL NOTES ABOVE AND LETTER THEM AT A NORMAL PACE MAKING NOTE OF HOW LONG IT TAKES TO LETTER EACH ENTRY. THEN REPEAT EACH ENTRY MAKING A SPECIAL EFFORT TO DO IT FASTER. MAKE NOTE OF THE TIME AND CALCULATE HOW MUCH FASTER YOU COMPLETED THE LETTERING ON THE SECOND TRY. SPEED IS IMPORTANT, BUT REMEMBER, YOU CANNOT SACRIFICE QUALITY TO ACHIEVE IT. WORK ON YOUR LETTERING UNTIL IT IS NEAT AND FAST!

Fig. 4-30 Lettering activity for Reinforcement Activity #1

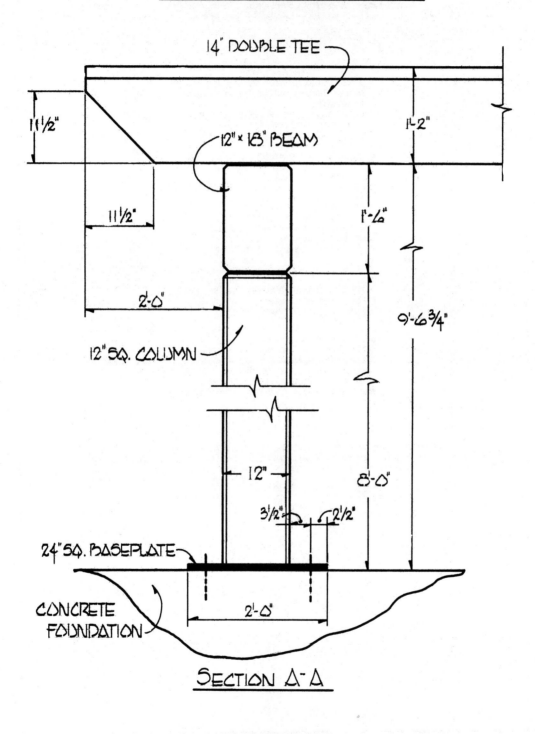

Fig. 4-31 Example for Reinforcement Activity #2

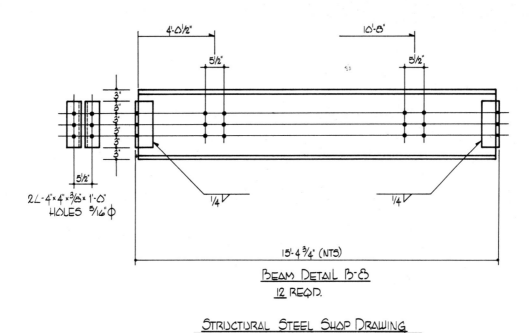

Fig. 4-32 Example for Reinforcement Activity #3

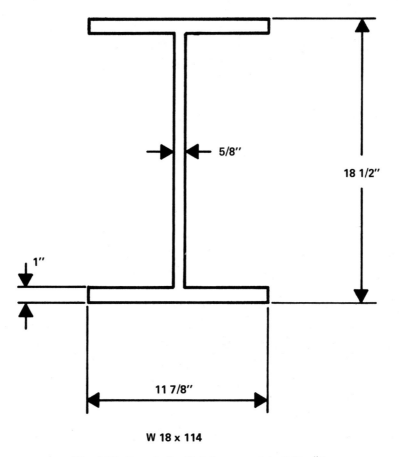

W 18 x 114

Fig. 4-33 Sample for Reinforcement Activity #4

Unit 5
Structural Connectors

OBJECTIVES

Upon completion of this unit, the student will be able to
- explain the applications of bolted, welded, riveted, split ring and shear plate connections in heavy construction.
- interpret common welding symbols.

CONNECTING STRUCTURAL MEMBERS

Heavy construction with steel and precast concrete members requires numerous field connections during the erection process. The reason for this is because the structural members are manufactured in a plant and shipped to the jobsite in pieces. Columns, beams, girders, joists, floor/roof members, and wall panels are all prepared for field connections during the fabrication process.

Structural members in heavy construction are connected by any one of three basic methods: bolting, welding, and riveting. Most large jobs will involve a combination of at least two of these methods. Designing connections is the job of an engineer or designer. However, drafters must also be familiar with structural connections in order to convert engineering calculations and sketches into finished connection details, Figures 5-1 and 5-2.

BOLTED CONNECTIONS

There are two categories of bolts used for field connections in heavy construction: common bolts and high-strength bolts. In addition, threaded rods of various diameters and lengths are used for bolting large, precast concrete members together, Figure 5-3.

Common bolts, identified by their square heads, are the least expensive type of bolt available for structural connections. However, their applications are limited. This is due to their low carbon-steel content and their inability to be tightened firmly enough for use in high-stress situations.

Most bolted connections in heavy construction require high-strength bolts. *High-strength bolts,* classified as A325 and A490, are made of a special steel. This special steel gives them a much greater tightening capacity and allows them to be used in places that would otherwise require hot-driven rivets. This is important because bolted connections have several advantages over riveted connections.

Due to the greater thicknesses of precast concrete members, threaded rods are sometimes used for making bolted connections. A *threaded rod* is a steel rod of specified diameter with one end threaded to accept a bolt, Figure 5-3. Threaded

Unit 5 Structural Connectors 61

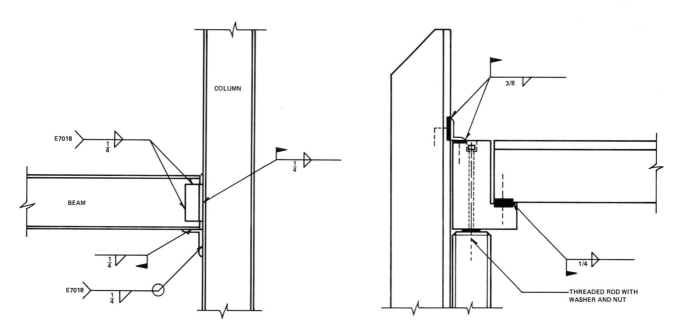

Fig. 5-1 Structural steel connection detail

Fig. 5-2 Structural concrete connection detail

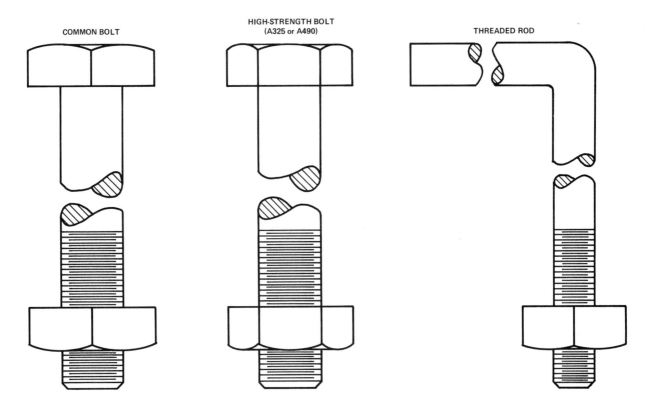

Fig. 5-3 Bolted connectors

rods are used most often in precast concrete beams to column connections. Figure 5-4 explains how to interpret thread notes commonly used with bolts and threaded rods in structural drafting.

SPLIT RING AND SHEAR PLATE CONNECTIONS

The applications of heavy timber and glued-laminated wood members are increased through the use of special metal connectors. The most common wooden fasteners are split ring connectors and shear plates. *Split ring connectors* are placed in grooves that have been cut in two members so they align when placed together for fastening. The joint between the two wood members is fastened with a bolt passed through a hole in the center of the split ring, Figure 5-5. In situations where a wooden joint may have to be disassembled and reassembled, shear plates are especially useful. *Shear plates* are commonly used in fabricating and erecting glued-laminated members. As in split ring connectors, bolts are the fastening devices used with shear plates, Figure 5-5.

WELDED CONNECTIONS

Arc welding, the fusion of metal by an electric arc, is the most common type of connection in heavy construction. Specifications for electrodes and standards of practice adopted by the American Society for Testing Materials (ASTM) and the American Welding Society (AWS) have eliminated problems that were associated with welding in the past and upgraded it to the most important type of structural connection process. Welding electrodes commonly used in heavy construction connections are of the E60 and E70 series.

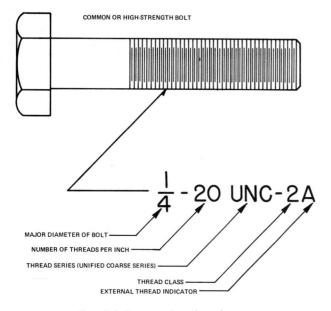

Fig. 5-4 Interpreting thread notes

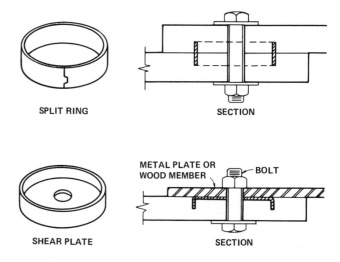

Fig. 5-5 Metal fasteners for wood connections

WELD TYPES AND SYMBOLS

Welds are classified according to the type of joint on which they are used. There are four common types of welds used in heavy construction connections. These are the back weld, fillet weld, plug or slot weld, and groove weld, Figure 5-6. Groove welds are subdivided into several subcategories. These subcategories are based on the shape of the groove to be welded: square groove, V groove, bevel groove, U groove, and flare bevel groove.

Welds are indicated on drawings by weld symbols. A completed weld symbol consists of the following.

- A horizontal reference line with a connected sloping line ending in an arrowhead
- A basic symbol indicating the type and/or size of weld
- A supplementary symbol further explaining the desired welding specification, Figure 5-7

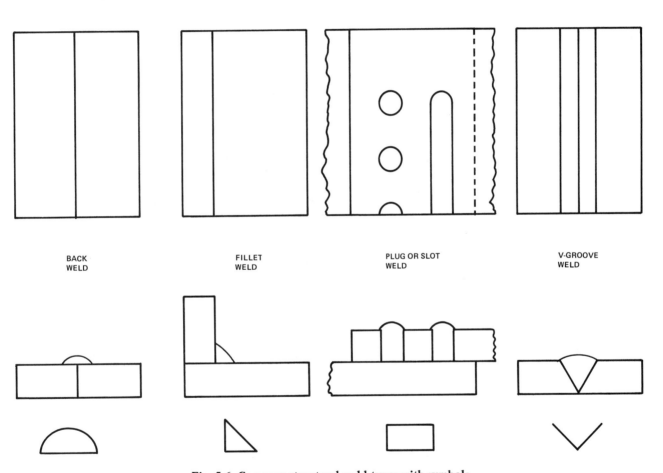

Fig. 5-6 Common structural weld types with symbols

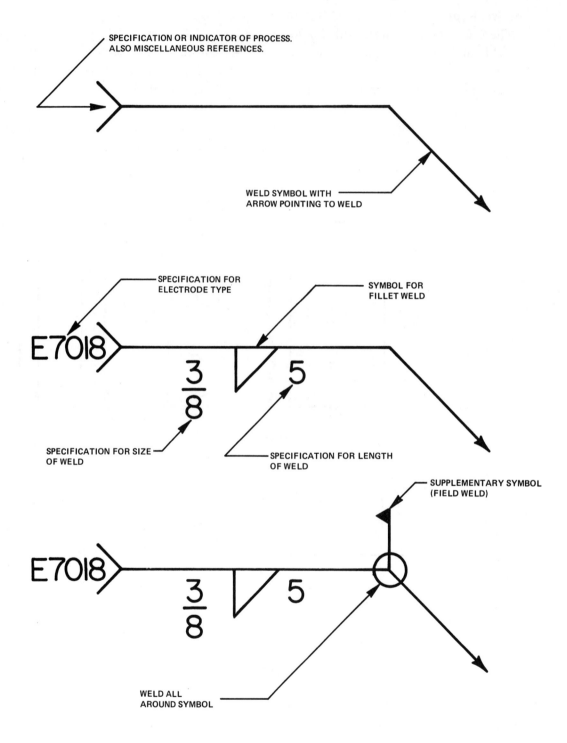

Fig. 5-7 Explanation of typical weld symbol

The basic symbols used for the various types of welds listed above are contained in Figure 5-8. Supplementary symbols that are commonly used by structural drafters, along with the basic symbols, are shown in Figure 5-9.

The structural drafter must be familiar not only with welding symbols, but certain rules governing their use. The arrow on the weld symbol shown in Figure 5-7 may point either to the right or to the left. However, this arrow must always form an angle with the horizontal reference line. The V-shaped tail of this same welding symbol is included when specifications concerning the weld are included at the tail of the symbol. If no specifications are included here, the V may be left off. When using a basic weld symbol that has a perpendicular leg, such as the fillet weld and certain groove welds (Figure 5-8), the leg is always drawn on the left-hand side as facing the symbol. Figure 5-10 illustrates several symbols, completely annotated to assist the student in learning to interpret welding symbols. Figure 5-11 shows examples of how welding symbols are used on structural drawings.

RIVETED CONNECTIONS

Riveting is no longer considered a major method of making structural connections in heavy construction. However, until around 1950, riveting was the only major connecting method. Because of its extensive applications in the past, the structural drafting student should be familiar with this connection process.

In heavy structural connections, hot rivets are driven through holes prepared in two structural steel members. The riveting process involves placing a heated rivet, with a button head on one end, into a prepared hole and hammering on the other end with a riveter or pneumatic hammer. This hammering secures the rivet firmly in the hole and fastens the structural members together.

Rivets are strong, long-lasting connectors. However, with the advent of high-strength bolts and vast improvements in welding materials and processes, riveted connections are no longer widely used.

BASIC WELD SYMBOLS FOR DRAFTING									
BACK	FILLET	PLUG OR SLOT	GROOVE						
			J	V	U	SQUARE	BEVEL	FLARE V	FLARE BEVEL
⌒	△	▢	⊢	∨	⊻	‖	⌶	⋎	⌒⌒

Fig. 5-8 Basic weld symbols for structural drafting

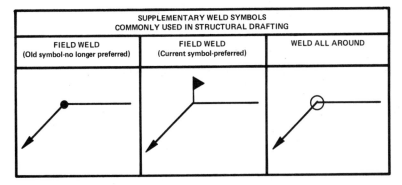

Fig. 5-9 Supplementary weld symbols

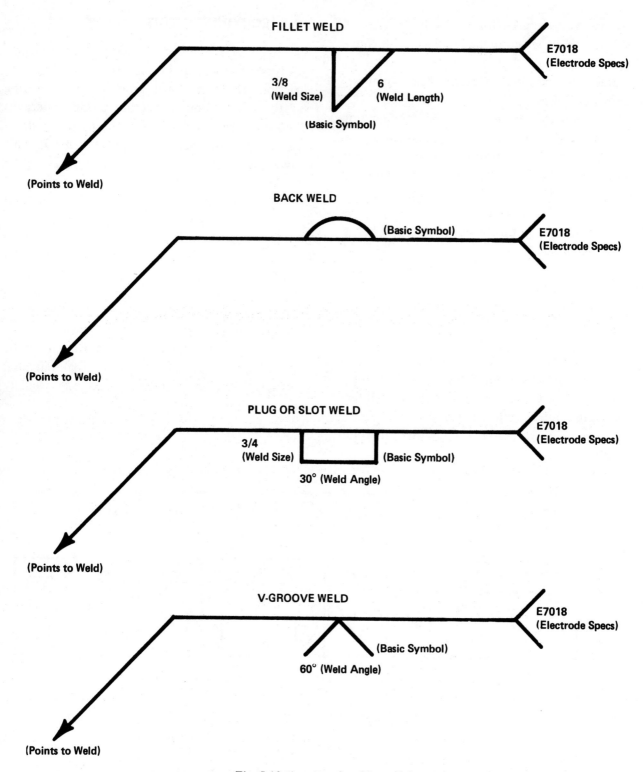

Fig. 5-10 Annotated weld symbols

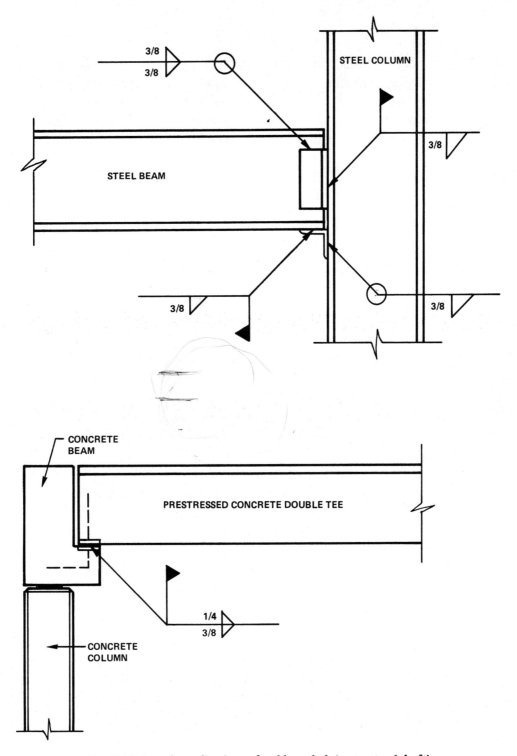

Fig. 5-11 Sample applications of weld symbols in structural drafting

SUMMARY

... Structural members in heavy construction are connected by bolting, welding, or riveting.
... Structural connections are designed by engineers or designers and detailed by drafters.
... There are two categories of bolts used for field connections in heavy construction: common bolts and high-strength bolts.
... Common bolts have limited applications and cannot be used in high-stress situations.
... High-strength bolts used in heavy construction connections are classified as A325 and A490.
... Structural wood connections are commonly made with split ring connectors and shear plates.
... Threaded rods are commonly used for precast concrete bolted connections.
... Welding is the most common type of structural connection used in heavy construction.
... The four most common types of welds in heavy construction connections are: back, fillet, plug or slot, and groove welds.
... Groove welds are subdivided into several categories: square groove, V groove, bevel groove, U groove, and flare bevel groove.
... Riveting is no longer considered a major structural connection method.

REVIEW QUESTIONS

1. List the three major types of structural connections used in heavy construction.
2. List the two categories of bolts used for field connections in heavy construction.
3. What type of bolts are most often used in heavy construction bolted connections?
4. List the two classifications of high-strength bolts used in structural connections.
5. How are thick, precast concrete columns and beams bolted together?
6. What is the most common type of structural connection used in heavy construction?
7. List the four most common types of welds used in heavy construction connections.
8. List five types of groove welds.
9. Sketch an example of a structural wood connection using a split ring connector and one using a shear plate.
10. Which of the following is no longer considered a major structural connection process?
    a. Welding
    b. Riveting
    c. Nailing
    d. Bolting
11. Which of the following groups is responsible for the design of structural connections?
    a. Junior drafters
    b. Contractors
    c. Architects
    d. Engineers

REINFORCEMENT ACTIVITIES

1. Complete your formal instruction in developing structural drafting lettering by doing Lettering Activity #5. This is the last in a series of five structured lettering activities designed to help develop an acceptable lettering style and speed. In completing this activity, make an extra effort to produce clear, neat lettering at an acceptable speed. Then as you apply what you have learned in completing the drafting activities that accompany the remaining units in this textbook, continue to make an effort to improve both your neatness, style, and speed in structural drafting lettering.
    a. Lettering Activity #5 Specifications
       On a B-size sheet of paper with borders 1/2" in from each side and no title block, copy the summary at the end of this unit in your best structural drafting lettering.

2. Begin to develop skills in the understanding and use of weld symbols by drawing weld symbols for each set of specifications below. Use a C-size sheet of paper with complete title block and borders.
    a. Back weld with E7018 electrodes to be field welded.
    b. Fillet weld with E7018 electrodes 1/4" x 6" long.
    c. Slot weld 3/4" with a 30° weld angle.
    d. V-groove weld at 60° with E7018 electrodes.
    e. Fillet weld to be field welded all around by 3/8".

3. Learn to apply weld symbols on structural drawings by completing the following activity. Prepare a C-size sheet of paper with a title block as shown in Figure 1-21 and 1/2" borders. At a scale of 1 1/2" = 1'-0", redraw the connection details in Figure 5-12. For each numbered situation requiring a weld, substitute the proper weld symbol. Also, size all weld plates and angles based on their proportions in Figure 5-12 and the needs of the situation.

70　Section 1　Overview of Structural Drafting

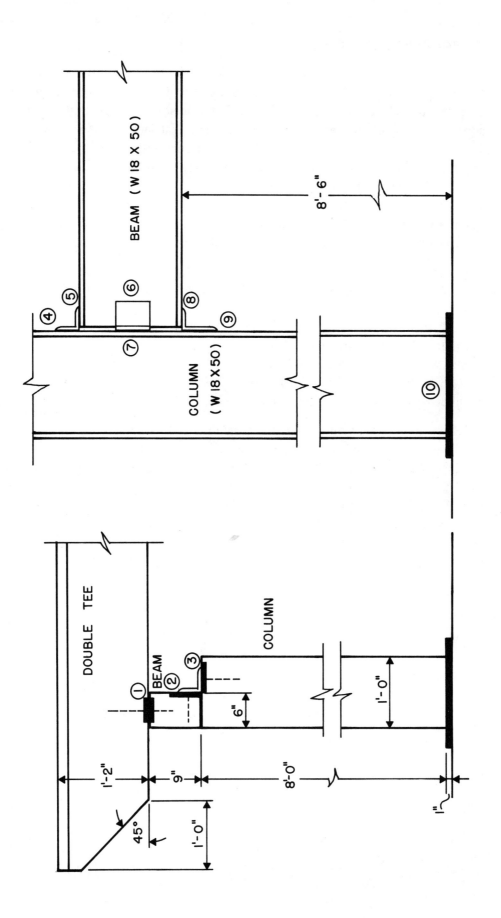

Fig. 5-12 Connection details for Reinforcement Activity #3

Section 2
Structural Steel Drafting

Unit 6
Structural Steel Framing Plans

OBJECTIVES

Upon completion of this unit, the student will be able to

- distinguish between engineering drawings and shop drawings.

- describe, designate, and illustrate the various structural steel products used in framing plans.

- properly use the American Institute of Steel Construction's Manual of Steel Construction for determining structural steel product designations and dimensions.

- properly construct structural steel framing plans according to engineering specifications.

STRUCTURAL STEEL DRAWINGS

In Unit 1, it was learned that structural drafters prepare two basic types of drawings: engineering drawings and shop drawings. Engineering drawings are used to provide general information for sales, marketing, engineering, and erection purposes. Shop drawings are used to provide more detailed information for fabrication purposes.

In structural steel drafting, engineering drawings are sometimes referred to as erection drawings depending on how they are to be used. Engineering drawings are prepared by drafters from sketches provided by structural engineers. The drawings include framing plans and sections which are symbolic representations of all steel members used in a structure, Figure 6-1.

This unit deals with the preparation of structural steel framing plans. In order to prepare structural steel framing plans, the drafter must first become familiar with the structural steel products used in framing plans.

STRUCTURAL STEEL FRAMING PRODUCTS

Rolled steel products are classified as being either a plate, a bar, or a shape. *Plates* are flat pieces of steel of various thicknesses. They are used as a framing product only for making changes to other framing members. Common uses are as stiffeners, gusset plates, and in making built-up girders. Plates are called out on drawings according to their thickness (in inches), their width (in inches), and their length (in feet and inches), Figure 6-2.

Bars are the smallest structural steel products and may have round, square, rectangular, or hexagonal cross-sectional configurations, Figure 6-3. Bars are not a structural steel framing product, but they can be used in modifying other steel framing products.

Shapes consist of W shapes, M shapes, S shapes, angles, channels, structural tees, structural tubing, and pipe and are the most important structural steel framing products.

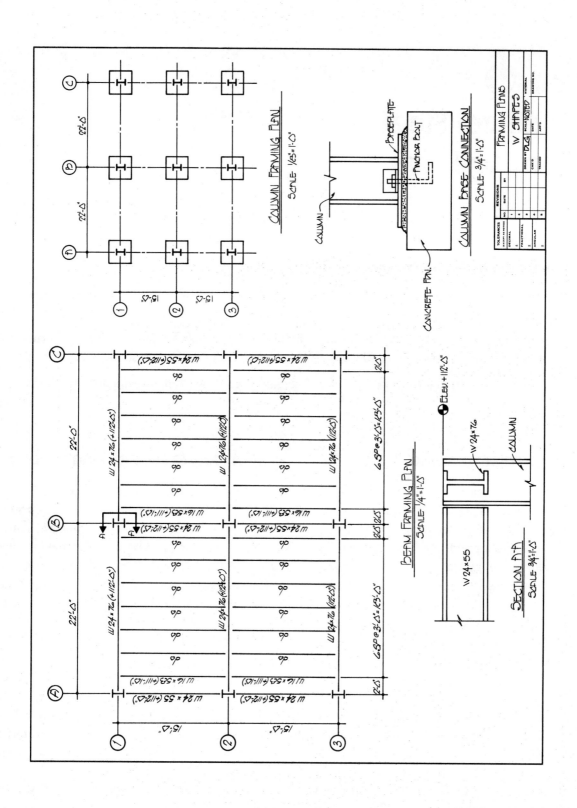

Fig. 6-1 Framing plans

Unit 6 Structural Steel Framing Plans 75

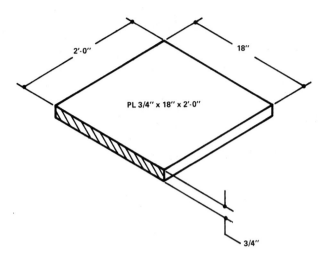

Fig. 6-2 Steel plate designations

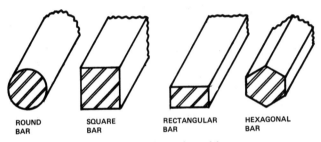

Fig. 6-3 Structural steel bars

W, S, and M Shapes

W, S, and M shapes are the new designations, set forth by the American Institute of Steel Construction (AISC), for shapes that previously were designated as WF, I, and M or Jr shapes. It is important to learn to use the new and proper designations when calling out structural steel shapes on drawings. The proper callout designations, dimensions for detailing, and properties for designing structural steel shapes are provided in the previously mentioned manual by the AISC. For the student's convenience, a portion of this manual has been reproduced in the Appendix of this book. Figure 6-4 contains a list of examples showing the old and the new designations for selected structural steel products.

S shapes in cross section have an I configuration with narrow flanges that slope in a manner similar to channel flanges, Figure 6-5. W shapes also have an I configuration in cross section, but their flanges are wider than those on S or M shapes and have a constant thickness, Figure 6-5.

New Designation	Type of Shape	Old Designation
W 24 x 76	W shape	24 WF 76
W 14 x 26		14 B 26
S 24 x 100	S shape	24 I 100
M 8 x 18.5	M shape	8 M 18.5
M 10 x 9		10 JR 9.0
M 8 x 34.3		8 x 8 M 34.3
C 12 x 20.7	American Standard Channel	12 [20.7
MC 12 x 45	Miscellaneous Channel	12 x 4 [45.0
MC 12 x 10.6		12 JR [10.6
HP 14 x 73	HP shape	14 BP 73
L 6 x 6 x 3/4	Equal Leg Angle	∠6 x 6 x 3/4
L 6 x 4 x 5/8	Unequal Leg Angle	∠6 x 4 x 5/8
WT 12 x 38	Structural Tee cut from W shape	ST 12 WF 38
WT 7 x 13		ST 7 B 13
ST 12 x 50	Structural Tee cut from S shape	ST 12 I 50
MT 4 x 9.25	Structural Tee cut from M shape	ST M 9.25
MT 5 x 4.5		ST 5 JR 4.5
MT 4 x 17.15		ST 4 M 17.15
PL 1/2 x 18	Plate	PL 18 x 1/2
Bar 1 ▢	Square Bar	Bar 1 ▢
Bar 1 1/4 ⌀	Round Bar	Bar 1 1/4 ⌀
Bar 2 1/2 x 1/2	Flat Bar	Bar 2 1/2 x 1/2
Pipe 4 Std.	Pipe	Pipe 4 Std.
Pipe 4 X-Strong		Pipe 4 X-Strong
Pipe 4 XX-Strong		Pipe 4 XX-Strong
TS 4 x 4 x .375	Structural Tubing: Square	Tube 4 x 4 x .375
TS 5 x 3 x .375	Structural Tubing: Rectangular	Tube 5 x 3 x .375
TS 3 OD x .250	Structural Tubing: Circular	Tube 3 OD x .250

Fig. 6-4 Structural steel designations

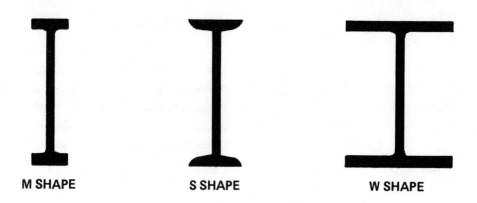

M SHAPE   **S SHAPE**   **W SHAPE**

NOTE
DIMENSIONS FOR DETAILING M, S, AND W SHAPES ARE PROVIDED IN SECTION I OF THE AISC *MANUAL OF STEEL CONSTRUCTION*. EXCERPTS ARE PROVIDED IN THE APPENDIX OF THIS BOOK.

Fig. 6-5 Structural steel shapes

M shapes are miscellaneous shapes and include all rolled shapes with an I cross-sectional configuration that cannot be classified as W or S shapes, Figure 6-5.

Angles

Angles are properly designated as L shapes and are of two types: equal angles and unequal angles, Figure 6-6. In both types, the legs have the same thickness, even though the legs of unequal angles differ in length. Examples of angles properly called out on drawings are: L 6 x 5 x 1/2 and L 4 x 4 x 1/4. The first example denotes an unequal angle in which one of the legs is 6" long, the other is 5" long, and both legs are 1/2" thick. The second example denotes an equal angle in which both legs are 4" long by 1/4" thick. The length of any given angle can also be designated by adding the required length at the end of the callout in feet and inches. If it was desired to have the angles called out as 6" long, as in the previous example, the designations would read L 6 x 5 x 1/2 x 0'-6" and L 4 x 4 x 1/4 x 0'-6".

Angle designations on older drawings read exactly as the previous examples with the exception of the uppercase L. The old symbol for angle was ∠. Because of its similarity to the letter L, drawings lettered freehand will show very little difference.

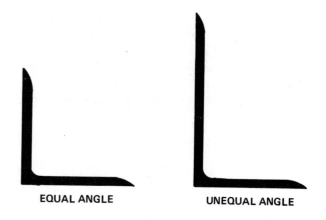

EQUAL ANGLE   **UNEQUAL ANGLE**

NOTE
DIMENSIONS FOR DETAILING L SHAPES ARE PROVIDED IN SECTION I OF THE AISC *MANUAL OF STEEL CONSTRUCTION*. EXCERPTS ARE PROVIDED IN THE APPENDIX OF THIS BOOK.

Fig. 6-6 L shapes

Channels

Channels are properly designated as C shapes. They have a squared C configuration with sloping flanges and a web of constant thickness, Figure 6-7. Channels are of two types: American Standard and Miscellaneous. Examples of properly called out channels are C 10 x 25, C 12 x 30, and C 15 x 33.9. All three of these examples denote American Standard channels. Some examples of Miscellaneous channel designations are MC 10 x 8.4, MC 12 x 10.6, and MC 18 x 42.7.

Unit 6 Structural Steel Framing Plans 77

Structural Tees

Structural tees are products cut from W, S, and M shapes by splitting the webs, Figure 6-8. Structural-tee designations are a modification of the structural shape designation from which they were cut, with the addition of an uppercase T. Sample structural-tee callouts are WT 6 x 95, MT 7 x 8.6, and ST 12 x 60. WT designations indicate that the structural tee was cut from a W shape, MT from an M shape, and ST from an S shape. For a more specific interpretation, the MT 7 x 8.6 above was cut from an M 14 x 17.2.

Structural Tubing

Structural tubing is manufactured in square, rectangular, and round cross-sectional configurations. It is often used as a structural column, Figure 6-9. Tubing designations on drawings provide

NOTE
DIMENSIONS FOR DETAILING C AND MC SHAPES ARE PROVIDED IN SECTION I OF THE AISC *MANUAL OF STEEL CONSTRUCTION*. EXCERPTS ARE PROVIDED IN THE APPENDIX OF THIS BOOK.

Fig. 6-7 C shapes

NOTE
DIMENSIONS FOR DETAILING MT, ST, AND WT SHAPES ARE PROVIDED IN SECTION I OF THE AISC *MANUAL OF STEEL CONSTRUCTION*. EXCERPTS ARE PROVIDED IN THE APPENDIX OF THIS BOOK.

Fig. 6-8 Structural tees

NOTE
DIMENSIONS FOR DETAILING STRUCTURAL TUBING SHAPES ARE PROVIDED IN SECTION I OF THE AISC *MANUAL OF STEEL CONSTRUCTION*. EXCERPTS ARE PROVIDED IN THE APPENDIX OF THIS BOOK.

Fig. 6-9 Structural tubing

the outside dimensions and the wall thickness dimensions. Sample structural tubing callouts are TS 5 x 5 x 0.375, TS 8 x 4 x 0.375, and TS 3 OD x 0.250. The first example denotes a square structural steel tube that is 5" square with walls 0.375 (3/8) of an inch thick. The second example denotes a rectangular structural steel tube that has sides 8" x 4" and walls that are 0.375 (3/8) of an inch thick. The final example denotes a round structural steel tube 3" in diameter with walls 0.250 (1/4) of an inch thick.

Structural Pipe

Structural pipe has a round, hollow cross-sectional configuration and is very effective for use as structural columns, Figure 6-10. Steel pipe is manufactured in three categories of strength: standard, X-strong, and XX-strong. The wall thickness determines the strength and the category. XX-strong pipe has thicker walls and is stronger than X-pipe, which has thicker walls and is stronger than standard pipe.

Examples of structural steel pipe designations are Pipe 3 Std, Pipe 3 X-strong, and Pipe 3 XX-strong. The first example denotes a 3" diameter pipe of the standard strength category. The second example denotes a 3" diameter pipe of the X-strong category. The final example denotes a 3" diameter pipe of the XX-strong category.

HEAVY-LOAD/LONG-SPAN FRAMING PRODUCTS

The structural steel shapes discussed in the preceding paragraphs are the most commonly used structural steel framing products. However, in certain heavy-load or long-span situations, standard rolled steel products do not meet the design requirements. When this is the case, a special built-up framing member can be designed that meets the requirements.

The most common type of built-up framing member is the built-up girder. A *built-up girder* is either a standard rolled shape that has been reinforced or a new shape made entirely of steel plates. In its publication *Structural Steel Detailing,* the AISC specifies five different types of built-up girders, Figure 6-11.

Figure 6-11A shows the first type of built-up girder in which an I configured rolled shape is reinforced by the addition of top and bottom plates permanently attached by welding. Figure 6-11B shows a similar built-up beam that substitutes a channel for the top plate. Figure 6-11C shows an example of a built-up girder that is used in composite construction with concrete. This type of girder would support a concrete floor resting on the shear connectors on top of it and be reinforced on the bottom with a steel plate. Figure 6-11D shows a deep, built-up girder made entirely of plates welded together. Figure 6-11E contains an example of a box girder which is composed of four welded plates.

STANDARD   X-STRONG   XX-STRONG

NOTE
DIMENSIONS FOR DETAILING STRUCTURAL PIPE ARE PROVIDED IN SECTION I OF THE AISC *MANUAL OF STEEL CONSTRUCTION.* EXCERPTS ARE PROVIDED IN THE APPENDIX OF THIS BOOK.

Fig. 6-10 Structural pipe

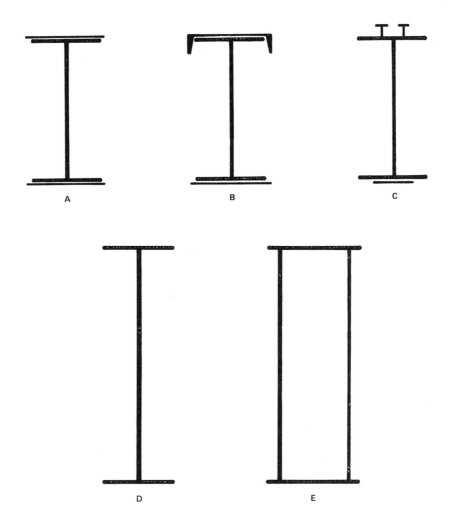

Fig. 6-11 Built-up girders

STRUCTURAL STEEL FRAMING PLANS

Structural steel framing plans are symbolic representations of columns, beams, built-up girders, and other framing members. They are used primarily for engineering and erection purposes. Structural steel framing plans are drawn by drafters from information provided by engineers through sketches. Figure 6-12 illustrates an engineer's sketch from which a structural steel drafter might prepare a framing plan. Figure 6-13 shows the framing plan that was prepared from this sketch.

A structural steel framing plan is a plan-view representation showing all columns, beams, girders, joists, bridging, etc., as they will appear when the framing for the structure being built is erected. Column centerlines are given number and letter designations and are completely dimensioned. Each structural member represented in the framing plan is given an identifying callout for easy reference. For example, beams may be labeled according to the designation given them in the AISC's *Manual of Steel Construction* or they may be given a mark number or both.

Drafting time in preparing structural steel framing plans is reduced through use of the symbol *do* which means ditto, or the same. When several members have the same designation or mark number and are located together on the framing plan, the *do* symbol may be used in place of the member designation. A note indicating the top of steel elevation is placed next to the member designation or on the drawing under the heading, *Notes*.

80 Section 2 Structural Steel Drafting

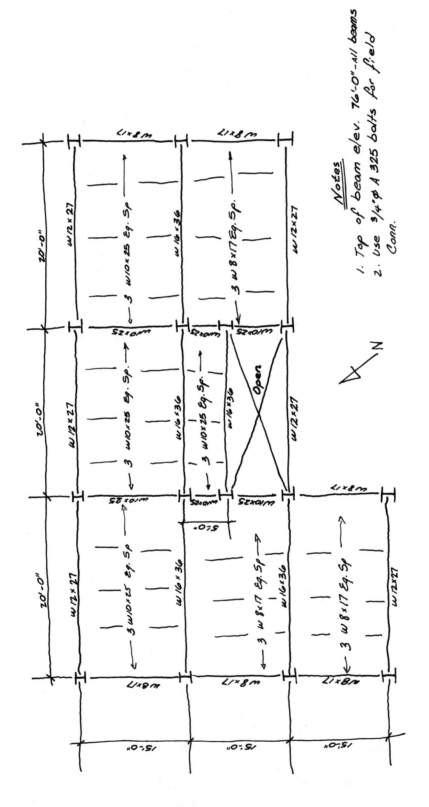

Fig. 6-12 Engineer's sketch

Unit 6 Structural Steel Framing Plans 81

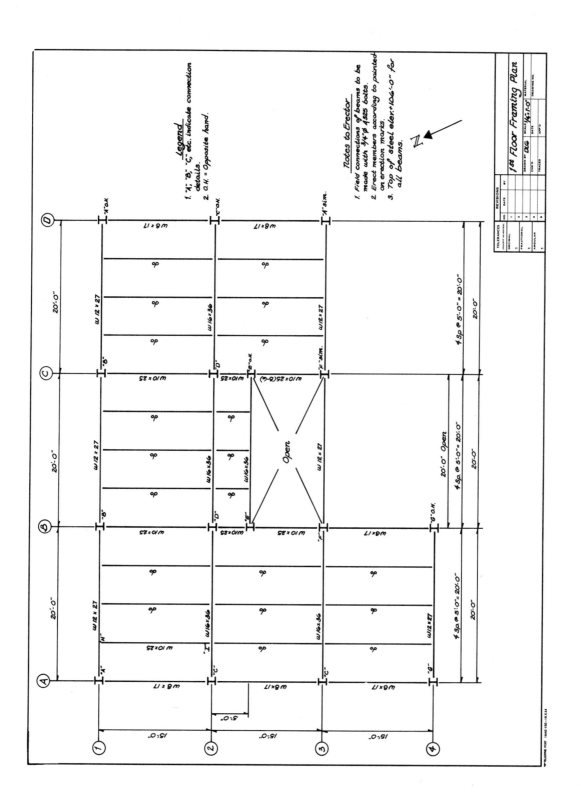

Fig. 6-13 Column-and-beam framing plan

Structural steel framing plans are of two types: column framing plans and beam framing plans. Figure 6-14 is an example of a column framing plan. An example of a beam framing plan is shown in Figure 6-13.

DRAWING STRUCTURAL STEEL FRAMING PLANS

Structural steel framing plans consist of column framing plans showing the vertical structural members (columns) and the foundations on which they bear and beam framing plans showing the horizontal structural members (beams, girders, joists). The procedures used by drafters in drawing column framing and beam framing plans differ. The column framing plan is usually prepared first.

Drawing Column Framing Plans

The column framing plan is a plan view of all columns used in a job and the foundations on which they bear. Accompanying the column framing plan is a column schedule, Figure 6-15.

The column schedule is drawn on the same sheet as the column framing plan if there is room. Column framing plans and schedules are prepared according to the following procedures:

1. Closely examine the engineer's sketches provided. Select an appropriate scale that will fit the framing plan comfortably on the sheet.
2. Lay out the centerline of column grid lines and draw in the plan view of each column, Figure 6-14.
3. Darken in column centerlines and label vertical centerlines with letter designations and horizontal centerlines with numbers, Figure 6-14.
4. Draw in the foundations that support the columns, Figure 6-14.
5. Complete the foundation plan by adding appropriate dimensions and a north arrow, Figure 6-14.
6. Add the column schedule as shown in Figure 6-15. Information required includes floor and roof elevations, top of foundation elevation, column designations, and splice points where applicable, Figure 6-15.

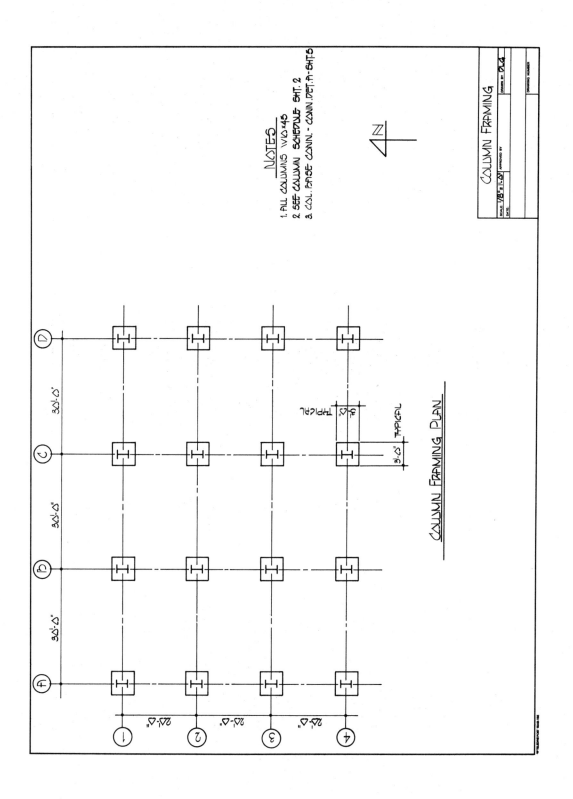

Fig. 6-14 Column framing plan

84 Section 2 Structural Steel Drafting

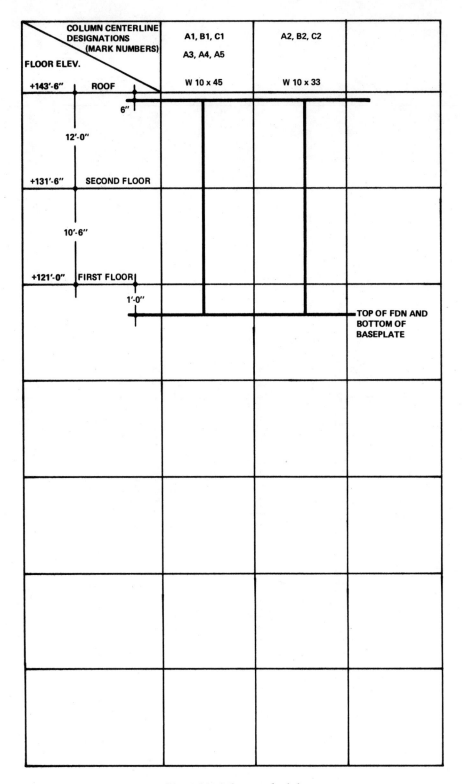

Fig. 6-15 Column schedule

Drawing Beam Framing Plans

Floor and roof framing plans are commonly referred to as *beam framing plans.* The beam framing plan repeats the plan view of the columns and centerline of column designations. In addition, every beam, girder, and/or joist used in the job is shown in the plan, Figure 6-13.

The beam framing plan is also coded to identify each structural connection that requires a connection detail to guide the erection crew in erecting the structural connection. Each individual connection situation is given a letter designation on the framing plan, Figure 6-13. To save drawing time, similar connection situations may share a common connection detail. The abbreviation *Sim* is used on the framing plan as a suffix to the letter designation to indicate two or more connections sharing the same detail. The abbreviation *Opp Hd,* meaning opposite hand, may be applied to connection situations that are exactly opposite so they may also share a common connection detail, Figure 6-13.

Beam framing plans are drawn according to the following procedures:

1. Closely examine the engineer's sketch(es) provided. Select an appropriate scale that will fit the framing plan comfortably on the sheet.
2. Lay out the column grid lines and draw in the columns. Label the centerlines of columns with number and letter designations as was done on the column framing plan, Figure 6-13.
3. Draw in all horizontal members (beams, girders, and joists). Label each member with the proper shape designation, Figure 6-13.
4. Label the top of steel elevation for each beam next to its shape designation or place a top of steel elevation indication on the drawing as a note, Figure 6-13.
5. Complete the beam framing plan by adding appropriate dimensions, notes, connection detail designations, and a north arrow, Figure 6-13.

SUMMARY

... Structural drafters prepare engineering drawings and shop drawings. Engineering drawings are sometimes called erection drawings.
... Framing plans are engineering drawings.
... Rolled steel products are classified as plates, bars, or shapes.
... Shapes are the most common structural steel framing products and include W shapes, S shapes, M shapes, angles, channels, structural tees, structural tubing, and pipe.
... W shape is the new designation for shapes that were designated WF in the past.
... S shape is the new designation for shapes that were designated I in the past.
... M shape is the new designation for shapes that were designated M or Jr in the past.
... Proper callout designations and dimensions for detailing steel shapes are provided in the *Manual of Steel Construction* published by the American Institute of Steel Construction (AISC).
... Angles are properly designated as L shapes; channels are properly designated as C or MC shapes; and structural tubing is properly designated as TS shapes.
... Structural tees are cut from W, M, and S shapes and are designated ST, MT, or WT depending on the shape from which they were cut.
... Structural pipe has three classifications depending on the wall thickness of the pipe: standard, X-strong, and XX-strong.
... The AISC, in its publication *Structural Steel Detailing,* specifies five types of built-up girders that are used in heavy-load or long-span situations.
... Structural steel framing plans are of two types: column framing plans and beam framing plans.
... Information for preparing structural steel framing plans is provided through engineering sketches.
... Drafters prepare structural steel framing plans from information contained in the engineering sketches and the AISC *Manual of Steel Construction.*

REVIEW QUESTIONS

1. What is another name for an engineering drawing?
2. What category of drawings are framing plans?
3. What are the three classifications of rolled steel products?
4. List the most common structural steel framing products.
5. What is the drafter's most reliable reference source for determining proper callout designations and dimensions for detailing?
6. List the old designations for the following:
   a. W shape
   b. M shape
   c. C shape
7. List the three designations for structural tees.
8. List the three classifications of structural pipe.
9. Sketch an example of the five types of built-up girders used in long-span or heavy-load situations.
10. Name the two types of structural steel framing plans.

REINFORCEMENT ACTIVITIES

All framing plans required in the following activities are to be drawn on C-size paper with a title block as in Figure 1-21 and 1/2" borders. All column schedules required are to be drawn on B-size paper with no title block and no border.

1. This activity is to be prepared in accordance with the information supplied in the engineer's sketch in Figure 6-16. At an appropriate scale, draw a complete column framing plan from the sketch provided. Refer to the AISC *Manual of Steel Construction* for answers to questions concerning dimensions of structural shapes.

2. Prepare a complete column schedule for the column framing plan in Reinforcement Activity #1.

3. Prepare a complete beam framing plan based on the information provided in the engineer's sketch in Figure 6-16. Refer to the AISC *Manual of Steel Construction* to insure that all beam designations are properly called out.

4. [This activity is to be p]repared in accordance with the information supplied in the engineer's sketch in Figure 6-17. At an appropriate scale, draw a complete column framing plan from the sketch provided. Refer to the AISC *Manual of Steel Construction* for answers to questions concerning dimensions of structural shapes.

5. Prepare a complete column schedule for the column framing plan in Reinforcement Activity #4.

6. Prepare a complete beam framing plan based on the information provided in the engineer's sketch in Figure 6-17. Refer to the AISC *Manual of Steel Construction* to insure that all beam designations are properly called out.

7. [This activity is to be p]repared in accordance with the information supplied in the engineer's sketch in Figure 6-18. At an appropriate scale, draw a complete column framing plan from the sketch provided. Refer to the AISC *Manual of Steel Construction* for answers to questions concerning dimensions of structural shapes.

8. Prepare a complete column schedule for the column framing plan in Reinforcement Activity #7.

9. Prepare a complete beam framing plan based on the information provided in the engineer's sketch in Figure 6-18. Refer to the AISC *Manual of Steel Construction* to insure that all beam designations are properly called out.

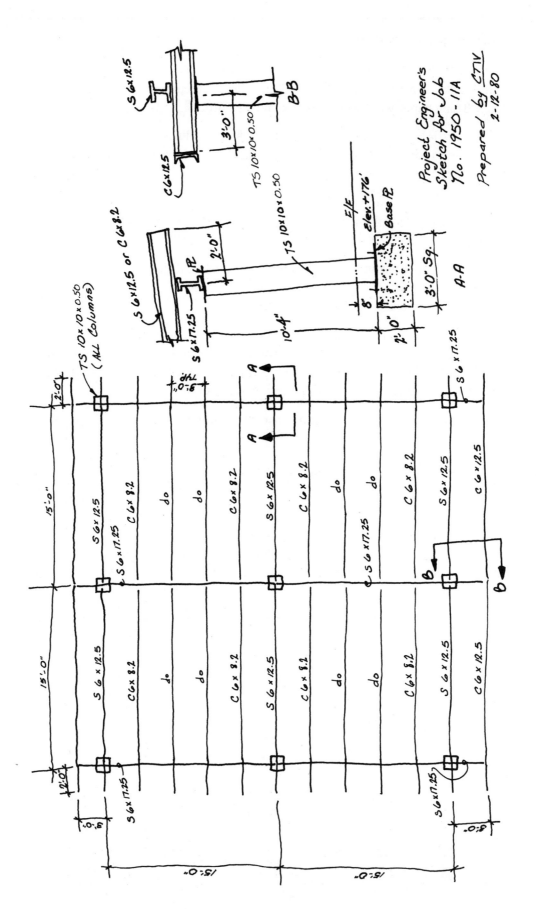

Fig. 6-16 Engineer's sketch

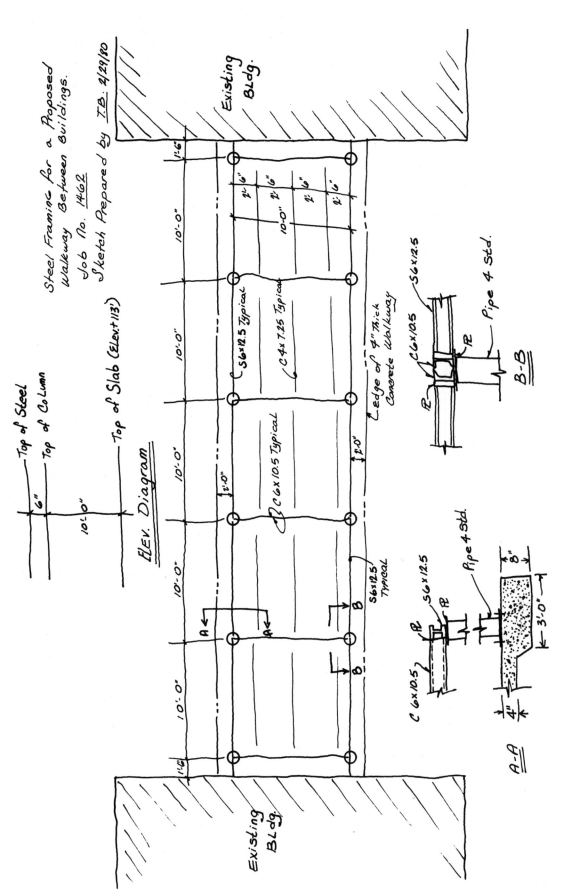

Fig. 6-17 Engineer's sketch

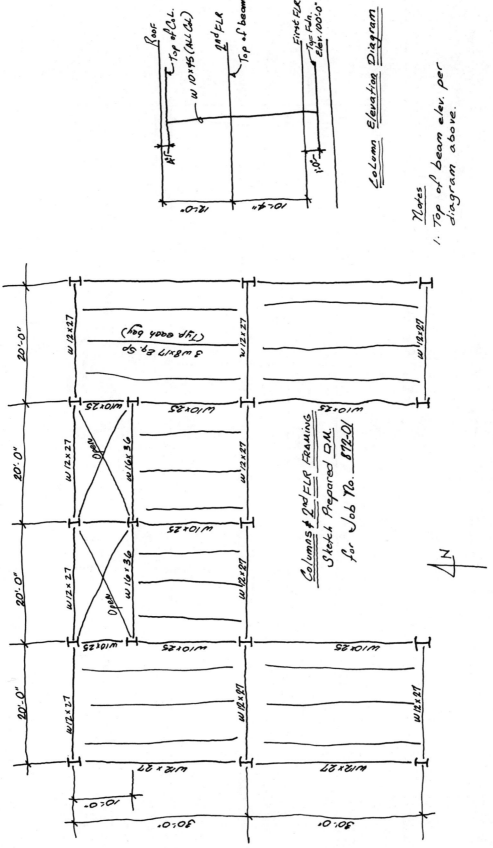

Fig. 6-18 Engineer's sketch

Unit 7
Structural Steel Sections

OBJECTIVES

Upon completion of this unit, the student will be able to
- define structural steel sections.
- prepare structural steel full, partial, and offset sections.

STRUCTURAL STEEL SECTIONS DEFINED

Sections are drawings provided to show the reader what materials are used in a job and how they fit together to form a completed structure. Sections in structural steel drafting may be prepared as single-line symbolic representations. They may also be prepared as scaled duplicates of the actual structural shapes being represented, Figures 7-1 and 7-2.

Sections are very important to the reader of a set of structural drawings. They clarify internal relationships in a structure that are not well defined on the framing plans. In addition to showing how a structure fits together, sections also

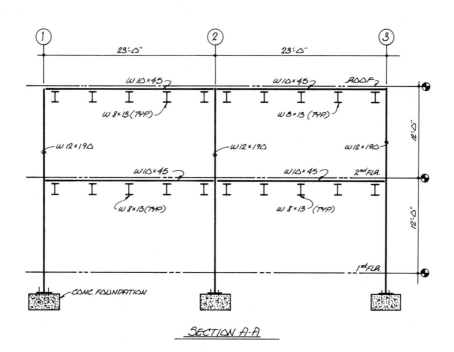

Fig. 7-1 Sample section

91

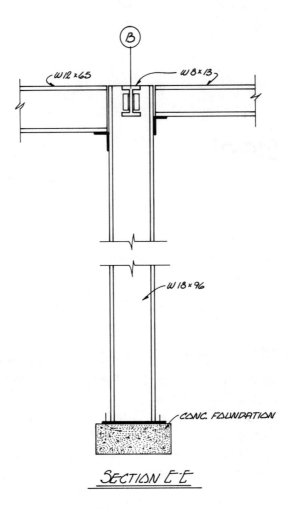

Fig. 7-2 Sample section

show height information. This height information includes such things as distances between floors, distances between floors and the roof, top of steel information, bottom of baseplate information, etc.

Sections are *cut* on framing plans, but drawn on separate section sheets. Figure 7-3 shows several ways in which sections are cut on framing plans. Figure 7-4 contains an explanation of several of the more commonly used section cutting symbols.

There are several different types of sections used in structural steel drafting. The most commonly used are full sections, partial sections, and offset sections.

FULL SECTIONS

Full sections cut through an entire building or structure. Full sections are of two basic types: longitudinal and cross sections. *Longitudinal sections* cut across the entire length of a building and may be represented as actual scaled drawings or as symbolic, single-line drawings. Figure 7-5 contains a scaled drawing and a symbolic representation of the same longitudinal section. *Cross sections* cut across the entire width of a building or structure and may also be represented as scaled drawings or symbolic representations. Figure 7-6 shows a scaled drawing and a symbolic representation of the same cross section.

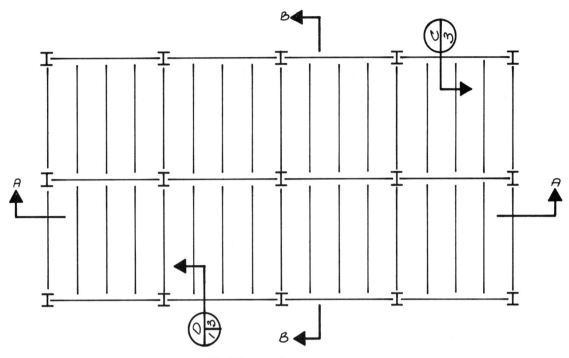

Fig. 7-3 Sample cut sections

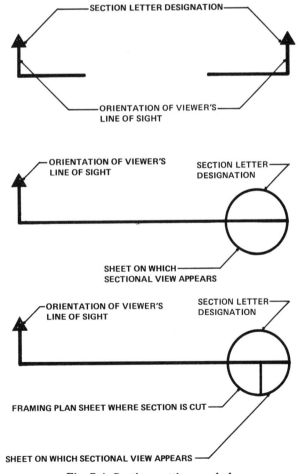

Fig. 7-4 Section cutting symbols

94 Section 2 Structural Steel Drafting

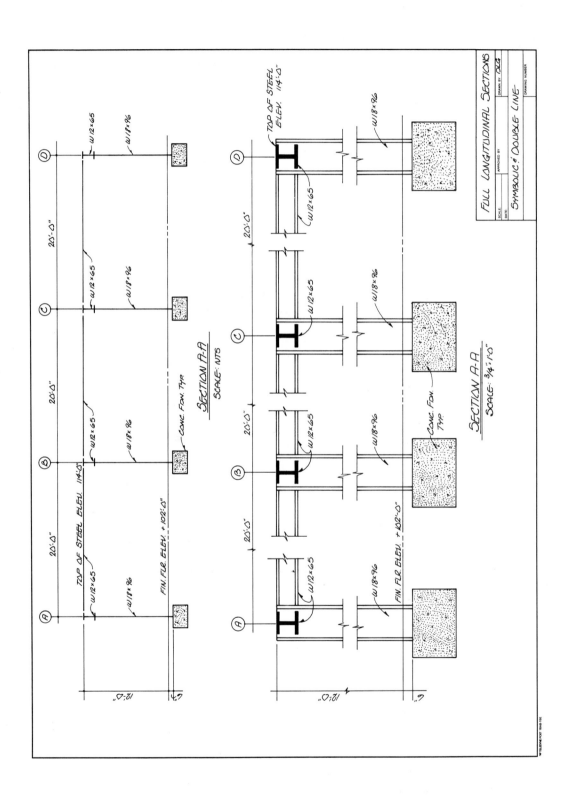

Fig. 7-5 Sample sections

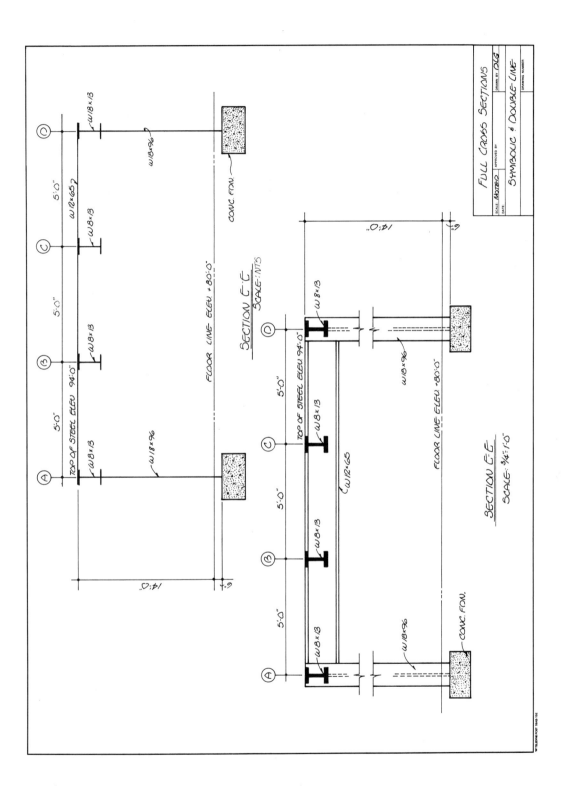

Fig. 7-6 Sample sections

How the drawings are to be used determines the extent of detail required on sectional drawings. If the drawings are to be used strictly for design and engineering purposes, single-line representations will usually suffice. If the drawings are to be used for guiding the erection crew or by other drafters in preparing fabrication details, actual scaled drawing representations are required.

All sections cut in structural steel drafting are structural sections. *Structural sections* show only the structural or load-bearing components of the building (foundation, floors, columns, and beams/girders). *Architectural sections,* on the other hand, show all of the interior features of the structure (carpet, flooring, paneling, suspended ceiling, baseboards, etc.). Since none of these architectural features is load bearing or structural, drafting time is shortened by excluding them.

PARTIAL SECTIONS

Isolated situations on a structural steel framing plan may be clarified without requiring full sections by using partial sections. *Partial sections* may be used to clarify internal relationships that are not cleared up by the longitudinal or cross sections. A framing plan may show both full and partial section cuts. Most framing plans have both because few framing plans are so simple that one longitudinal and one cross section will clarify all situations. Figure 7-7 has two examples of partial sections as they are used in structural steel drafting.

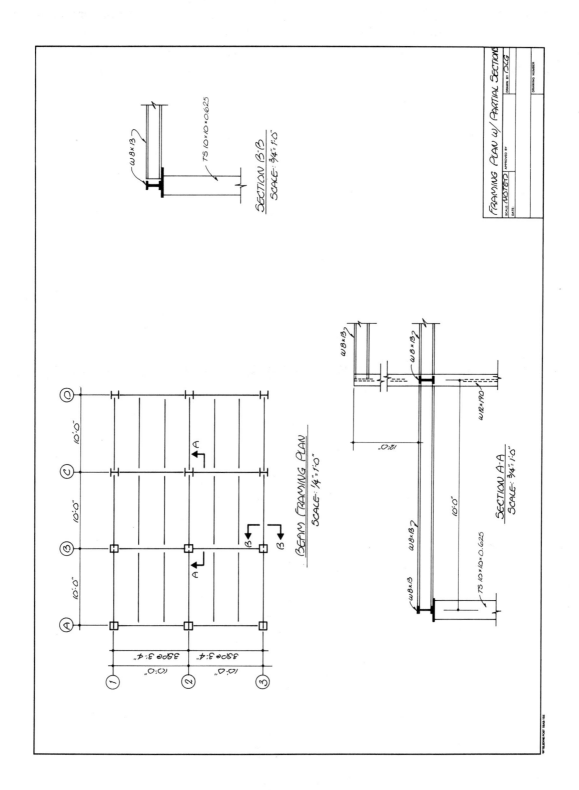

Fig. 7-7 Framing plan with partial sections

OFFSET SECTIONS

All section cutting plane lines do not run straight across the framing plan. Often, interior situations requiring clarification do not fall along a straight extension of the cutting plane line. When this is the case, an *offset section* may be cut. Figure 7-8 illustrates how an offset section cutting plan appears on a framing plan. The sectional drawings for offset sections are drawn as if the offset interior portion of the structure aligned with a straight extension of the cutting plane line from its point of origin, Figure 7-9.

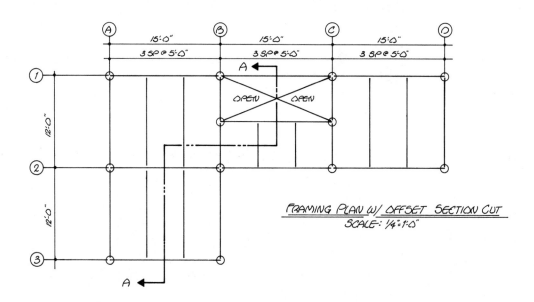

Fig. 7-8 Offset section cut

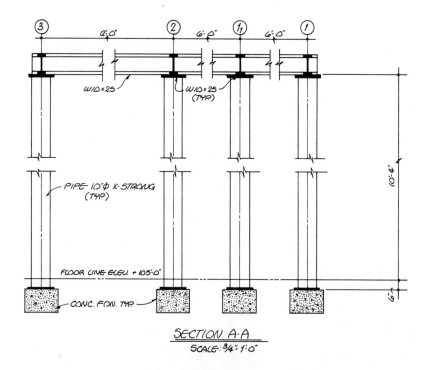

Fig. 7-9 Sample section

SECTION CONVENTIONS

One of the reasons for cutting sections is to show the reader of a set of plans what materials the structure is made of. All materials used in a structure have standard symbols that are used in sectional drawings. In concrete and wood construction, there is an extensive list of materials that might be used in a structure. In steel construction, however, the list can be abbreviated to include only concrete and steel. Figure 7-10 gives an example of the sectioning symbols for concrete and steel as well as examples of how they are applied on sectional drawings. It should be noted that section conventions are used infrequently in structural steel drafting.

DRAWING STRUCTURAL STEEL SECTIONS

Structural steel drafters obtain the information needed to prepare sectional drawings from the following two sources:

- The framing plans on which the sections were cut
- Engineer's sketches from which the framing plans were drawn

The AISC *Manual of Steel Construction* is also a valuable reference source in preparing sectional drawings. It is helpful in clarifying any ambiguities relative to proper callouts and dimensions.

The framing plans indicate what sections must actually be drawn. The engineer's sketch(es) provide information on height relationships and how the structure fits together. Structural steel sectional drawings are prepared according to the following procedures:

1. Examine the framing plan(s) to determine what sectional drawings must be prepared.
2. Examine each sectioned situation individually and determine the following information:

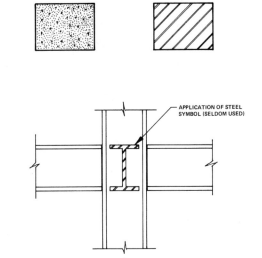

Fig. 7-10 Section and symbols

- What materials are used in the section?
- How do the structural components fit together?
- What are the important height distances that should appear on the sectional drawing? These will include such distances as: bottom of baseplate elevation, top of foundation elevation, floor elevations, roof elevations, column heights, and top of steel elevations.

3. Determine whether scaled drawings or symbolic representations are required and draw the sectional drawings at any appropriate scale. When scaled drawings are required, a scale of 3/4" = 1'-0" or larger should be used. Symbolic representations may be drawn at smaller scales or not to scale.

SUMMARY

... Sections are drawings that show the materials used in a job and how they fit together to form a completed structure.

... Sections clarify internal relationships in a job that are not clear or do not appear on the framing plan.

... Sections are important sources of height information for readers of structural steel drawings.

... The most commonly used types of sections in structural steel drafting are full, partial, and offset sections.

... Full sections cut through an entire building or structure along an imaginary cutting plane.

... Full sections are of two types: longitudinal and cross section.

... Longitudinal sections cut across the entire length while cross sections cut across the entire width of a structure.

... Sections in structural steel drafting may be drawn as scaled duplicates of the actual structural shapes or as single-line symbolic representations.

... All sections drawn in structural steel drafting are structural sections and only show load-bearing features such as the foundation, floor, columns, and beams/girders.

... Partial sections are used to clarify isolated situations not cleared up by full sections.

... Offset sections are used to clarify detailed interior situations in a structure that do not fall along a straight extension of the cutting plane line from its point of origin.

REVIEW QUESTIONS

1. Define *sections* as used in structural steel drafting.
2. List the three most common types of sections used in structural steel drafting.
3. Define *full section*.
4. List the two types of full sections.
5. Distinguish between the two types of sections in question 4.
6. All sections in structural steel drafting are classified as what type of section? What does this mean?
7. Sketch an example of a partial section.
8. Explain what an offset section is and how one is used in structural steel drafting.

REINFORCEMENT ACTIVITIES

The title block as shown in Figure 1-21 and 1/2" borders are required for all activities. The information required to complete the activities that follow is contained in Figure 7-11. The student may also wish to refer to the AISC *Manual of Steel Construction* for dimensional information.

1. Prepare a single-line representation of the view indicated by Section AA. Refer to Figures 7-1 and 7-5 for examples.

2. Prepare a double-line drawing of the representation completed in Activity #1. Refer to Figure 7-5 for an example.

3. Prepare a single-line representation of the view indicated by Section BB. Refer to Figure 7-6 for an example.

4. Prepare a double-line drawing of the representation completed in Activity #3. Refer to Figure 7-5 for an example.

5. Prepare a single-line representation of the offset section indicated by Section CC. Refer to Figures 7-1, 7-5, and 7-6 for an example.

6. Prepare a double-line drawing of the representation completed in Activity #5. Refer to Figure 7-9 for an example.

7. Prepare a double-line drawing of the view indicated by Section DD. Refer to Figure 7-2 for an example.

8. Prepare a double-line drawing of the view indicated by Section EE. Refer to Figure 7-2 for an example.

9. Prepare a double-line drawing of the view indicated by Section FF. Refer to Figure 7-2 for an example.

10. Prepare a double-line drawing of the view indicated by Section GG. Refer to Figure 7-2 for an example.

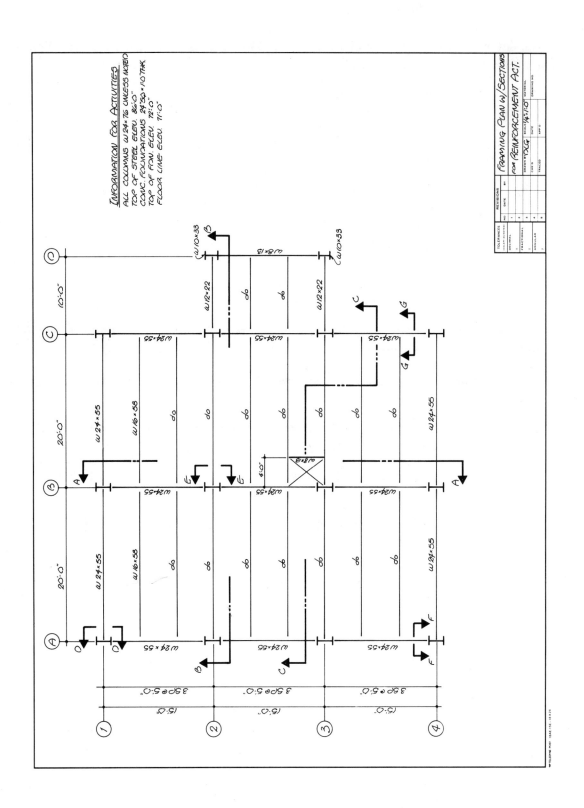

Fig. 7-11 Framing plan for drafting activities

Unit 8
Structural Steel Connection Details

OBJECTIVE

Upon completion of this unit, the student will be able to
- prepare complete structural steel baseplate, framed, and seated connections.

STRUCTURAL STEEL CONNECTION DETAILS

Connection details in structural drafting are used to show exactly how all structural members are to be connected during erection. They are very similar to sections except they are much more detailed and they isolate on the area in which the connection is located. For example, a column baseplate connection shows only that portion of the column immediately around the baseplate and the foundation on which it rests. Beam-to-column and beam-to-beam connections also isolate on the connection. Figure 8-1 shows several examples of structural steel connection details.

Connection details are used in a number of different ways and are an important part of a set of structural steel working drawings. Drafters use the details, along with sections and framing plans, in preparing fabrication details of structural members. Erection crews use connection details to guide them in putting a structure together. Designing connection details is the responsibility of an engineer or designer. Drawing them is the responsibility of the drafter.

Structural steel connections may be either bolted or welded. Riveting, once widely used, is no longer considered a major steel-connecting process. Welded and/or bolted connections are used for the following:

- Connect column baseplates to a foundation
- Make column splices
- Connect beams to columns
- Connect beams to beams

Column baseplate connections are usually bolted. Beam-to-column and beam-to-beam connections may be either bolted or welded and they may be either framed or seated or a combination of both.

Baseplate Connection Details

Structural steel columns are fitted with a steel baseplate designed to fit over anchor bolts cast into a concrete foundation. The detail showing exactly how this connection is made and specifying all necessary information about the connection is called a *baseplate connection detail*.

A baseplate connection detail must be provided for every different connection situation. This means that baseplate connection situations in which all vital information (baseplate size, anchor bolt size, and connector specifications) is the same, may share a common connection detail. If any of this information or any other information relating to the connection differs, a separate connection detail must be provided.

A common practice is to call out the various connection details that are to be drawn on the framing plan. Letter designations such as A, B, C, etc., are used to do this. Baseplate connection details are called out on the column framing plan. Figure 8-2 illustrates commonly used baseplate connection details for W, S, and M-shaped columns.

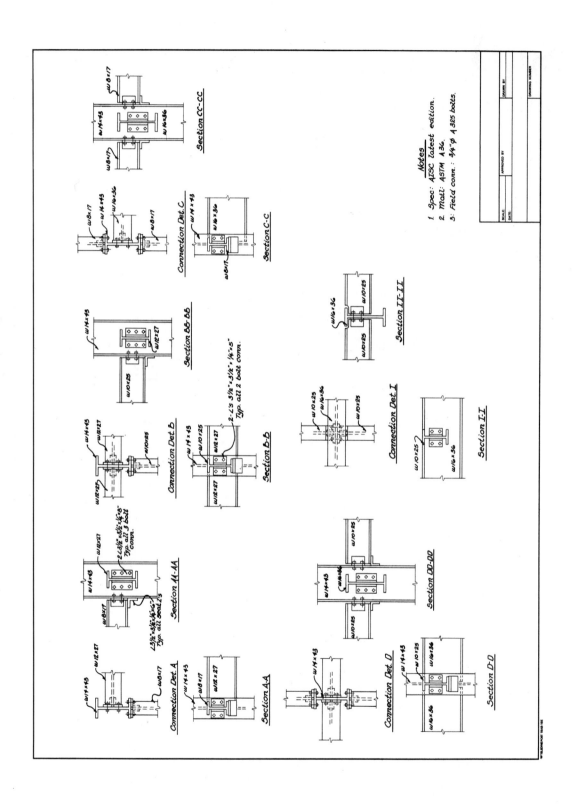

Fig. 8-1 Connection details

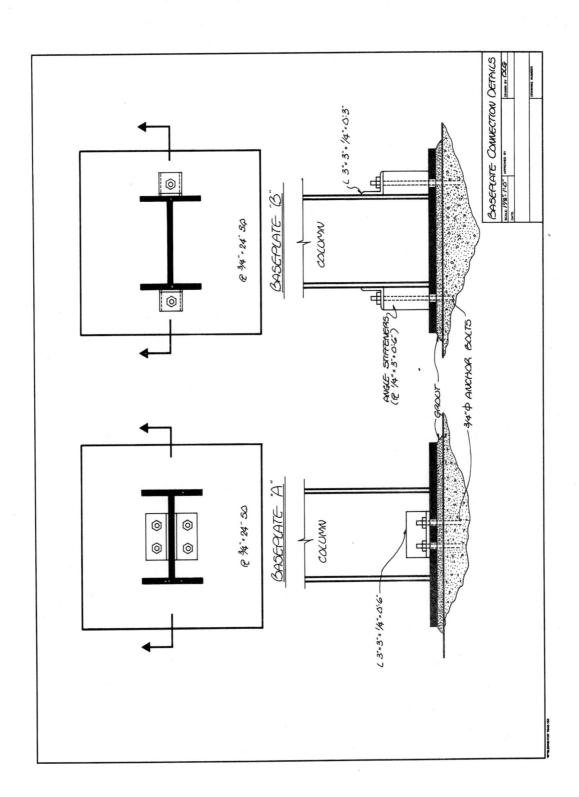

Fig. 8-2 Baseplate connection details

Figure 8-3 shows baseplate connections for tubular and pipe columns. The sizes and grades of plates, angles, bolts, etc., in these details were specified by an engineer. The details themselves were drawn by a drafter.

Framed Connections and Seated Connections

A common connection method for joining beams to columns and beams to beams is framed connection. A *framed connection* involves connecting one member to another member at the webs. This is accomplished by attaching angles to the webs of one member and connecting these angles to the web of the other member. All connections in a framed connection may be either welded or bolted.

It is common practice to connect angles or any other connectors that are to be attached in the shop by welding and then make field connections by bolting. Figures 8-4, 8-5, and 8-6 show examples of framed and seated connections. Note that a connection detail is actually a plan view of the connection with sections cut through it to show elevations of the connection.

Figure 8-4 contains two connection details, A and B, which were properly prepared. Connection Detail A shows a W6 x 15.5 resting on a plate that has been welded to the top of a 10" square tubular column. A framed connection has been provided for connecting an MC 6 x 12 to the web of the W6 x 15.5. This connection was accomplished by preparing two angles with one hole apiece to accept a bolt. These angles were then welded to opposite sides of the web of the W6 x 15.5 beam in the shop.

Upon erection of the structure, a bolt was passed through aligning holes in the angles and the channel to be connected. Connection Detail B in Figure 8-4 is very similar to Connection Detail A. All of the remaining connection details in Figures 8-5 and 8-6 were accomplished in much the same manner as Connection Detail A.

Seated connections occur when a beam rests on top of a column or another beam, or when a beam must be attached to the flange of a column. Seated connections are often used together with framed connections. By doing this, erection is made easier. The erectors are given a surface to place the beam on while connections are being made.

All of the connection details contained in Figures 8-4, 8-5, and 8-6 combine seated and framed connections. In Connection Detail A, Figure 8-4, the W6 x 15.5 beam resting on the TS 10 x 10 x 1/4 represents a seated connection. The beam actually sits on the column and is connected to it by welding. The connection of the beam and the channel is a framed connection.

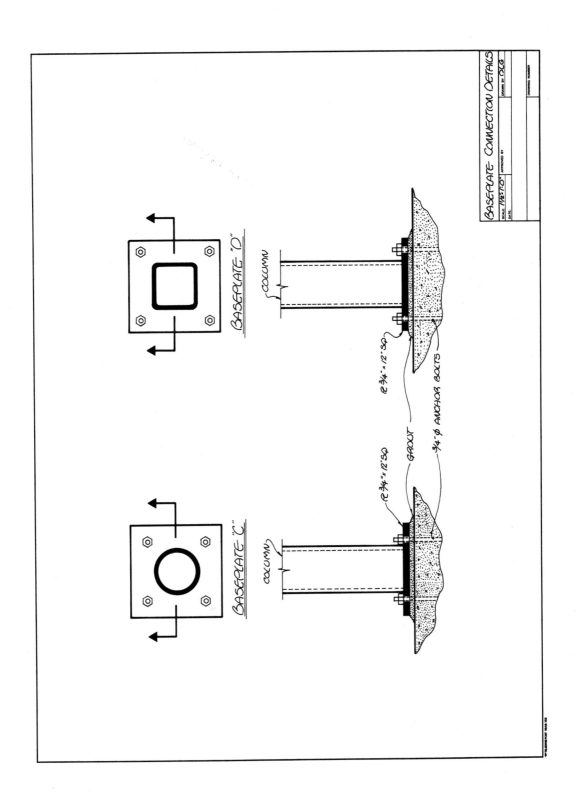

Fig. 8-3 Baseplate connection details

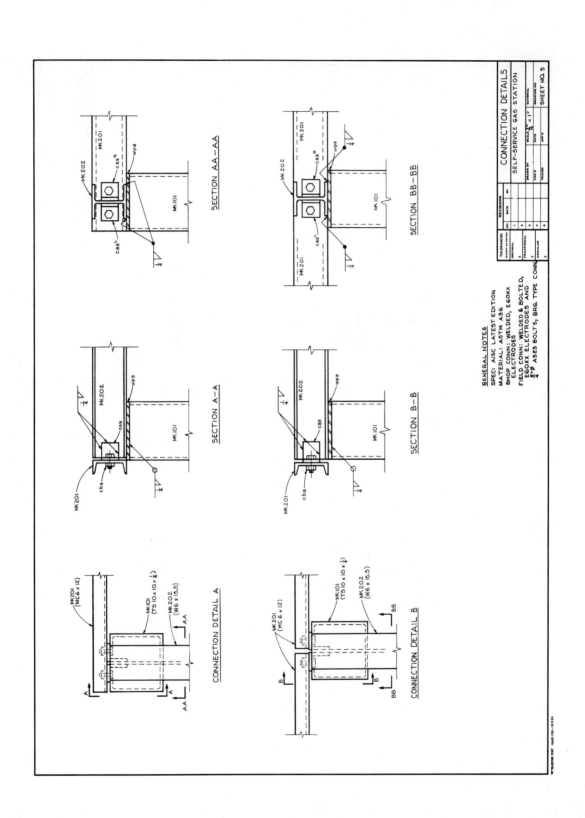

Fig. 8-4 Connection details

Unit 8 Structural Steel Connection Details 109

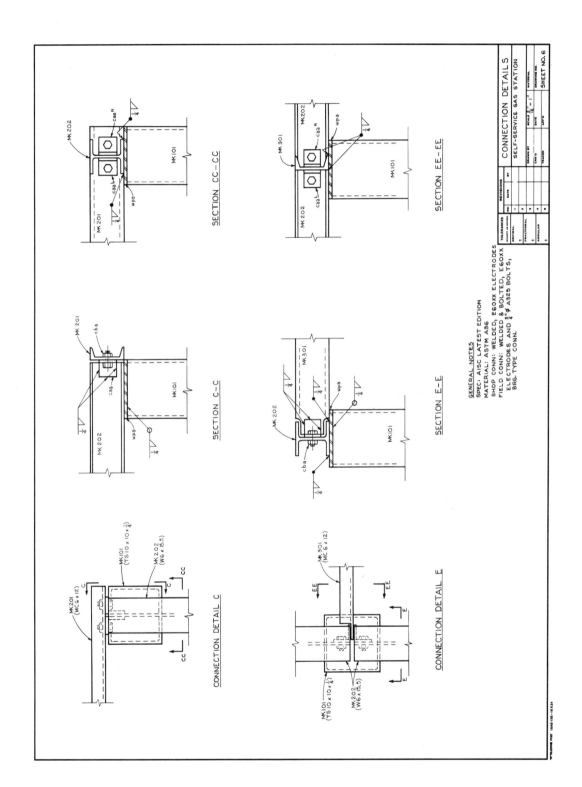

Fig. 8-5 Connection details

110 Section 2 Structural Steel Drafting

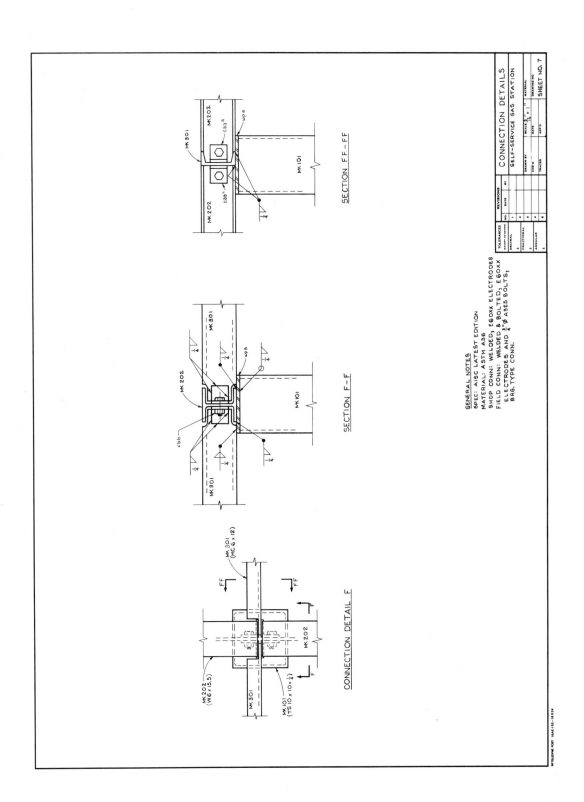

Fig. 8-6 Connection details

A more common type of seated connection involves a beam framing into the flange of W, S, or M-shaped column. In this case, an angle is attached to the column during fabrication and the beam is seated on the angle during erection, Figure 8-7.

Connection details for beam-to-beam and beam-to-column connections are called out by letter designations on the beam framing plan, Figure 6-13. All connector specifications are provided by an engineer or designer.

Drawing Structural Steel Connection Details

Designing structural connections is the job of an engineer or experienced designer. Drawing connection details is the job of the drafter. In order to draw connection details, the drafter must have the following items:

- All pertinent framing plans
- All pertinent sections
- Engineering specifications for all connections

When these things are in hand, the drafter proceeds as follows:

1. Examine the column framing plan(s) to determine what and how many baseplate details are required. Examine the beam framing plan(s) to determine what and how many connection details are required.
2. Using the framing plan(s) and the sections together, construct plan views for each connection detail required. Connection details are drawn to a scale of 3/4'' = 1'-0'' or larger.
3. Cut necessary sections on the connection detail plan views and draw the required elevation views.
4. Using the engineering specifications as a guide, label all connectors and connecting material. Add any additional notes that may be needed for clarification.

SUMMARY

... Structural steel connection details are used to show exactly how all structural members in a job are connected during erection.

... Connection details are similar to sections except they are more detailed and isolate on the connection area only.

... Connection details are used by drafters in preparing fabrication details of structural members and by the erection crew in putting the structure together.

... Designing structural connections is the job of the engineer or designer. Drawing them is the job of the drafter.

... Structural connections may be either bolted or welded. Riveting is no longer considered a major steel-connecting process.

... Typical structural steel connection situations that must be detailed are column baseplate connections, beam-to-column connections, and beam-to-beam connections.

... Baseplate connection details are called out on the column framing plan and must be provided for every different column connection situation.

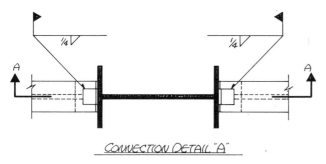

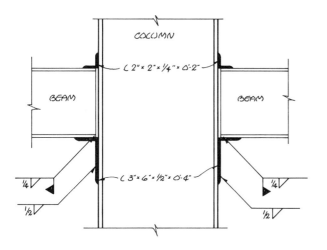

Fig. 8-7 Connection detail

... Beam-to-column connection details may be either framed or seated. Beam-to-beam details are usually framed.

... A framed connection joins two structural members at the webs. A seated connection provides a bearing surface for one member to sit on.

... A connection detail contains a plan view of the connection with appropriate sections cut to provide elevations of the connection.

... In order to draw connection details for a job, the structural drafter must have the framing plan(s), the sections, and engineering data on connector specifications.

REVIEW QUESTIONS

1. Why are connection details included in a set of structural steel working drawings?
2. How do connection details differ from sections?
3. How and by whom are connection details used?
4. In preparing connection details, what is the division of labor between engineering and drafting?
5. Name the two most common connecting processes used in structural steel connections.
6. Name three structural steel connection situations that must be detailed.
7. Explain how and where connection details are called out.
8. Explain the difference between a framed and a seated connection by constructing a simple sketch of each.
9. Provide a simple sketch that illustrates all of the components of a beam-to-column connection detail.
10. What three things are needed by the structural drafter in preparing connection details for a job?

REINFORCEMENT ACTIVITIES

The following activities should be done on C-size paper with 1/2" borders and a title block as shown in Figure 1-21. Column-and-beam framing information for the activities is contained in Figure 8-8. Sectional information and engineering data for the activities is contained in Figure 8-9. Several of the activities listed may be drawn on one sheet.

1. At a scale of 1" = 1'-0", prepare baseplate connection detail "A."
2. At a scale of 1" = 1'-0", prepare baseplate connection detail "B."
3. At a scale of 1" = 1'-0", prepare baseplate connection detail "C."
4. At an appropriate scale, prepare connection detail "A" (beam to column).
5. At an appropriate scale, prepare connection detail "B" (beam to column).
6. At an appropriate scale, prepare connection detail "C" (beam to column).
7. At an appropriate scale, prepare connection detail "D" (beam to column).
8. At an appropriate scale, prepare connection detail "E" (beam to column).
9. At an appropriate scale, prepare connection detail "F" (beam to column).
10. At an appropriate scale, prepare connection detail "G" (beam to column).
11. At an appropriate scale, prepare connection detail "H" (beam to column).
12. At an appropriate scale, prepare connection detail "I" (beam to beam).
13. At an appropriate scale, prepare connection detail "J" (beam to beam).

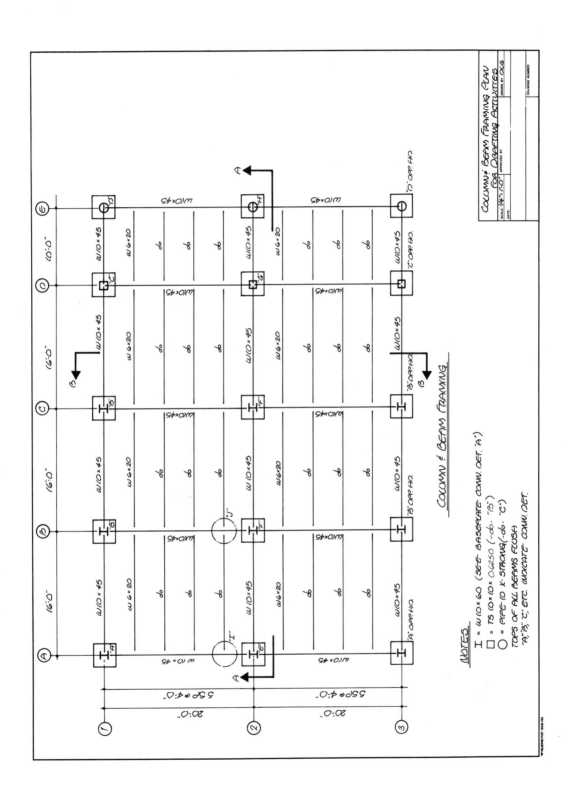

Fig. 8-8 Column-and-beam framing plan for drafting activities

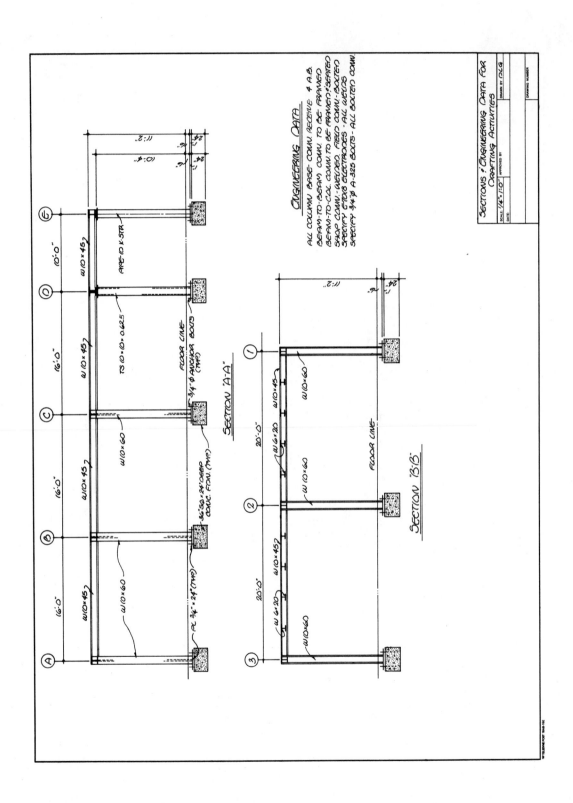

Fig. 8-9 Sections for drafting activities

Unit 9
Structural Steel Fabrication Details

OBJECTIVES

Upon completion of this unit, the student will be able to
- define structural steel shop drawings.
- define structural steel fabrication details.
- construct fabrication details for structural steel columns and beams.

STRUCTURAL STEEL SHOP DRAWINGS DEFINED

In Unit 1, it was learned that a complete set of structural working drawings consists of engineering drawings for engineering purposes and shop drawings for fabrication purposes. Engineering drawings in structural steel drafting include framing plans, sections, and connection details. These were covered in Units 6, 7, and 8. Shop drawings include fabrication details and bills of material. This unit will deal with fabrication details. Bills of material will be covered in Unit 10.

Shop drawings are comprehensive, precisely detailed drawings accompanied by bills of material. Shop drawings are prepared for use by shop workers in fabricating structural steel members for a job. The primary component of shop drawings is the fabrication detail. Fabrication details are prepared by structural steel drafters. The information required in constructing the fabrication details for a job is contained in the framing plan(s), sections, and connection details.

STRUCTURAL STEEL FABRICATION DETAILS DEFINED

Fabrication details are orthographic drawings of structural steel columns and beams. They contain all of the information required by shop workers in fabricating the structural members used in any given job. Each different column and beam used in the framing of a job must have an individual fabrication detail. This fabrication detail shows length information, locations of plates and angles, positions of holes that are to be drilled, and any other information relating to the fabrication process.

CONSTRUCTING STRUCTURAL STEEL FABRICATION DETAILS

In order to construct the fabrication details for a job, the drafter must have the framing plan(s) for the job, the sections, and the connection details. The framing plan(s) provides such information as:

- How many columns must be detailed?
- What is the structural shape of each column?
- How many of a certain column designation are required?

The framing plan provides this same information for beams.

Sections provide height information valuable in calculating the height of columns. Connection details provide the information necessary for converting centerline of column distances into actual beam lengths. The details also provide information needed for locating holes, plates, angles, etc. Connection specifications also appear on the connection details.

Figure 9-1 shows a removed portion of a framing plan, a section through the removed portion, and a connection detail. The fabrication detail shown was developed from information provided on the framing plan, section, and connection detail. Examine Figure 9-1 closely to determine how the actual member length of the beam and hole locations were arrived at.

Column Fabrication Details

Structural steel fabrication details contain the following information:

- At least two views (plan and elevation) of the column
- Complete locational dimensions for holes, plates, and angles
- The baseplate
- Connection specifications
- Miscellaneous notes for the fabricator, Figure 9-2

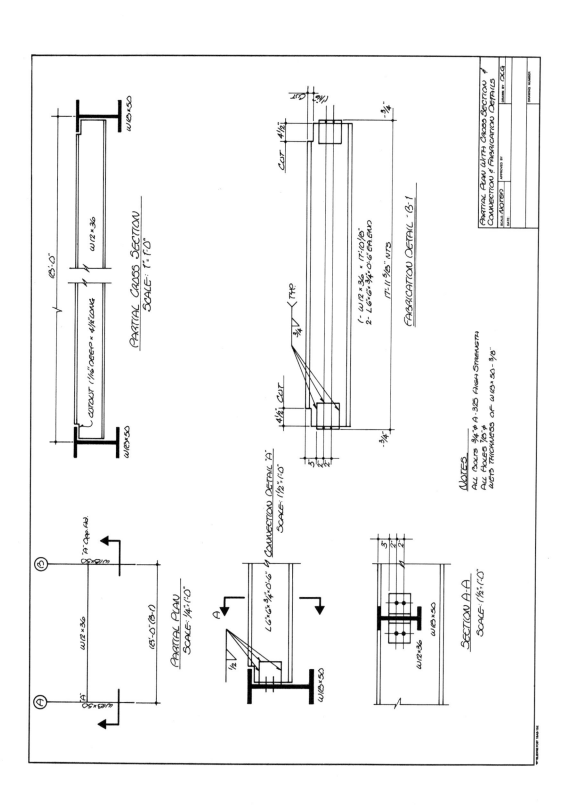

Fig. 9-1 Partial plan with section and details

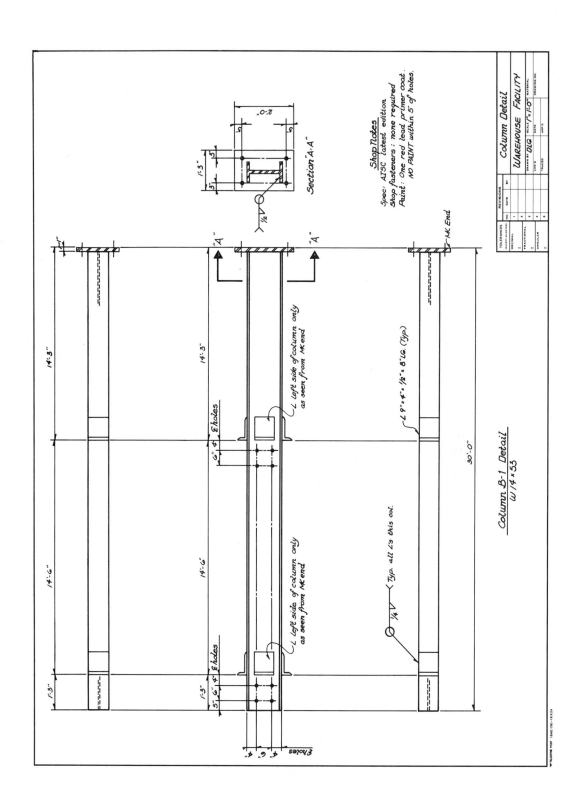

Fig. 9-2 Sample column detail

Column fabrication details are drawn according to the following procedures:

1. The column length is determined. Column length calculations are illustrated in Figures 9-3 and 9-4.
2. The required orthographic views of the column are drawn to the proper length and the overall length dimension is placed on the drawing. It is common practice in structural steel drafting to draw width and depth dimensions to scale, but the length not to scale (NTS). Scaled dimensions should be drawn to a scale of 3/4" = 1'-0" or larger, Figure 9-2.
3. Locations of holes, plates, angles, etc., are determined from the connection details and placed on the drawing with accompanying dimensions, Figure 9-2.
4. A section is cut through the column facing the baseplate. The sectional view is drawn to show how the baseplate is attached and its configuration, Figure 9-2.
5. Any connection specifications or other information that should be noted is taken from the connection details and entered on the drawing, Figure 9-2.

Beam Fabrication Details

Structural steel beam fabrication details contain the following information:

- An elevation of the beam with end views or sections when required for clarity
- Complete locational dimensions for holes, plates, and angles
- Length dimensions
- Connection specifications
- Cutouts
- Miscellaneous notes for the fabricator, Figure 9-5

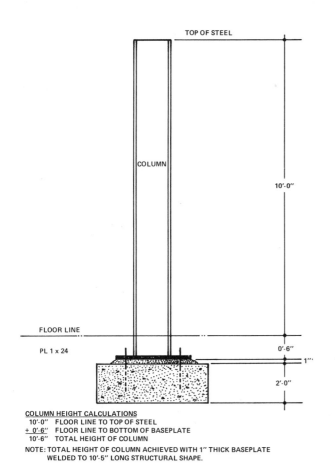

Fig. 9-3 Column height calculations

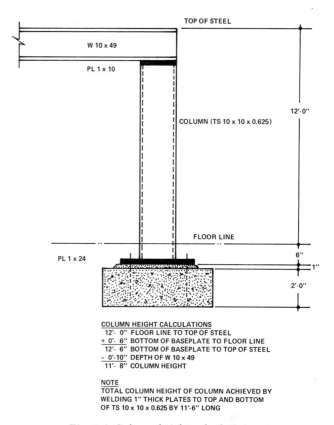

Fig. 9-4 Column height calculations

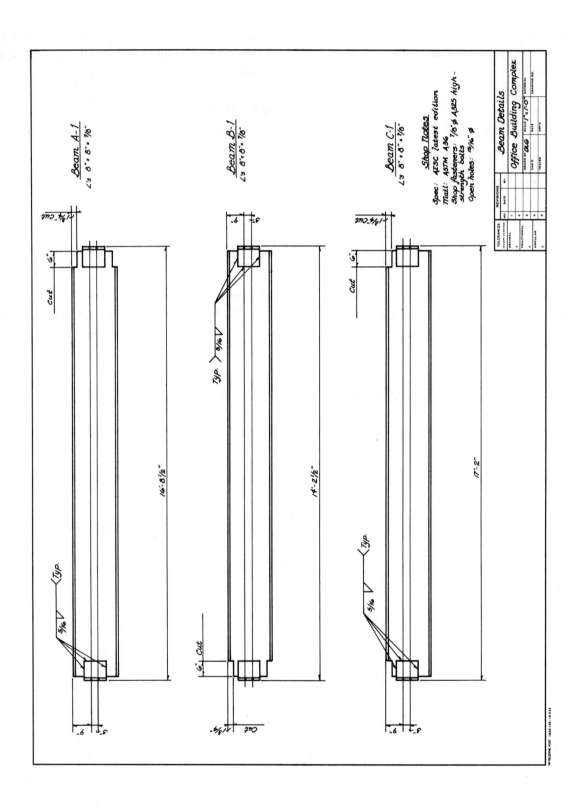

Fig. 9-5 Sample beam details

Beam fabrication details are drawn according to the following procedures:

1. The beam length is determined. Beam length calculations are illustrated in Figures 9-6 and 9-7.
2. The required orthographic views of the beam are drawn to the proper length and the overall dimension is placed on the drawing. It is common practice in structural steel drafting to show depth and width dimensions to scale, but the length not to scale (NTS). Scaled dimensions should be drawn to a scale of 3/4" = 1'-0" or larger, Figure 9-5.
3. Locations of holes, plates, angles, etc., are determined from the connection details and placed on the drawing with accompanying dimensions, Figure 9-5.
4. Any connection specifications or other information that should be noted is taken from the connection details and entered on the drawing, Figure 9-5.

SUMMARY

... A complete set of structural steel working drawings consists of engineering drawings for engineering purposes and shop drawings for fabrication purposes.
... Shop drawings include fabrication details and bills of material.
... The primary component of shop drawings is the fabrication detail.
... Fabrication details are orthographic drawings of structural steel columns and beams showing all information necessary for fabrication.
... Fabrication details are prepared by structural drafters from information found on the framing plan(s), sections, and connection details.
... Column fabrication details contain: orthographic views of the column; locational dimensions for plates, angles, and holes; length dimensions; the baseplate configuration; connection specifications; and miscellaneous notes for the fabricator.
... Beam fabrication details contain: orthographic views of the beam; sections or end views when required for clarity; length dimensions; locational dimensions for holes, angles and plates; cutouts; connection specifications; and miscellaneous notes for the fabricator.

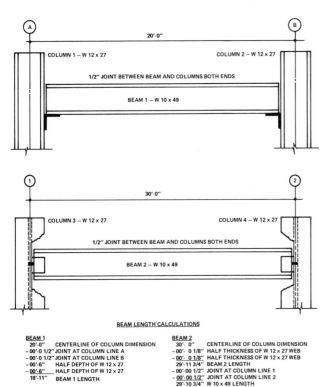

Fig. 9-6 Beam length calculations

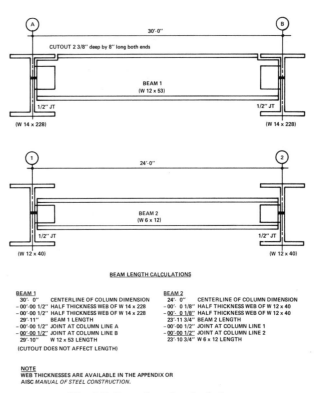

Fig. 9-7 Beam length calculations

REVIEW QUESTIONS

1. Define the term *shop drawing* as used in structural steel drafting.
2. Define the term *fabrication detail* as used in structural steel drafting.
3. What is the primary component of shop drawings?
4. What must the structural steel drafter have in order to prepare the fabrication details for a job?
5. What information is contained in a column fabrication detail?
6. What information is contained in a beam fabrication detail?
7. Construct a simple sketch that illustrates how to calculate column heights for I-shaped columns.
8. Construct a simple sketch that illustrates how to calculate the beam length for a beam that frames into the webs of two I-shaped columns.
9. Construct a simple sketch that illustrates how to calculate the beam length for a beam that frames into the flanges of two I-shaped columns.
10. Construct a simple sketch that illustrates how to calculate the beam length for a beam that frames into the web of two other beams.

REINFORCEMENT ACTIVITIES

All activities are to be completed on either B-size or C-size paper with 1/2" borders and a title block as in Figure 1-21. In all activities the student must refer to Figures 9-9 through 9-12. All information necessary for completing the activities is contained in these figures and Figure 9-8.

1. Construct a complete column fabrication detail for column A2, Figure 9-8.
2. Construct a complete column fabrication detail for column B1, Figure 9-8.
3. Construct a complete column fabrication detail for column C1, Figure 9-8.
4. Construct a complete column fabrication detail for column E2, Figure 9-8.
5. Construct a complete beam fabrication detail for beam B-1, Figure 9-8.
6. Construct a complete beam fabrication detail for beam B-2, Figure 9-8.
7. Construct a complete beam fabrication detail for beam B-3, Figure 9-8.
8. Construct a complete beam fabrication detail for beam B-4, Figure 9-8.
9. Construct a complete beam fabrication detail for beam C-1, Figure 9-8.

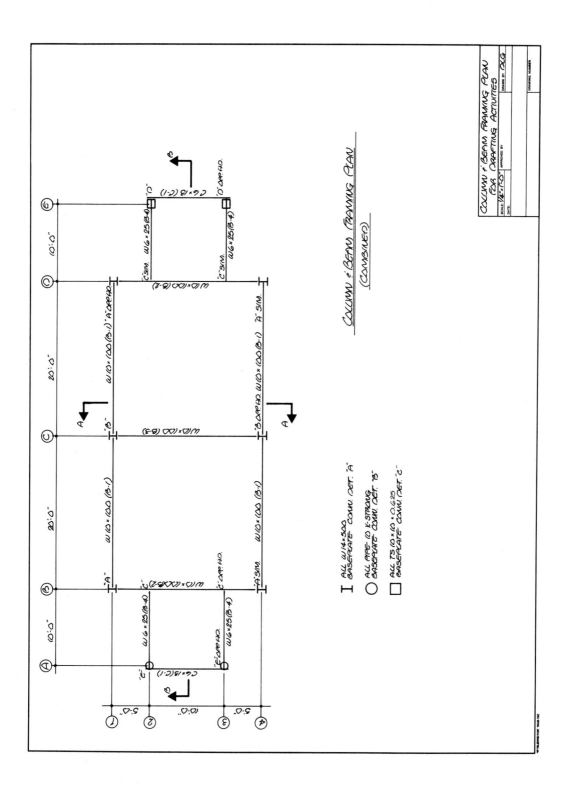

Fig. 9-8 Column-and-beam framing plan for drafting activities

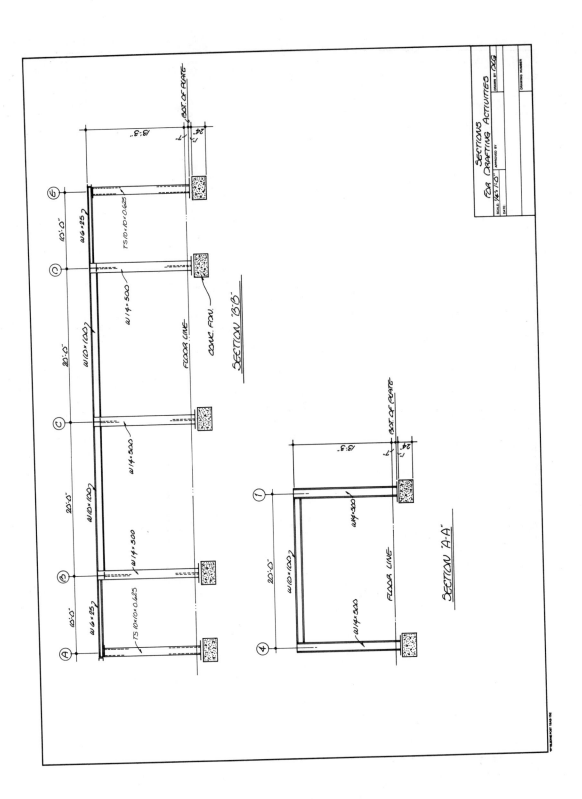

Fig. 9-9 Sections for drafting activities

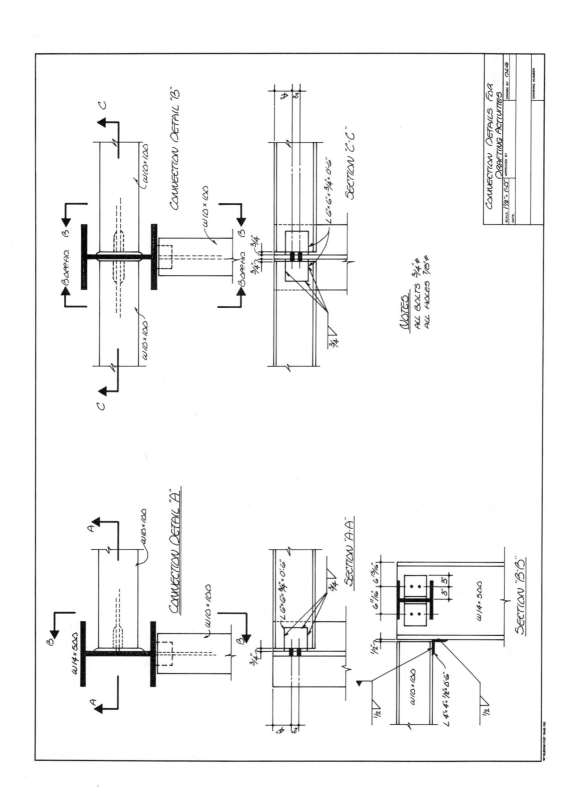

Fig. 9-10 Details for drafting activities

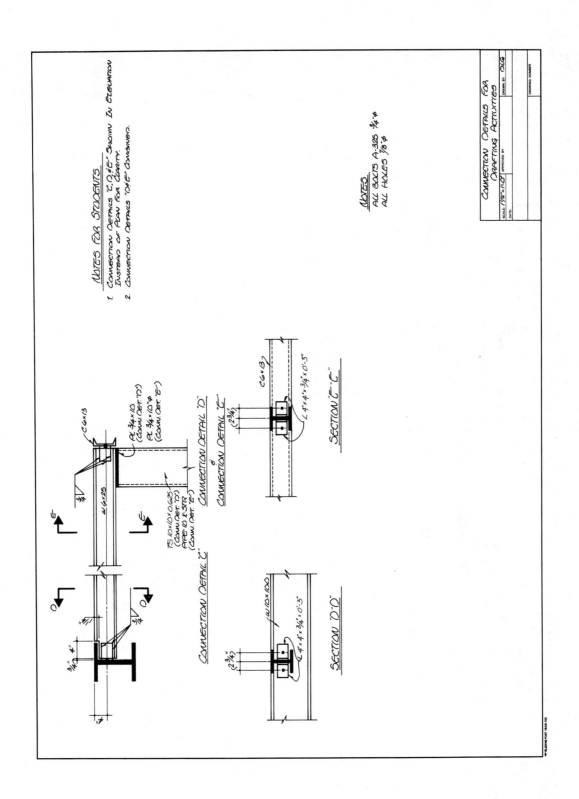

Fig. 9-11 Connection details for drafting activities

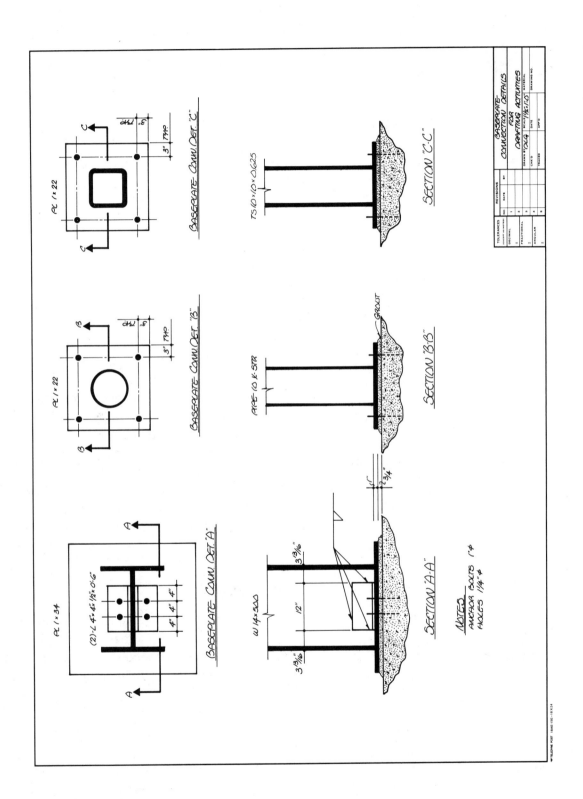

Fig. 9-12 Baseplate connection details

Unit 10
Structural Steel Bills of Materials

OBJECTIVES

Upon completion of this unit, the student will be able to

- define the terms advance bill and shop bill.
- distinguish between advance bills and shop bills.
- prepare advance bills and shop bills for structural steel jobs.

BILLS OF MATERIALS

In addition to preparing engineering drawings and shop drawings, structural steel drafters are also called upon to prepare bills of materials. *Bill of materials* is a general term including other more specific terms such as the following:

- Advance bill of materials
- Advance bill
- Materials list
- Preliminary bill
- Shop bill of materials
- Shop bill

There are actually two types of bills of materials that the drafter must be concerned with. All of the terms above are different names for these two types.

The first type involves drafting and the purchasing department. It is used to order all of the steel shapes and pieces required to fabricate an entire job. It is called alternately *advance bill of materials, material lists, preliminary bill,* and *advance bill,* with the latter term being the most commonly used.

The second type involves drafting and the shop or fabrication workers. It is used to identify all of the steel pieces required in the fabrication of each individual structural member for a job and accompanies the shop drawings. This type of bill of materials is alternately called the *shop bill of materials* or more commonly the *shop bill.* A shop bill may be placed right on the shop drawing or may be attached to it as a separate sheet.

Advance Bills

Most structural steel fabrication companies do not stockpile large quantities of steel shapes and materials. Rather, the amounts of steel needed for a job are ordered on a job-by-job basis from rolling mills and steel warehousing firms. An *advance bill* is a comprehensive listing of all steel shapes and materials that will be required to complete a given job, Figure 10-1.

The advance bill is prepared immediately following completion of initial drafting on the fabrication details, often before they have been checked. This is why the advance bill is sometimes referred to as the preliminary bill.

ADVANCE BILL

JOB NUMBER __2632__   PREPARED BY __OLG__   SHEET NUMBER __1__ OF __1__

ITEM NUMBER	NUMBER REQUIRED	SHAPE	DESCRIPTION	LENGTH	REMARKS
			COLUMNS		
1	1	W	12×190	15'-6"	B4
2	2	W	12×72	15'-6"	B5, B6
3	4	W	8×40	15'-6"	A4, A5, A6, A7
4	2	W	8×24	15'-6"	C4, C7
5	2	PIPE	10 X-STRONG	12'-6½"	C5, C6
6	2	TS	10×10×0.625	12'-6½"	C1, C3
			BEAMS		
7	3	W	16×40	22'-1½"	
8	6	W	16×40	23'-0"	
9	2	W	10×45	16'-11"	
10	5	S	7×20	10'-6"	
11	2	S	7×20	9'-5½"	
			PLATES		
12	2	PL	¾×24	18'-0"	CUT FOR BASEPLATES
13	1	PL	¾×22	8'-0"	CUT FOR BASEPLATES
			ANGLES		
14	1	L	6×6×5/8	46'-0"	CUT FOR CONNECTIONS
15	1	L	4×6×5/8	32'-0"	CUT FOR CONNECTIONS

Fig. 10-1 Sample advance bill

The exact makeup of advance bill forms varies from company to company. However, all advance bill forms contain the following information at a minimum:

- A job number that corresponds with the job number on the shop drawings.
- A sheet number that is used for the bill of materials sheet.
- A space in which the person preparing the bill is identified.
- A list of numbers so that every piece listed on the bill has an item number.
- A space beside each item number in which the number required of each piece is listed.
- A space in which the structural shape for each piece is identified.
- A space in which the length of each piece is listed.
- An additional space for remarks pertaining to each piece.

Figure 10-2 contains a sample bill form that might be used in industry. This form has been generalized from numerous other forms designed for specific companies. However, it serves the purpose of familiarizing the drafting student with advance bill forms.

Shop Bills

A *shop bill* is used to draw from the material ordered that which is needed in the fabrication of a specific member(s) in a job. Methods of preparation vary from company to company. A common practice is to prepare a shop bill to accompany each sheet of fabrication details in a set of shop drawings. The shop bill may be placed on the drawing or attached to it as a separate form.

The information contained in a shop bill is the same as that contained in the advance bill in terms of the form itself. Item numbers, the number of each piece required, item lengths, etc., are also all listed on the shop bill. The difference between the two is twofold. First, the advance bill covers an entire job, while the shop bill is geared toward individual members. Second, the advance bill lists all of the columns for a job together, all of the beams for a job together, all of the plates together, etc. The shop bill lists all of the materials required for each specific member in a job together.

Like the advance bill form, shop bill forms vary from company to company. Within a specific company, however, the two forms are often the same or very similar. Figure 10-3 shows a completed shop bill that was prepared on a form that was generalized from several used in industry.

BILL OF MATERIALS FORM

JONES STEEL FABRICATION COMPANY, INC.
201 COUNT RD
STEELTOWN, FLORIDA

JOB NUMBER_____ PREPARED BY_____ BILL OF MATERIALS SHEET_____ OF_____

ITEM NUMBER	NUMBER REQUIRED	SHAPE	DESCRIPTION	LENGTH	REMARKS

Fig. 10-2 Sample bill of materials form

SHOP BILL

JOB NUMBER __768__   PREPARED BY __DLG__   SHEET NUMBER __1__ OF __1__

ITEM NUMBER	NUMBER REQUIRED	SHAPE	DESCRIPTION	LENGTH	REMARKS
			2 BEAMS	1-B-1 & 1 B-2	
1	1	W	10 × 49	20'-10¾"	
2	1	W	10 × 49	19'-10¾"	
3	8	L	6 × 6 × 1/2	6"	
4	16	BOLTS	3/4"φ	–	A325 HIGH STRENGTH
			1 BEAM	B-3	
5	1	W	16 × 40	12'-10½"	
6	4	L	6 × 5 × 1/2	10"	
7	12	BOLTS	3/4 φ	–	A325 HIGH STRENGTH
			2 CHANNELS C-1 &	C-2	
8	1	C	10 × 15.3	16'-0"	
9	1	C	10 × 15.3	15'-8¾"	
			1 CHANNEL	C-3	
10	1	C	15 × 40	19'-5"	
11	2	L	7 × 7 × 1/2	9"	
12	1	PLATE	15 × 1/2	3¾"	
13	6	BOLTS	7/8"φ	–	A307

Fig. 10-3 Sample shop bill

PREPARING BILLS OF MATERIALS

A number of different people are involved in the preparation of a bill of materials. If it is an advance bill, a drafter prepares it, a checker checks it, and then it is passed along to the purchasing department for their input. If it is a shop bill, a drafter prepares it, a checker checks it, and it is sent to fabrication along with the shop drawings.

The forms for the advance bill and the shop bill may be the same or very similar. However, the procedures for preparing the two different types of bills of materials vary due to their differing uses.

Preparing Advance Bills

A drafter assigned to prepare the advance bill for a job needs prints of all of the fabrication details for the job and an advance bill form. With these things in hand, the drafter proceeds as follows:

1. The fabrication details are arranged into columns and beams and listed on the form. The columns are listed first and the beams second. Next, all miscellaneous pieces such as angles, plates, bars, etc., are grouped together and listed, Figure 10-2.
2. All pertinent information such as the number required of each piece, shape, description, length, and remarks when applicable, etc., are entered on the form, Figure 10-2.
3. The entire bill is given a thorough examination by the drafter preparing it. This is done to insure that all necessary information has been included and is correct.

Preparing Shop Bills

A drafter assigned to prepare the shop bill(s) for a job or portion of a job needs prints of all fabrication details to be billed and a shop bill form. With these things in hand, the drafter proceeds as follows:

1. Each individual member to be billed is isolated and taken separately. The main structural shape and all pertinent data about it is listed first. This is followed by a listing of all other items such as plates, angles, bars, bolts, etc., required in the fabrication of the member, Figure 10-3.
2. Step 1 is repeated for every structural member that has been detailed, Figure 10-3.
3. The entire bill is given a thorough examination by the drafter preparing it. This is done to insure that all necessary information has been included and is correct.

SUMMARY

... The term bill of materials is a general term that includes a number of terms that are alternately used in structural steel drafting departments.
... Common alternative terms for bill of materials are materials list, preliminary bill, advance bill, shop bill of materials, and shop bill.
... All of the various terms for bill of materials are simply other names for the two types of bills that are used in structural steel drafting. These two types are most commonly referred to as advance bill and shop bill.
... An advance bill is a comprehensive listing of all steel shapes and materials that are needed to fabricate an entire job. In lay terms it might be considered a grocery list for a structural steel job.
... A shop bill is a listing of the steel shapes and materials that are needed to fabricate one or several individual structural members in a job. In lay terms the shop bill can be viewed as a recipe for an individual item included in the grocery list.
... Advance bill and shop bill forms may be exactly alike or very similar. The makeup of the forms varies from company to company. However, all forms contain the following information at a minimum: a job number, a sheet number, item numbers, shape descriptions, lengths, and remarks.
... Advance bills are prepared immediately following completion of the fabrication details, often before they have been checked.
... Advance bills go from drafting to the purchasing department for further preparation.
... Shop bills are prepared after the fabrication details have been checked and approved.
... Shop bills accompany the shop drawings to fabrication.

REVIEW QUESTIONS

1. The term *bill of materials* is a general term that includes several other terms. List four of these terms.
2. There are two types of bills of materials used in structural steel drafting departments. List the most common names for the two.
3. Define the term *advance bill*.
4. Define the term *shop bill*.
5. Distinguish between advance bill and shop bill.
6. Sketch an example of a bill form that includes all of the information minimally required for an advance bill or a shop bill.

REINFORCEMENT ACTIVITIES

1. Using the dimensions set forth in Figure 10-4 and the format shown in Figure 10-1, prepare a blank advance bill form on a B-size sheet of paper.
2. Using the dimensions set forth in Figure 10-4 and the format shown in Figure 10-3, prepare a blank shop bill form on a B-size sheet of paper.
3. Using the blank advance bill form prepared in Activity #1, prepare a complete advance bill for the fabrication details contained in Figures 10-5 and 10-6.
4. Activities #4 through #12 are to be completed on the Shop Bill Form prepared in Activity #2. Enter the necessary information for the column in Figure 10-5 onto the shop bill form.
5. Prepare the shop bill for beam 1, Figure 10-6.
6. Prepare the shop bill for beam 2, Figure 10-6.
7. Prepare the shop bill for beam 3, Figure 10-6.
8. Prepare the shop bill for beam 4, Figure 10-6.
9. Prepare the shop bill for beam 5, Figure 10-6.
10. Prepare the shop bill for beam 6, Figure 10-6.
11. Prepare the shop bill for beam 7, Figure 10-6.
12. Prepare the shop bill for beam 8, Figure 10-6.

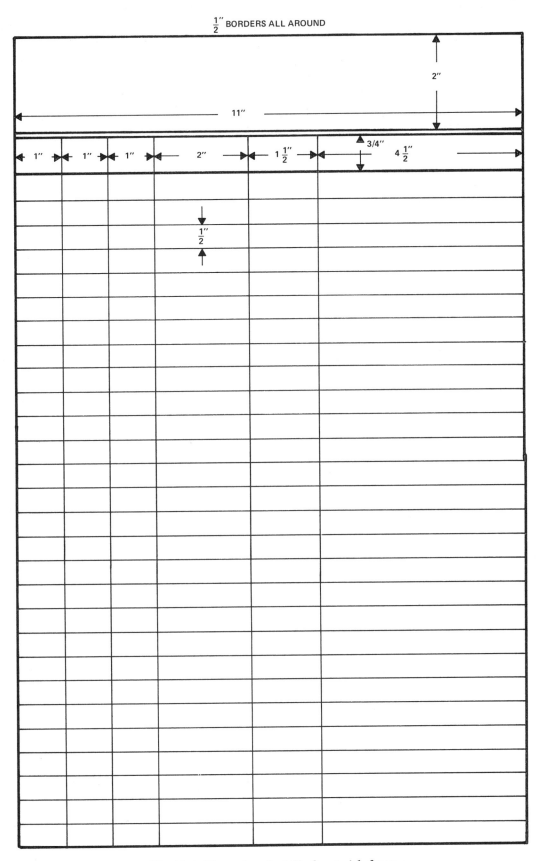

Fig. 10-4 Dimensions for bill of materials forms

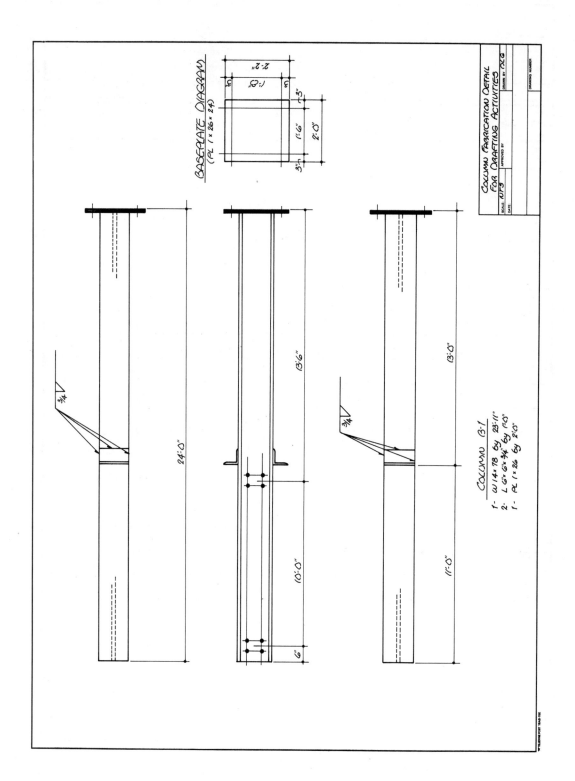

Fig. 10-5 Sample column fabrication detail

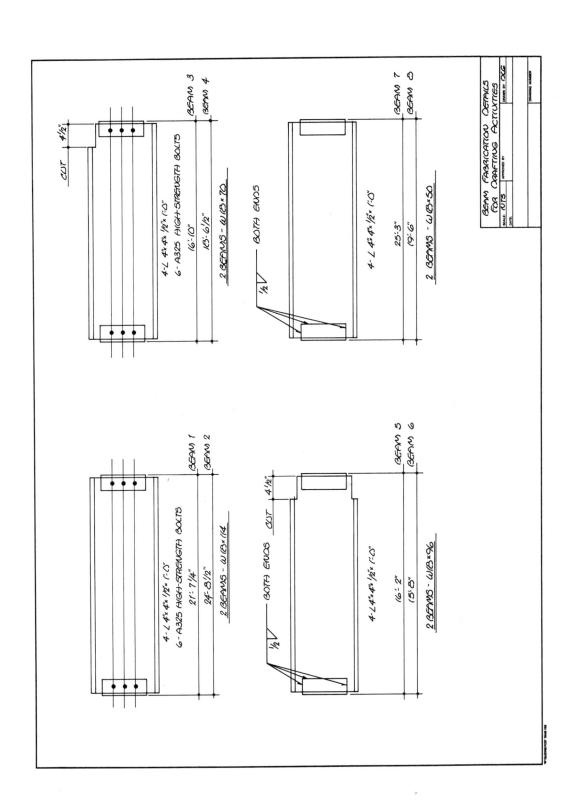

Fig. 10-6 Sample beam fabrication detail

Section 3
Structural Precast Concrete Drafting

Unit 11
Precast Concrete Framing Plans

OBJECTIVE

Upon completion of this unit, the student will be able to
- construct precast concrete column framing plans, beam framing plans, floor/roof framing plans, and wall framing plans from raw data available.

PRECAST CONCRETE FRAMING PLANS

Structural precast concrete drafters prepare two types of drawings: engineering drawings and shop drawings. Framing plans are engineering drawings. Taken together as a set, the engineering drawings and the shop drawings make up a set of structural working drawings.

The four basic types of framing plans in precast concrete drafting are column, beam, floor/roof, and wall. Figures 11-1 through 11-6 show examples of these types of framing plans. The examples are a complete set of framing plans for a warehouse facility. They were prepared from guidelines set forth in architectural plans.

Framing plans are the heart of a set of structural precast concrete working drawings and serve the same purposes as the floor plan in a set of architectural plans. Most companies require separate column, beam, floor, wall, and roof framing plans. Information required on framing plans includes the following:

- A plan view layout of the structural members
- Mark numbers for each structural member
- Product schedule, general notes, and legend
- North arrow
- Complete dimensions
- Centerline of column designations (column framing plan only)

Each of these items is illustrated in Figures 11-1 through 11-6. Several of these require further explanation before proceeding to the various steps in preparing precast concrete framing plans.

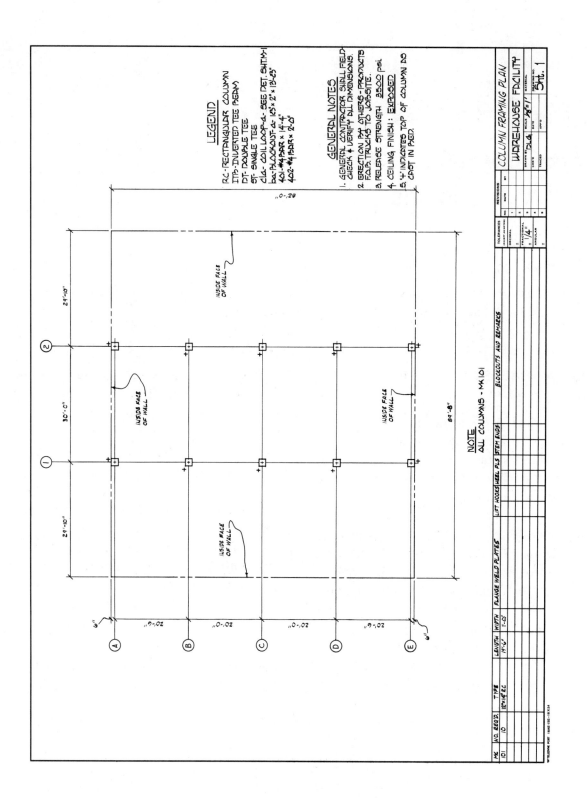

Fig. 11-1 Column framing plan

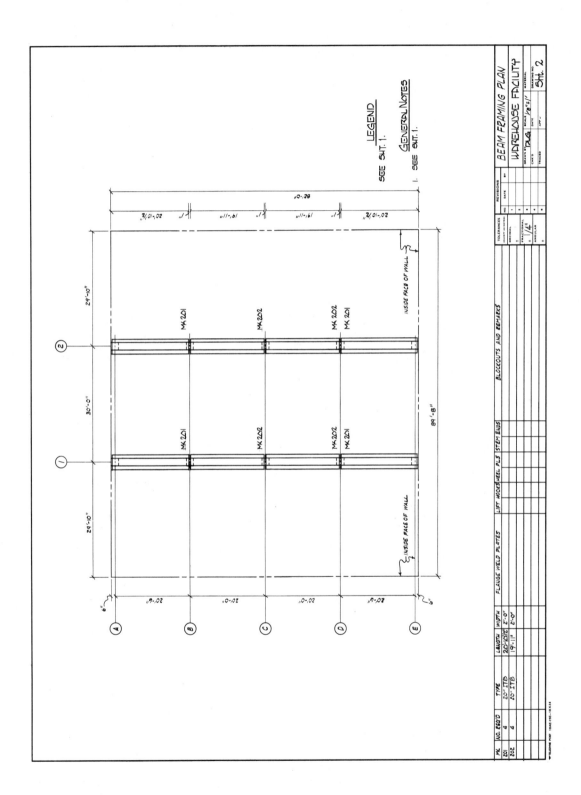

Fig. 11-2 Beam framing plan

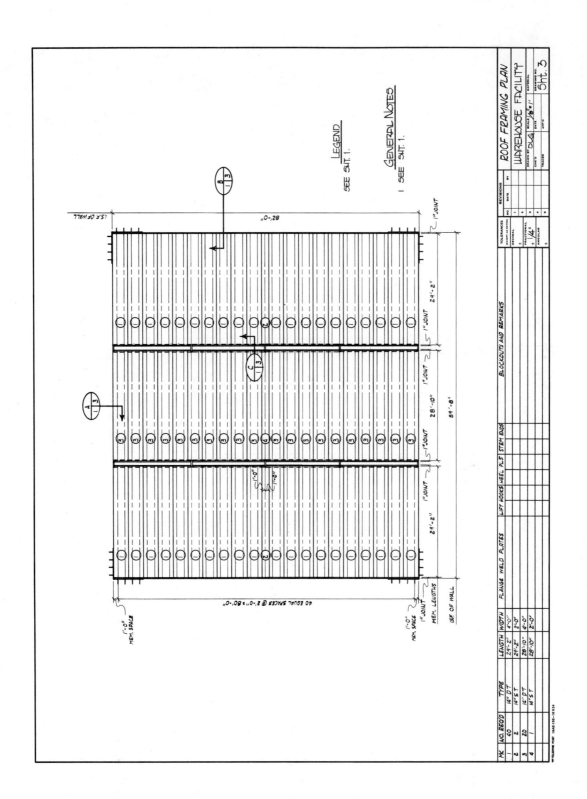

Fig. 11-3 Roof framing plan

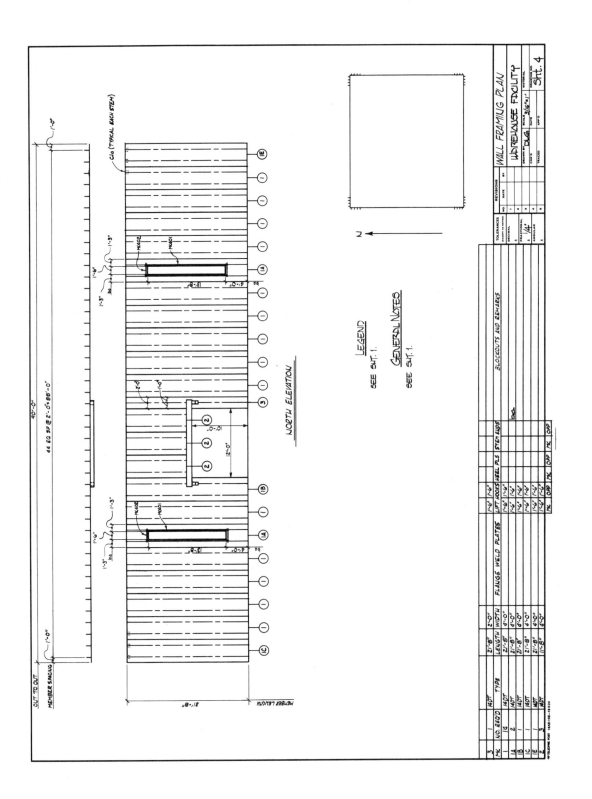

Fig. 11-4 Double-tee wall panel framing plan (north elevation)

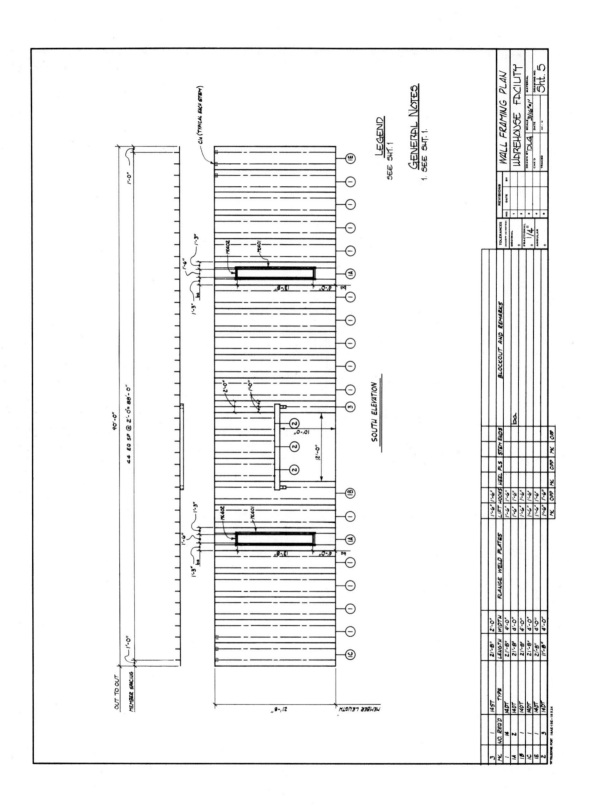

Fig. 11-5 Double-tee wall panel framing plan (south elevation)

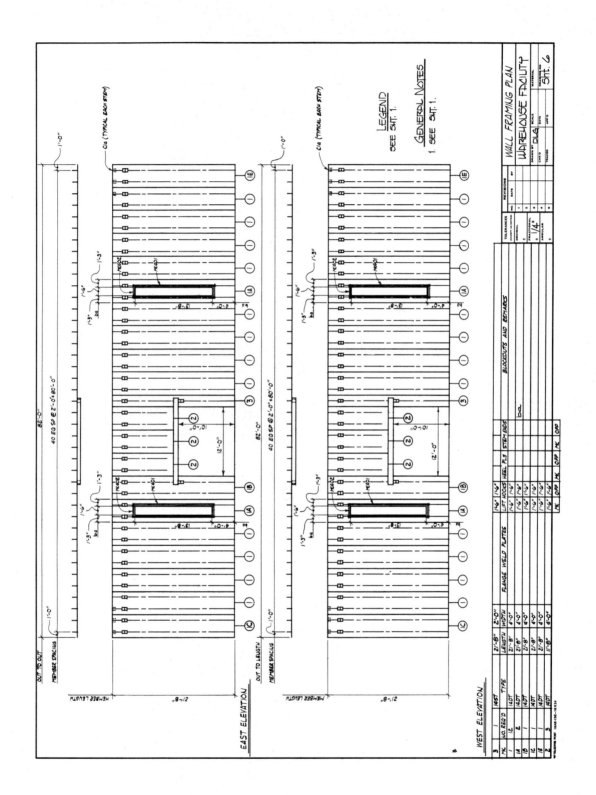

Fig. 11-6 Double-tee wall panel framing plan (east and west elevations)

Mark Numbers

Precast concrete structures are composed of numerous individual structural members. These individual members are fabricated in a manufacturing plant, shipped to the jobsite, and erected. In order to manage the process, a system of mark numbers is used by most companies. Assigning mark numbers to each individual structural member is much like numbering the pieces of a jigsaw puzzle so that it can be easily assembled.

In order to distinguish structural members within a job, it is common practice to assign mark numbers from different series. For example, columns are usually numbered within the 100 series. This means that the numbers assigned to columns fall between 100 and 199. Beams are numbered in the 200 series, while floor, wall, and roof members are numbered from 1 to 99.

The rules for assigning mark numbers are:

- All members that have the same length, width, thickness, shape, reinforcing, and stranding receive the same mark number.
- Any member that differs in even one of these ways receives a different mark number.
- Members that are alike in all of these ways, but still differ in some other way, receive a mark number suffix (i.e. mark numbers 100, 100A, 100B, 100C, etc).

Examine the mark numbers in Figures 11-1 through 11-6 and try to determine the reasons for the differences in numbers. These reasons are not always clear from the framing plan. Sometimes, the drafter is required to examine sections and details to find an answer. Sections and details are covered in later units.

Product Schedules

Product schedules provide a summary of all structural products contained in a framing plan. These schedules serve as an easy reference for information such as the following:

- Types of products used
- Length and width of all products
- How many of each product are used on a given framing plan
- Various other items of miscellaneous information

Figure 11-7 shows a product schedule that might be found on a framing plan, usually along the bottom border line. Individual companies differ slightly in the makeup of their product schedules, but all companies use them. The sample schedule in Figure 11-7 can be used for all drafting activities in this unit.

Unit 11 Precast Concrete Framing Plans 149

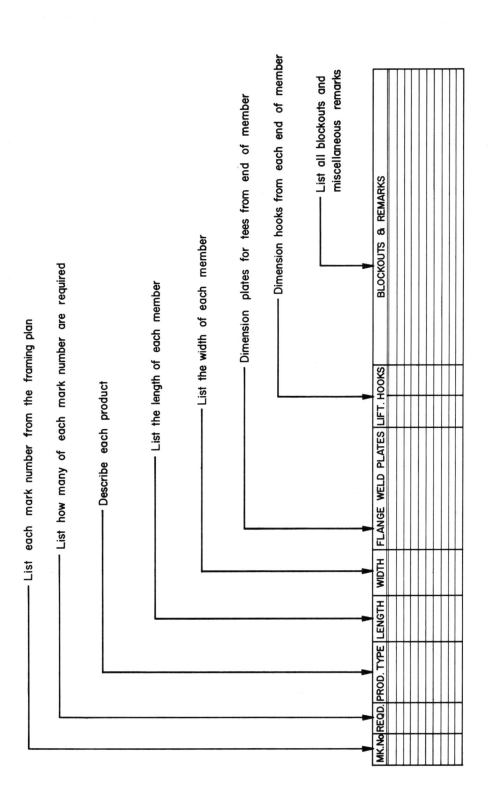

Fig. 11-7 Sample product schedule

General Notes

Notes are an important part of framing plans. Notes provide information to further clarify the drawings. There are various types of notes used on framing plans, but the most common are *general notes*.

These are notes that apply to the entire job as opposed to some notes that apply only to a specific part of the job. Figure 11-8 has a sample of the general notes that might be found on a framing plan. To save time, it is common practice to place the general notes for a job on a cover sheet or on the first sheet of the job only, Figure 11-1. On successive sheets, only the notes that specifically apply to each sheet are listed. The reader's attention is directed to the first sheet or the cover sheet for the job-wide general notes, Figure 11-9. Miscellaneous notes to the fabricator, contractor, subcontractors, and architect are also placed on the drawings when required.

Legend

An important part of the framing plans and all sheets in a set of precast concrete working drawings is the legend. The *legend* is a key for readers that explains symbols and abbreviations used on the drawings.

Legends are handled much like the general notes. A master legend for the entire job is compiled and placed on the cover sheet or first sheet of the job rather than on each individual sheet, Figure 11-1. This is common practice; however, some companies do still place a legend on every sheet that contains symbols or abbreviations. Individual company differences relative to standard operating procedures are usually explained in a company's drafting practices manual.

Figure 11-10 lists a number of common abbreviations and symbols used in precast concrete drafting. Anytime a symbol or abbreviation is used it must be explained in a legend. Legends are particularly important in precast concrete drafting because abbreviations and symbols often vary from company to company. Figure 11-11 shows a sample legend that might be found in a set of precast concrete working drawings.

One of the most common entries in a precast concrete legend is the blockout. A *blockout* is an opening left in a precast member for a window, door, skylight, conduit, ductwork, piping, or for numerous other reasons. Blockouts are formed by placing a wooden, styrofoam, or cardboard object the size and shape of the desired opening in the fabrication bed before pouring in the concrete. Once the concrete is dry, the blockout material is removed leaving the desired opening.

Figures 11-4 and 11-5 show examples of blockouts in precast concrete members. Blockouts are labeled in the legend by letter designation, width in inches, depth in inches, and length in feet and inches, Figure 11-11.

GENERAL NOTES

1. GENERAL CONTRACTOR SHALL FIELD-CHECK AND VERIFY ALL DIMENSIONS AND CONDITIONS AT JOBSITE.
2. ERECTION INCLUDES PLACING OUR PRODUCTS AND MAKING OUR MEMBER CONNECTIONS ONLY.
3. ERECTION BY OTHERS, PRODUCTS F.O.B. TRUCKS JOBSITE.
4. NO GROUTING, POINTING, CAULKING, OR FIELD-POURED CONCRETE.
5. RELEASE STRENGTH: _____ psi (UNLESS OTHERWISE NOTED).
6. CEILING FINISH.
7. BLOCKOUTS SMALLER THAN 10" x 10" (EXCEPT AS NOTED ON SHOP DRAWING) TO BE FIELD-CUT BY PROPER TRADES.
8. WALL PANEL FINISHES.

Fig. 11-8 Sample general notes that apply to an entire job

General Notes

1. These notes apply to sht. 6 only.
2. See sht. 1 for job-wide general notes.
3. Shipped loose items reqd. this sht.:
   32 nbpa   14 nbpb   18 nbpc

Fig. 11-9 General notes that apply to a specific framing plan

PRODUCT ABBREVIATIONS

BM	—	Beam
CF	—	Core floor
COL	—	Column
DT	—	Double tee
FS	—	Flat slab
ITB	—	Inverted tee beam
LB	—	L beam or ledger beam
LT	—	Lin tee
RB	—	Rectangular beam
RC	—	Rectangular column
SqB	—	Square beam
SqCol	—	Square column
ST	—	Single tee
WP	—	Wall panel

MISCELLANEOUS ABBREVIATIONS AND SYMBOLS

ab — anchor bolt (aba = anchor bolt-a)
 b — blockout (ba = blockout-a)
bp — baseplate (bpa = baseplate-a)
     bearing pad (bpa = bearing pad-a)
cb — connection bolt (cba = conn. bolt-a)
cp — connection plate (cpa = conn. plate-a)
hp — heel plate (hpa = heel plate-a)
tr — threaded rod (tra = threaded rod-a)
wa — weld angle (waa = weld angle-a)
wp — weld plate (wpa = weld plate-a)
#3 Bar — 3/8" ∅ Reinforcing bar (#301, 302, 303, etc.)
#4 Bar — 4/8" ∅ Reinforcing bar (#401, 402, 403, etc.)
#5 Bar — 5/8" ∅ Reinforcing bar (#501, 502, 503, etc.)
#6 Bar — 6/8" ∅ Reinforcing bar (#601, 602, 603, etc.)
#7 Bar — 7/8" ∅ Reinforcing bar (#701, 702, 703, etc.)
#8 Bar — 8/8" ∅ Reinforcing bar (#801, 802, 803, etc.)

Fig. 11-10 Common abbreviations and symbols in precast concrete drafting

Legend

ba - Blockout - a - 5" x 6" x 1'-2"
bb -   "   - b - 6" x 3" x 1'-0"
bc -   "   - c - 10" x 2" x 0'-9"
bd -   "   - d - 3" x 3" x 0'-3"
be -   "   - e - 6" x 9" x 2'-1"
#301 - #3 Bar x 6'-0" lg. straight
#302 - #3 Bar x 1'-2" lg. Bt. See Detail
#401 - #4 Bar x 4'-2"   "   "   "
#501 - #5 Bar x 18'-0" lg. straight
#502 - #5 Bar x 1'-9"   "   "   "
aba - Anchor bolt - a - 1/2"∅ x 1'-6" lg.
bpa - Baseplate - a - See Detail
cpa - Connection Plate - a - See Detail
wpa - Weld Plate - a - See Detail

Fig. 11-11 Sample legend

DRAWING FRAMING PLANS

Precast concrete drafters obtain the information they need to prepare framing plans from architectural plans, engineer's sketches, or contractor's sketches. The framing plans may either be prepared by one drafter or by a team of drafters. If one person is assigned all of the framing plans, he or she will proceed in the following order: 1) column framing plan, 2) beam framing plans, 3) floor framing plan (if applicable), 4) wall framing plans, and 5) roof framing plan.

If a team of drafters prepares the framing plans simultaneously, they must coordinate their efforts very closely. The procedures for preparing the various types of framing plans vary slightly and should be confronted individually.

Figure 11-12 shows an architect's rough drawing of the structural plans for a small department store. The job calls for precast columns, beams, walls, and roof members. Figures 11-13 through 11-16 illustrate precast concrete framing plans that were prepared according to the architect's rough drawing. These plans were drawn according to the procedures that follow.

Column Framing Plan

1. From the raw data supplied (architect's drawings), determine the centerline of column dimensions, select an appropriate scale, and lay out construction lines for the centerlines of the columns.
2. Determine the depth and width of the columns and draw the plan view of each column locating it on the proper centerline, Figure 11-13.
3. Add centerline of column number and letter designations and dimensions as shown in Figure 11-13.
4. Assign mark numbers to each column based on information available. Mark number changes or suffixes are subject to occur as the drawings progress.
5. Add the products schedule, fill in the title block, add general notes for the entire job, and add a north arrow. North arrows are presumed to be straight up on the sheet unless otherwise specified, Figure 11-13.
6. On a separate pad of paper, begin compiling a master legend for the entire job. Once the job, including all engineering and shop drawings, is complete, the legend is entered on the first page or on a cover sheet. If the legend is lengthy, it is common practice to save drafting time and space on the sheet by having it typed rather than lettered by hand, Figure 11-13.

Unit 11 Precast Concrete Framing Plans 153

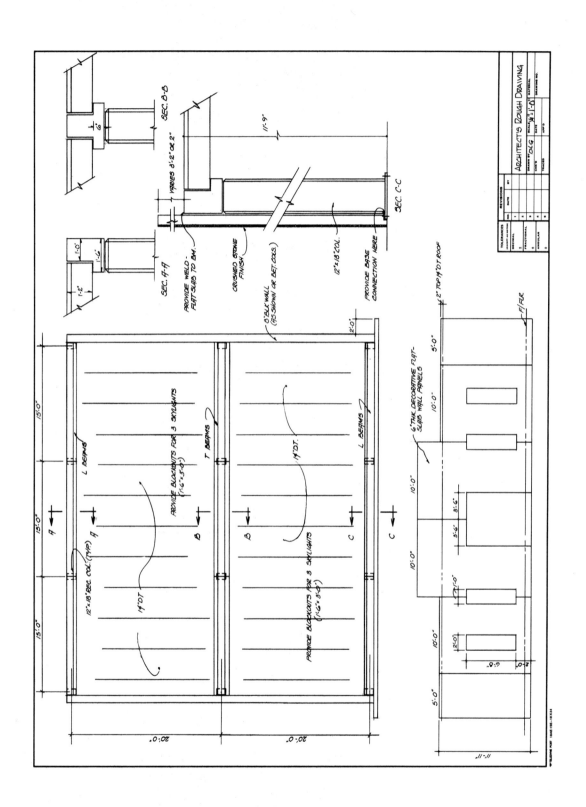

Fig. 11-12 Architect's rough drawing of the structural plans for a small department store

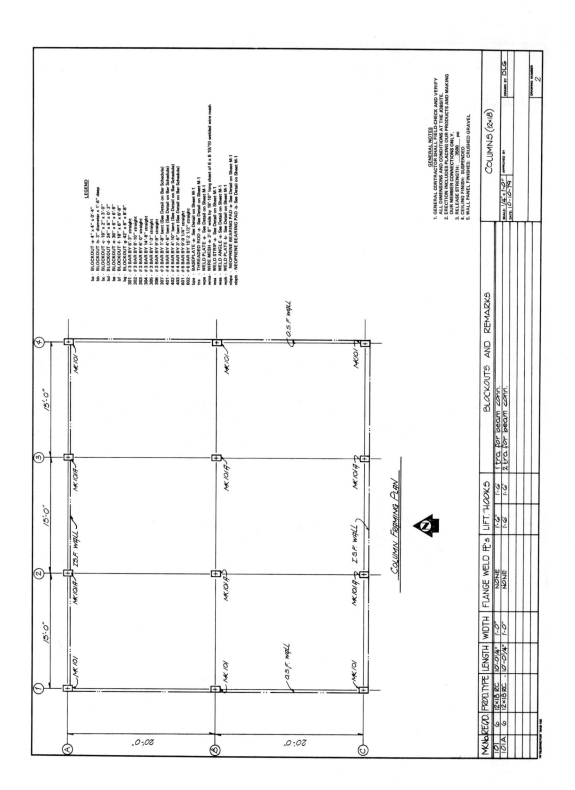

Fig. 11-13 Column framing plan prepared from architect's rough drawing

Beam Framing Plan

1. Determine the types and sizes of beams to be used.
2. Basing your calculations on the centerline of column dimensions, determine the beam lengths. It is common practice to leave a 1″ joint between abutting beams, Figure 11-14. A sample beam length calculation would proceed as follows: (For Mark Number 202, Figure 11-14)
   a. Record the centerline of column dimension for the columns that support Mark Number 202: 15′-0″.
   b. From this dimension, subtract one-half of the 1″ joint from each end of Mark Number 202 or a total of 1″. This leaves a total beam length of 14′-11″.
3. Place a new sheet over the completed column framing plan and trace each column using hidden lines. Draw the beams resting on top of the columns and show the appropriate joint between abutting beams. Widths of beams and the joints between them may be exaggerated for clarity, Figure 11-14.
4. Add the appropriate dimensions as shown in Figure 11-14. Centerline of column dimensions are optional on the beam framing plan and are often omitted.
5. Assign mark numbers to each beam based on information available. Mark number changes and suffixes are subject to occur as the drawings progress.
6. Complete the products schedule and title block; add any notes that apply specifically to the beam framing plan (see the Fabricator's note in Figure 11-14); refer readers to the first sheet for the legend and add the north arrow, Figure 11-14. Note that the blockout section of the products schedule is not filled in until later when the beam-to-column connection details have been drawn.
7. Examine each place where a beam rests on a column and assign letter designations for each different situation. Situations that are very close to being the same receive the same letter with the abbreviation *Sim*, meaning similar. Situations that are a mirror image of another situation receive the same letter with the abbreviation *Opp Hd*, meaning opposite hand.

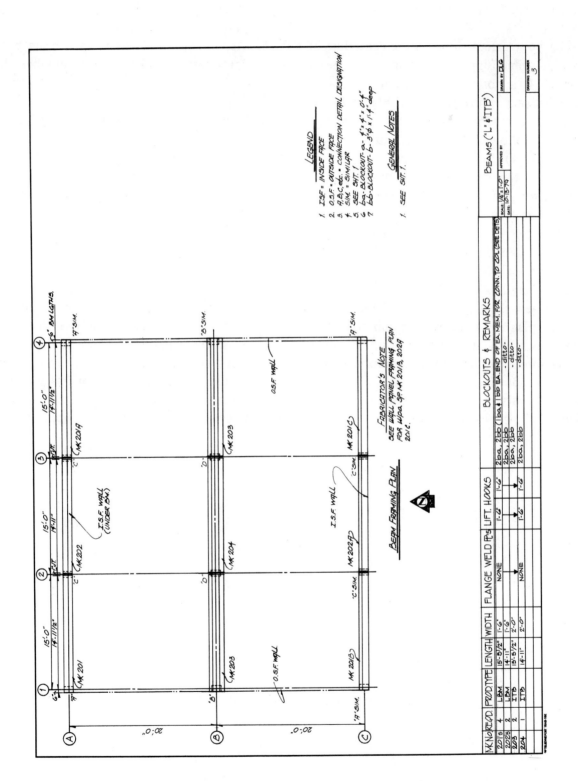

Fig. 11-14 Beam framing plan prepared from architect's rough drawing

Floor/Roof Framing Plans

1. Determine what product is to be used for framing the roof or floor and then look up the width of the framing product.
2. Place a new sheet over the beam framing plan and trace each beam. The bearing surfaces of the beams should be drawn with hidden lines.
3. Calculate the required amount of framing. In Figure 11-12, the amount of framing can be determined to be 46'-0". This amount is arrived at by first adding together the three 15'-0" centerline of column dimensions for a total of 45'-0". An additional 6" of framing is required to cover from the centerline of the outside columns to the edge of the walls, requiring an additional 1'-0" to be added to the initial 45'-0". This makes a total of 46'-0" of framing required. Since 14" double tees are 4'-0" wide, the roof may be framed with 11 full double tees and 1 cut in half and used as a 2'-0" wide single tee, Figure 11-15.
4. Based on the calculations from Step 3, above, draw the roof members and add necessary dimensions, Figure 11-15.
5. Examine the raw data closely to determine if openings (blockouts) are required in the roof. In Figure 11-12, the architect has requested 3, 1'-6" x 3'-0" openings in each of the two bays of framing for skylights. The requested blockouts are placed in the roof members, dimensioned, and assigned a letter designation, Figure 11-15.
6. Cut sections to clarify internal relationships as shown in Figure 11-15. The actual section breakouts are drawn after the framing plans have been completed.
7. Assign a mark number to each roof member based on information available. Mark number changes or suffixes are subject to occur as the drawings progress.
8. Complete the products schedule and title block; add any notes that apply specifically to the roof framing plan; refer readers to the first sheet for the legend and add the north arrow, Figure 11-15. Note that the information for the Flange Weld Plates section of the products schedule is used for floor or roof framing plans when double or single are the framing product. However, the information that is entered in this section is supplied by the engineer and not computed by the drafter. See Figure 11-17.

158 Section 3 Structural Precast Concrete Drafting

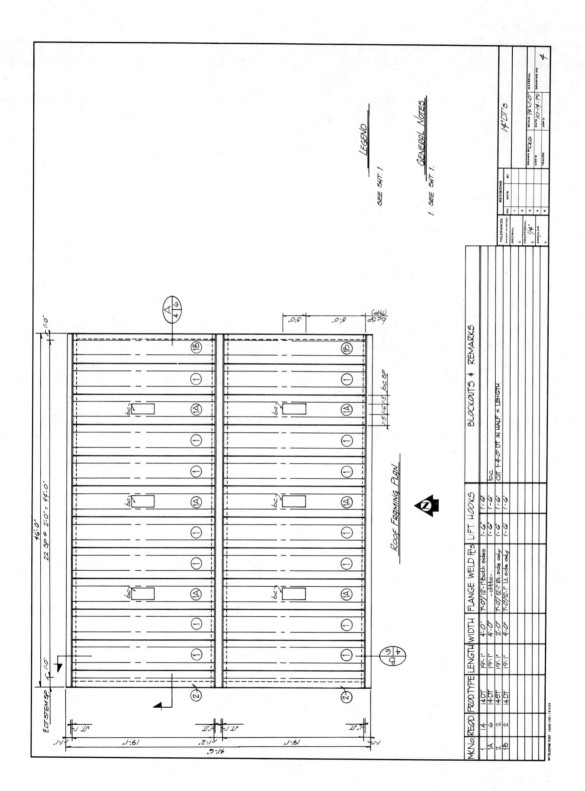

Fig. 11-15 Roof framing plan prepared from architect's rough drawing

Wall Panel Framing Plans

1. From the raw data, determine the size and type of product used to frame the walls. In Figure 11-12, 6″ thick, flat-slab wall panels with a decorative coating of crushed gravel are specified.
2. Determine the total amount of framing, the widths of framing members, and the heights of framing members. From Figure 11-12, it can be determined that the total amount of framing required is 50′-0″. To frame this, the architect has requested 4 full, 10′-0″ wide panels and 2 panels cut to 5′-0″ wide.
3. Using the information determined in Step 2 above, lay out the wall panel framing. Wall panel framing plans consist of a plan view and an elevation of each wall panel drawn as if the viewer is on the inside of the building looking out. The columns and beams are superimposed on the wall panel elevations with phantom lines, Figure 11-16.
4. Locate all required openings for doors, windows, etc., in the wall panel elevations, assign each opening a blockout designation, and add dimensions, Figure 11-16.
5. Identify any connections that are required involving the wall panels, obtain the engineer's design for the connections, and draw the connectors in the appropriate wall panels as indicated in the calculations. Figure 11-17 has sample engineer's connection calculations.
6. Add necessary dimensions as shown in Figure 11-16. Assign mark numbers for each wall panel based on information available. Mark numbers and suffixes are subject to occur as the drawings progress.
7. Complete the products schedule and title block; add any notes that apply specifically to the wall panel framing plan; add a key plan and code it to show which wall is being drawn (Figure 11-16); refer readers to the first sheet for the legend and add the north arrow next to the key plan.

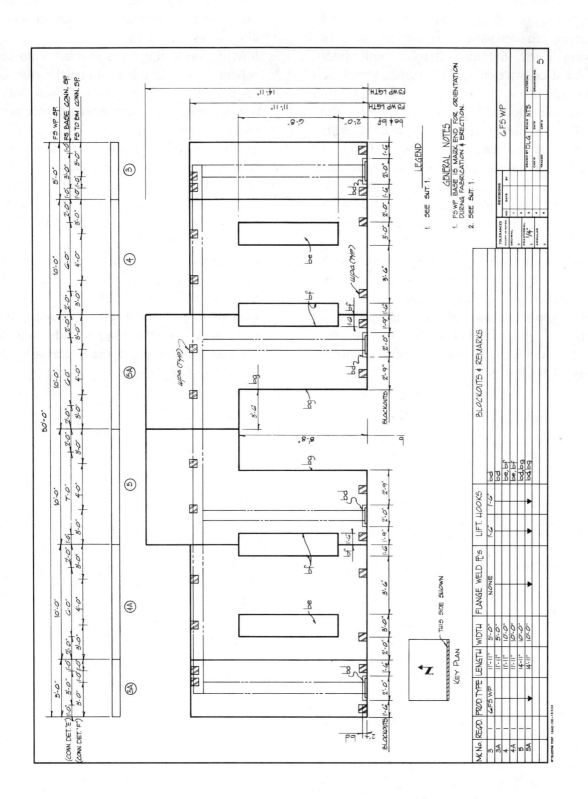

Fig. 11-16 Wall framing plan prepared from architect's rough drawing

SUMMARY

... Precast concrete drafters prepare two types of drawings: engineering drawings and shop drawings. Framing plans are engineering drawings.
... There are four basic types of framing plans in precast concrete drafting: column, beam, floor/roof, and wall.
... Information required on framing plans includes: a plan view layout of all structural members; mark numbers for each member; product schedule, general notes, and legend; north arrow; complete dimensions, and centerline of column designations (required on column framing plans, optional on beam framing plans, and not required on floor, roof, or wall panel framing plans).
... All members in a precast concrete job must be assigned mark numbers.
... Floor, roof, and wall members receive mark numbers between 1 and 99, columns receive mark numbers between 100 and 199, and beams receive mark numbers between 200 and 299.
... All precast members in a job that have the same length, width, thickness, shape, reinforcing, and stranding receive the same mark number.
... Any member that meets the above criteria but still differs in some way receives a mark number suffix.
... Any member that does not have the same length, width, thickness, shape, reinforcing, or stranding receives a different mark number.
... Product schedules are provided on framing plans as summaries of the products used in the plan.
... General notes pertain to an entire job and are usually placed on the first sheet of the job or on a cover sheet.
... Only notes that apply specifically to a particular sheet and are not contained in the general notes are placed on successive sheets of a job.
... A legend is a guide to readers of a set of plans that explains abbreviations and symbols.
... A master legend for a job is usually compiled and placed on the first sheet or a cover sheet rather than on each sheet in a job.
... Blockouts in precast concrete members are given a letter designation and placed in the legend by width in inches, depth in inches, and length in feet and inches.
... Precast concrete drafters obtain the information they need to prepare framing plans from architect's drawings, engineer's sketches, or contractor's sketches.

REVIEW QUESTIONS

1. Name the two types of drawings that are prepared by precast concrete drafters.
2. List the four basic types of precast concrete framing plans.
3. List the information that is required on a framing plan.
4. What is the purpose of a product schedule on a framing plan?
5. Why are general notes usually placed on the first sheet of a set of plans rather than repeated on each individual sheet?
6. Explain the purpose of a legend on precast concrete drawings.
7. Provide a legend entry for an opening in a precast concrete member that is to be 3'-0" long, 6" deep, and 1'-0" wide.
8. From what group of numbers would a precast concrete beam be assigned a mark number?
    a. 1-99
    b. 100-199
    c. 200-299
    d. 300-399

9. From what group of numbers would a precast concrete floor, wall, or roof member be assigned a mark number?
   a. 1–99
   b. 100–199
   c. 200–299
   d. 300–399
10. From what group of numbers would a precast concrete column be assigned a mark number?
    a. 1–99
    b. 100–199
    c. 200–299
    d. 300–399

REINFORCEMENT ACTIVITIES

The raw data for the following drafting exercises is provided in the architect's drawing contained in Figure 11-18 and the engineer's calculations contained in Figure 11-17.

1. On a C-size sheet of paper, prepare a column framing plan.
2. On a C-size sheet of paper, prepare a beam framing plan.
3. On a C-size sheet of paper, prepare a wall panel framing plan for the north wall.
4. On a C-size sheet of paper, prepare a wall panel framing plan for the south wall.
5. On a C-size sheet of paper, prepare a wall panel framing plan for the east wall.
6. On a C-size sheet of paper, prepare a wall panel framing plan for the west wall.
7. On a C-size sheet of paper, prepare a roof framing plan.

The raw data for the following drafting activities is provided in the architect's drawing contained in Figure 11-19 and the sample roof framing plan provided in Figure 11-20.

8. The architect's drawing in Figure 11-19 calls for a precast concrete roof of either prestressed flat slabs or cored flat slabs. The sample roof framing plan in Figure 11-20 is an example of how the three-unit apartment complex might be framed in 4" thick, flat-slab members. Using Figure 11-20 as an example, construct another roof framing plan for the three-unit apartment complex in Figure 11-19 using 8" thick x 4' wide cored flat slabs. The drawing should be done on a C-size sheet of paper.

9. Using the architect's drawing of a three-unit apartment complex in Figure 11-19 as a starting point, change the overall length of the building to 66'-0" and the overall width to 42'-0". Then construct a roof framing plan for the building using 4" thick x 10' wide prestressed flat slabs. Use Figure 11-20 as an example and put the drawing on C-size sheet of paper.

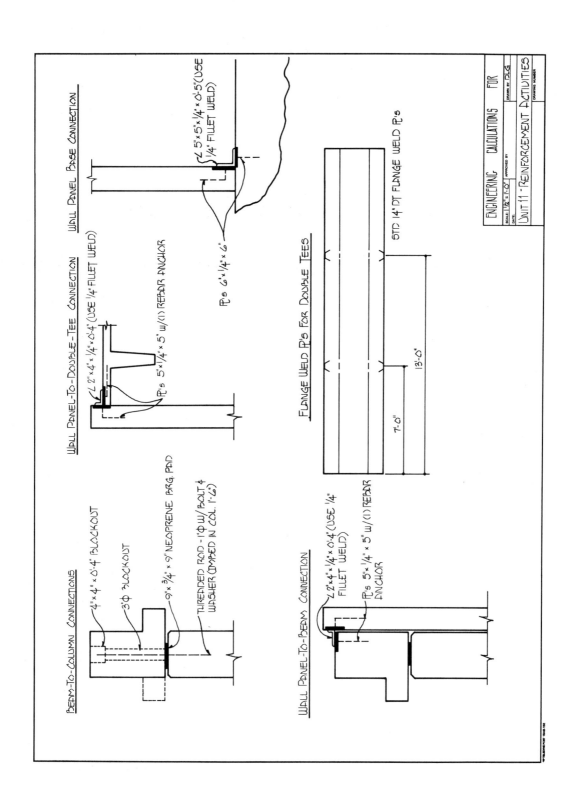

Fig. 11-17 Engineering calculations for Reinforcement Activities

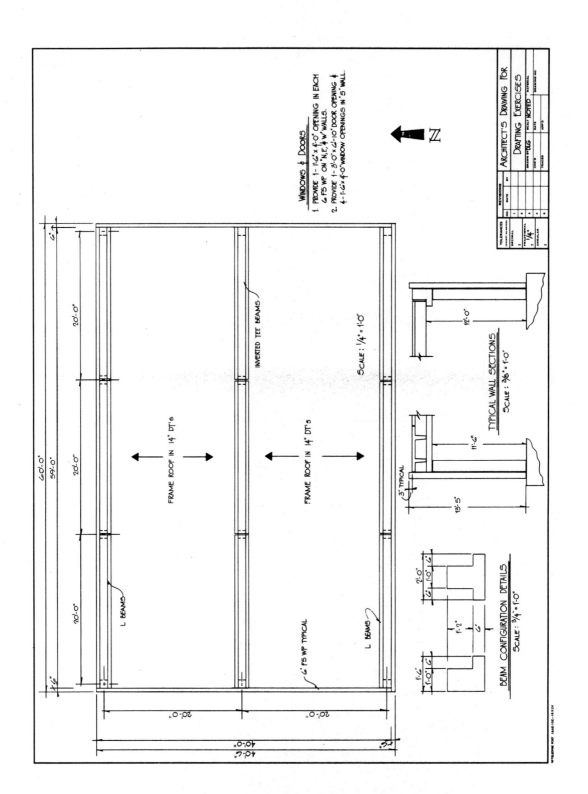

Fig. 11-18 Raw data for Reinforcement Activities

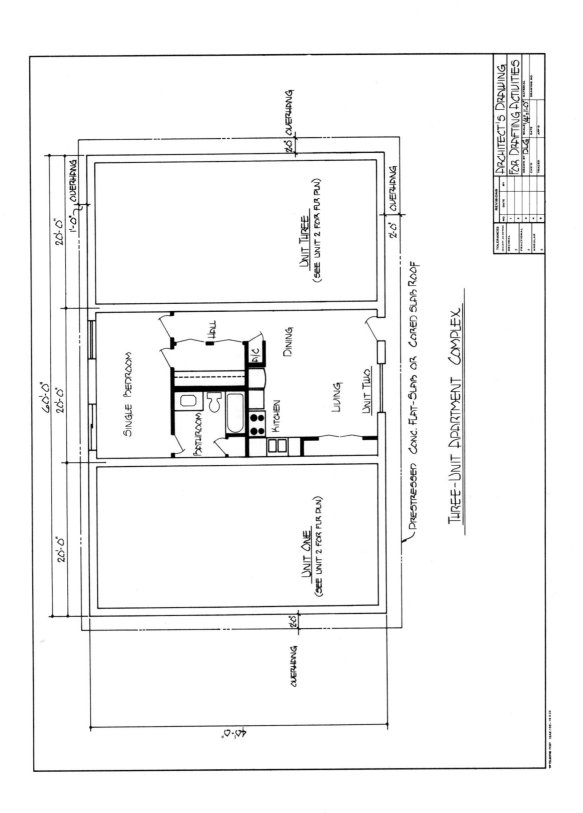

Fig. 11-19 Architect's drawing for drafting activities

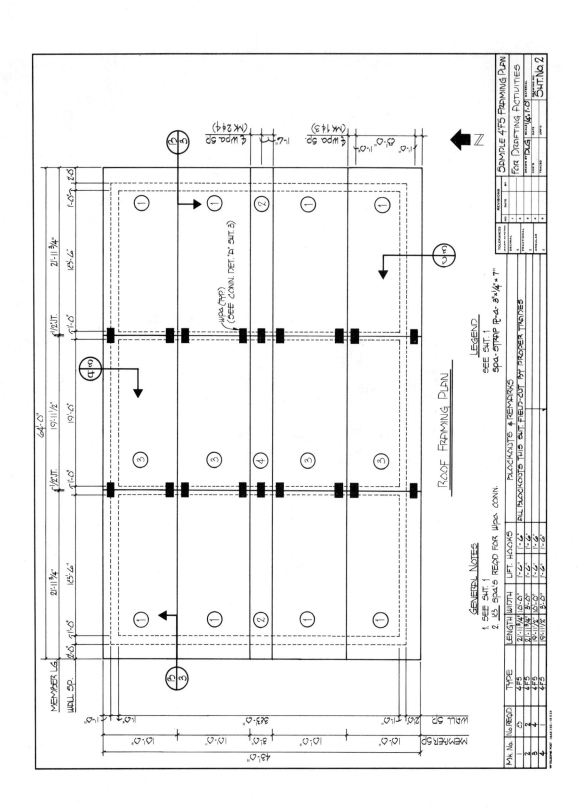

Fig. 11-20 Sample roof framing plan for drafting activities

Unit 12
Precast Concrete Sections

OBJECTIVES

Upon completion of this unit, the student will be able to

- define precast sections.
- prepare precast concrete full, partial, and offset sections.
- illustrate examples of structural section conventions.

PRECAST CONCRETE SECTIONS DEFINED

Sections are drawings that show the materials used in a job and how they fit together to form a completed structure. They are cut on framing plans with a section cutting symbol along imaginary planes through the plan. Figure 12-1 shows an annotated example of a commonly used section cutting symbol. Refer to Figure 7-4 for more examples.

Sections are a very important part in a set of precast concrete working drawings. Sections clarify internal relationships in a structure that either do not appear on the framing plans or appear as hidden lines. In addition to showing how a structure fits together, sections also show height information. This information includes such items as the distance from the finished floor to the ceiling or other important height information. Figure 12-2 shows two full sections that were cut on the framing plan found in Figure 11-15. Examine these two figures closely. Make note of the close relationship that exists between framing plans and sections.

There are several different types of sections used in precast concrete drafting. The most commonly used sections are full, partial, and offset sections.

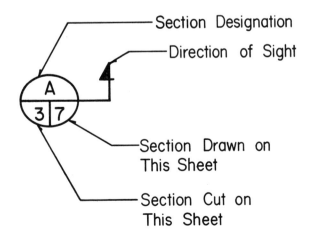

Fig. 12-1 Annotated section cutting symbol

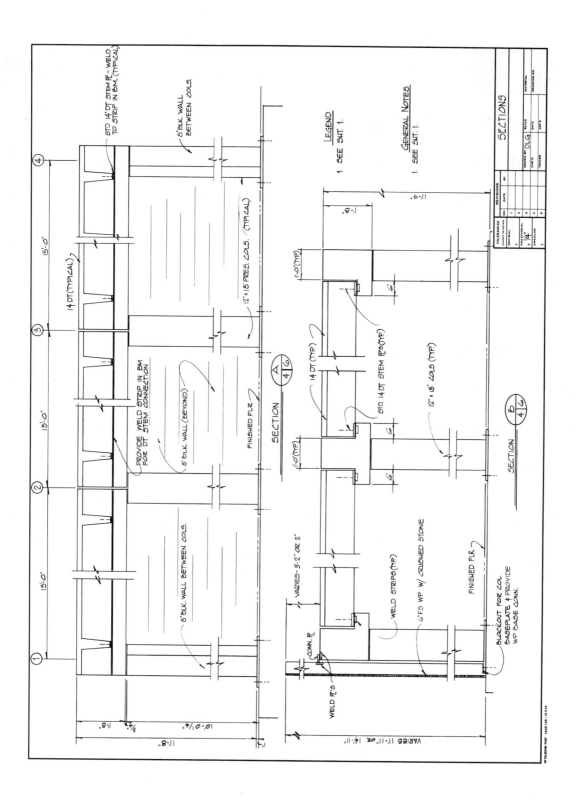

Fig. 12-2 Precast concrete sections

FULL SECTIONS

Full sections are sections that cut through an entire building or structure. Full sections are of two basic types: longitudinal and cross sections. Longitudinal sections cut across the entire length of a building, Figure 12-3. Cross sections cut across the entire width of a building, Figure 12-4. Both types of full sections fall into the broader category, *structural sections*.

All sections cut in precast concrete drafting are structural sections. This means that the sectional drawings show only the structural components of a structure (foundation, floors, walls, columns, beams, and roof), the materials of which they are made, and how they are connected. Architectural sections, on the other hand, show all of the interior details of the structure such as carpet, paneling, suspended ceiling, and baseboards, etc. Since these interior details are not part of the structural makeup of a structure, drafting time is shortened by excluding them.

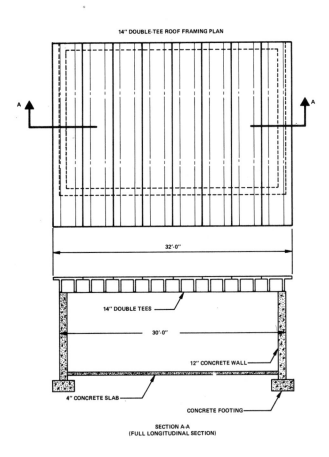

Fig. 12-3 Full longitudinal section

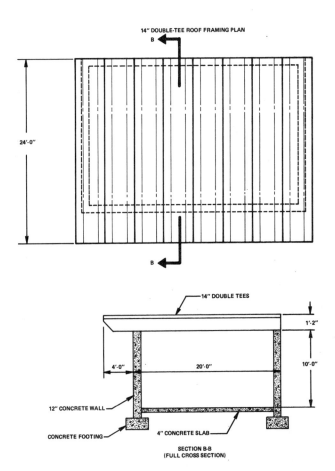

Fig. 12-4 Full cross section

PARTIAL SECTIONS

Partial sections are often used in precast concrete drafting to clarify isolated portions of a structure. Full sections are very involved and time-consuming. If the desired clarification can be provided with less than a full section, a partial section is cut and drawn. A complete set of precast concrete working drawings may contain full sections as well as several partial sections. Several examples of partial sections as used in precast concrete drafting are shown in Figure 12-5.

OFFSET SECTIONS

When cutting a section through a framing plan or some portion of a structure, the cutting plane line does not always run straight. To clarify detailed interior situations that may not fall along a straight extension of the cutting plane line, the line may offset, Figure 12-6. When the cutting plane line is offset, the sectional drawing is drawn as if the offset interior portion of the structure aligned with the straight extension of the cutting plane line from its point of origin, Figure 12-7.

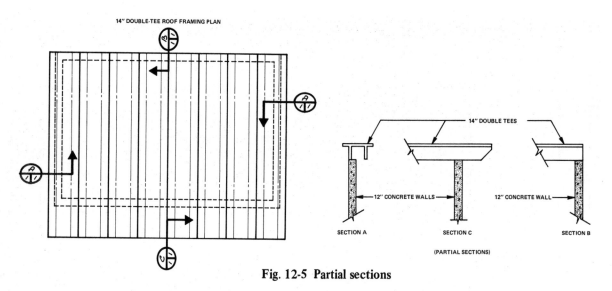

Fig. 12-5 Partial sections

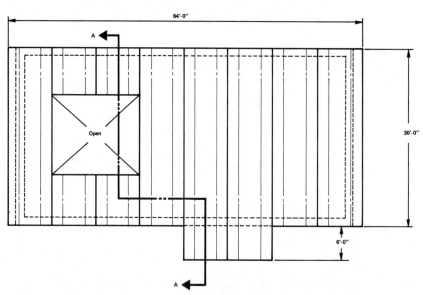

Fig. 12-6 Framing plan with offset section cut

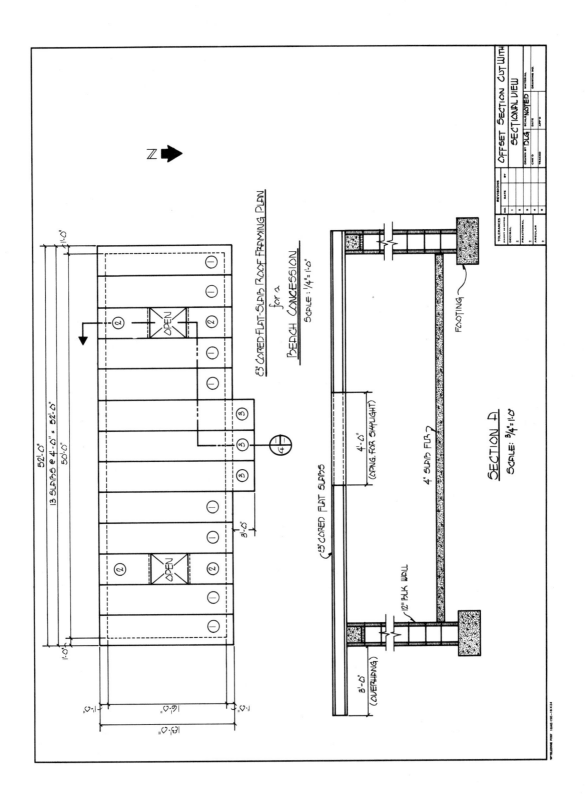

Fig. 12-7 Offset section cut with sectional view

172 Section 3 Structural Precast Concrete Drafting

SECTION CONVENTIONS

One of the primary purposes of sections is to show the reader of a set of plans what materials the structure is made of. All materials used in building have standard symbols that may be used in sectional drawings, or an individual company may use its own materials symbols and explain them in a legend. Examples of building materials symbols commonly used in precast concrete drafting are shown in Figure 12-8. Although these symbols are used in general, many companies cut down on drafting time by minimizing their use. It is common practice for a company that specializes in the fabrication of precast concrete products to make sectional drawings without using the standard symbols for steel and concrete. This can be done without confusion because only the structural components of a structure are included in the sections and these will be constructed of concrete with steel connectors. The use of sectioning symbols is definitely required only where a structure is composed of several different structural materials.

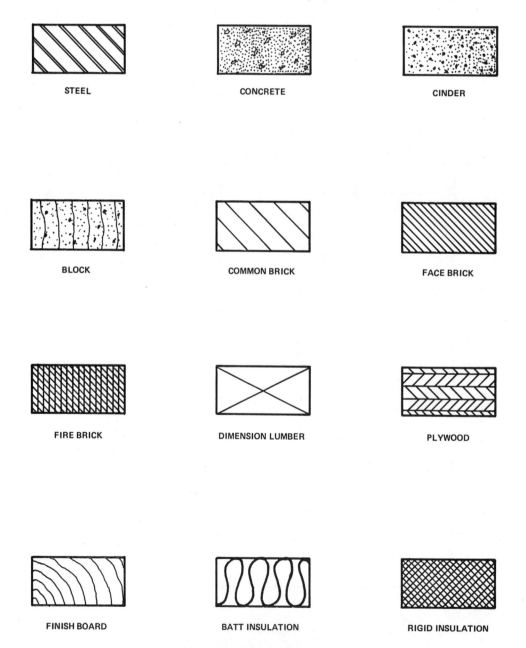

Fig. 12-8 Common sectioning symbols

DRAWING PRECAST CONCRETE SECTIONS

Precast concrete drafters obtain the information they need to prepare sectional drawings from two sources:

- The framing plans where the sections were cut
- The raw data from which the framing plans were prepared

The framing plans indicate what sections actually must be drawn. The raw data from which the framing plans were prepared provides information on how the structural members fit together, how they are to be connected, and height information. In Figures 12-9 and 12-10, common precast concrete sectional drawings are illustrated. These sectional drawings were prepared according to the following procedures:

1. From the framing plan(s), determine what sectional drawings must be prepared.
2. Determine whether the sectional drawings are to be full sections, partial sections, or offset sections. This indicates how many sheets or how much space on one sheet is required to accommodate the sectional drawings.
3. Examine each sectioned situation individually and answer the following questions for each:

   - What materials are used in the section?
   - How do the structural components fit together?
   - What are the important height distances in the section?

4. Based on the information obtained in step three, select an appropriate scale and draw the sectional drawing. Sections may be drawn to a number of different scales, depending on the degree of detail desired, the complexity of the section, and the amount of space available for drawing. Sections in precast concrete drafting are usually drawn to a scale of 3/4" = 1'-0" or larger.

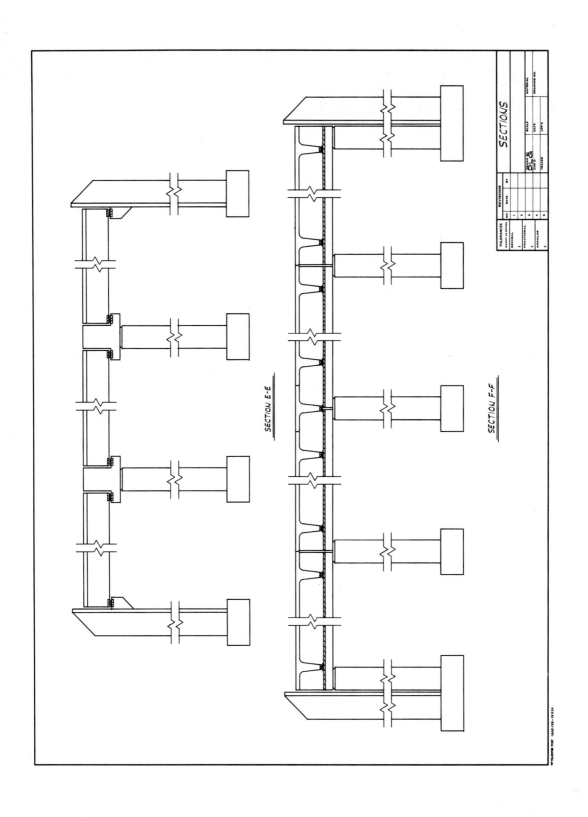

Fig. 12-9 Precast concrete full sections

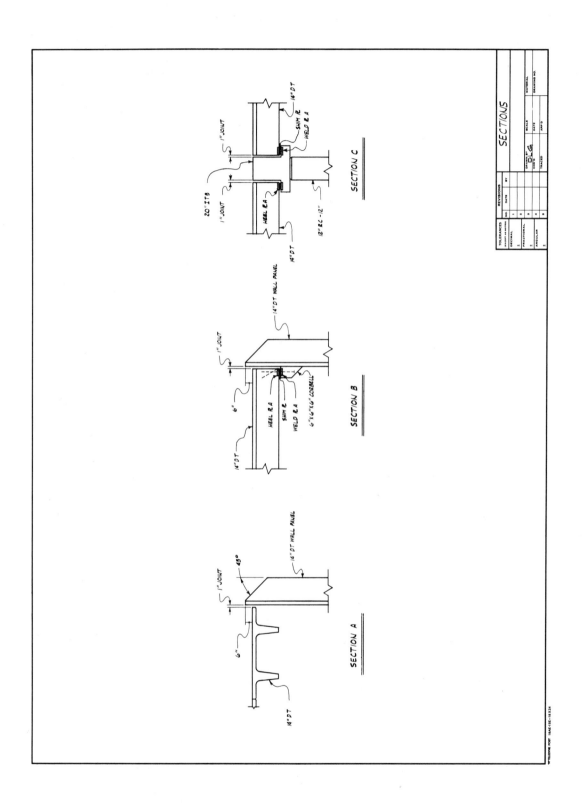

Fig. 12-10 Precast concrete partial sections

SUMMARY

... Sections are drawings that show the materials used in a job and how they fit together to form a completed structure.
... Sections clarify internal relationships in a job that do not appear on a framing plan or appear hidden.
... Sections are also important sources of height information in precast concrete working drawings.
... The most commonly used types of sections in precast concrete drafting are full, partial, and offset sections.
... Full sections cut through an entire building or structure along an imaginary cutting plane.
... Full sections are of two basic types: longitudinal and cross sections.
... Longitudinal sections cut across the entire length of a structure while cross sections cut across the entire width of a structure.
... All sections used in precast concrete drafting are considered structural sections.
... Structural sections differ from architectural sections in that they contain only the structural components (foundation, floors, walls, columns, beams, and roof) and exclude interior details such as carpeting, paneling, baseboards, suspended ceilings, etc.
... Partial sections are used to clarify isolated portions of a structure and should be used anytime when a full section is not required.
... Offset sections are used to clarify detailed interior situations in a structure that do not fall along a straight extension of the cutting plane line from its point of origin.
... When drawing precast concrete sections, materials are symbolized by standard section conventions or by individual company symbols that are explained in a legend.

REVIEW QUESTIONS

1. In a single sentence, define *sections*.
2. What is the purpose of a section?
3. List the three most common types of sections in precast concrete drafting.
4. Define *full section*.
5. List the two types of full sections.
6. Distinguish between longitudinal and cross sections.
7. All sections in precast concrete drafting fall into what broad category?
8. How do structural sections differ from architectural sections?
9. Sketch an example of a partial section.
10. Explain how an offset section is used.

REINFORCEMENT ACTIVITIES

The title block as shown in Figure 1-21 and 1/2" borders are required for all activities.

1. Using the roof framing plan in Figure 12-11 as the raw data, construct a full longitudinal section. The section should be drawn on a C-size sheet of paper at a scale no smaller than 3/4" = 1'-0".

2. Using the roof framing plan in Figure 12-11 as the raw data, construct a full cross section. The section should be drawn on a C-size sheet of paper at a scale no smaller than 3/4" = 1'-0".

3. Figure 12-12 contains a roof framing plan of 8" cored flat slabs for an auto repair shop and office. On a C-size sheet of paper, at a scale of 1" = 1'-0", construct a full longitudinal section for the structure.

4. Figure 12-12 contains a roof framing plan of 8" cored flat slabs for an auto repair shop and office. On a C-size sheet of paper, at a scale of 1" = 1'-0", construct a full cross section through the garage area.

5. Figure 12-12 contains a roof framing plan of 8" cored flat slabs for an auto repair shop and office. On a C-size sheet of paper, at a scale of 1" = 1'-0", construct a full cross section through the office area.

The following activities are taken from Figure 12-13 and can all be placed on one, C-size sheet of paper.

6. Construct the partial section "A" at a scale of 3/4" = 1'-0" showing only the structural situation at the roof.

7. Construct the partial section "B" at a scale of 3/4" = 1'-0" showing the structural situation from the bottom of the footing to the top of the flat-slab wall panels.

8. Construct the partial section "C" at a scale of 3/4" = 1'-0" showing the structural situation from the bottom of the footing to the top of the double-tee roof.

9. Construct the partial section "D" at a scale of 3/4" = 1'-0" showing the structural situation from the bottom of the footing to the top of the double-tee roof.

The following activity involves major alterations to Figure 12-13 and should be placed on a separate sheet of C-size paper.

10. Construct a full cross section of the structure in Figure 12-13 at a scale of 1" = 1'-0" with the following alterations:
    a. Convert the L beam above the back wall to a rectangular beam, 12" wide x 18" deep. Extend the double tees to have full bearing on the rectangular beam.
    b. Convert the ITB to a rectangular beam, 12" wide x 18" deep and extend both bays of double tees to a half bearing on the beam less 1/2" each to provide a 1" joint between abutting double tees.
    c. Leave the L beam at the front wall and all other structural situations intact.

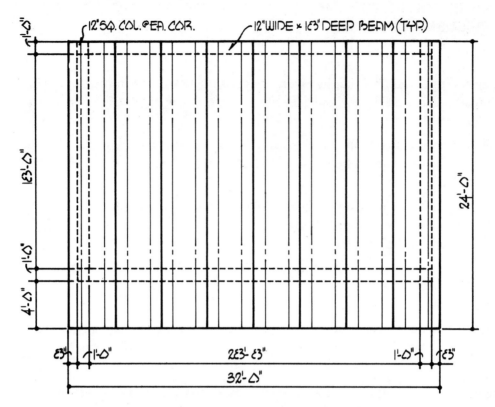

14" DT ROOF FRAMING PLAN

ADDITIONAL INFORMATION FOR SECTIONAL VIEWS
1. Columns rest on a footing of concrete 2'-0" wide x 1'-0" deep.
2. Columns are 9'-6" high.
3. Beams rest directly on top of columns.
4. 14" double tees rest directly on top of the beams.
5. The bottom of the 4" concrete slab floor is 8" above the top of footing.

Fig. 12-11 Roof framing plan for Reinforcement Activities

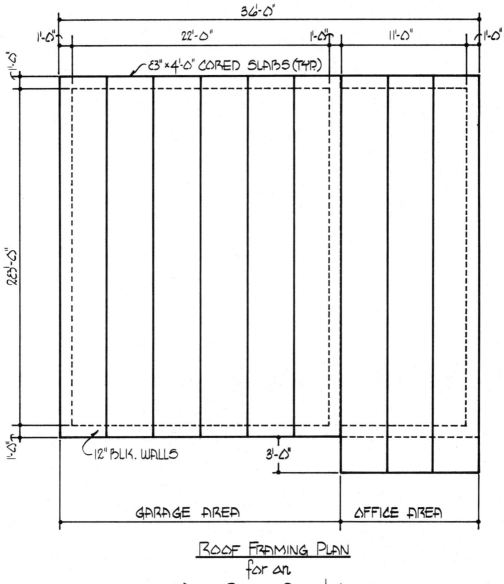

Fig. 12-12 Roof framing plan for Reinforcement Activities

ADDITIONAL INFORMATION FOR SECTIONAL VIEWS
1. 12" block walls rest on 2'-0" wide x 1'-0" deep concrete footings.
2. Cored slabs rest directly on block lintel beams filled with concrete (See Section A – Figure 12-7).
3. The bottom of the 4" concrete slab floor is 8" above the top of footing.
4. The office area has an 8'-0" high ceiling. The garage area has a 12'-0" high ceiling.

180 Section 3 Structural Precast Concrete Drafting

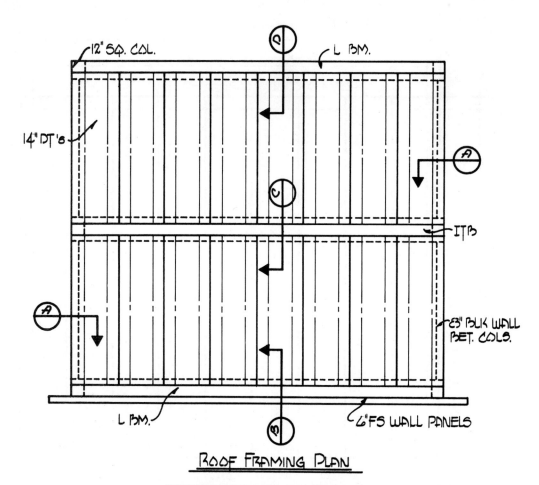

ADDITIONAL INFORMATION FOR SECTIONAL VIEWS
1. Left-side, right-side, and back walls are 8" block built-up between the columns from the concrete slab floor to the bottom of the beams or double tees.
2. The 6" flat-slab wall panels extend from the top of footing to a distance of 3'-0" above the top of the double-tee roof.
3. Columns are 12" square and rest on concrete footings 2'-0" wide x 1'-0" in depth.
4. L and ITB beam configuration dimensions may be obtained from Figure 12-2.
5. Concrete slab floor 4" thick is located 8" above the top of footing.
6. Bottom of double-tee elevation is 10'-0" above the top of the concrete slab floor.

Fig. 12-13 Roof framing plan for Reinforcement Activities

Unit 13
Precast Concrete Connection Details

OBJECTIVE

Upon completion of this unit, the student will be able to

- construct each of the following: precast concrete baseplate connection details; precast concrete bolted beam-to-column connection details; precast concrete welded connection details; and precast concrete haunch connection details.

PRECAST CONCRETE CONNECTION DETAILS DEFINED

Precast concrete products are fabricated in a manufacturing plant and shipped to the jobsite by truck and train to be erected. Lumber that is used in building a wooden structure must be fastened together by nails, clips, or other means. Precast concrete members must also be fastened or connected. Precast concrete connections at the jobsite are achieved by bolting, welding, or a combination of both.

In order to convey the intended connection designs to readers of the plans, connection details are drawn. A *connection detail* is a plan view of a connection with a section cut through it to show an elevation view of the actual connection. Typical connections requiring details in precast concrete drafting are:

- Column baseplate connections
- Bolted beam-to-column connections
- Welded connections
- Haunch connections

COLUMN BASEPLATE CONNECTIONS

Columns are one of the most frequently used precast concrete products. So that they can be attached to a footing, columns are cast with a heavy steel baseplate in one end. This baseplate is provided with drilled holes at specific locations that fit over anchor bolts cast in the footing.

Each column is erected by placing it over its corresponding anchor bolts which have been fitted with combination washers and leveling nuts. Once the column is up, its height and degree of level can be adjusted by turning the leveling nuts. Once leveled and the top of the column is at the proper elevation, another set of washers and nuts is tightened down on each anchor bolt. This is done until tne column fits firmly against the top of the baseplate permanently securing the column in place, Figures 13-1 and 13-2.

Baseplates cast in columns are of two basic types. The first type is a flat plate that extends beyond the perimeter of the column. It contains holes drilled through it to fit over anchor bolts cast in the footing, Figure 13-1. The second type of baseplate fits flush with the side of the column. It contains holes for anchor bolts and has small, three-sided compartments attached to it to accommodate the anchor bolt connections, Figure 13-2.

The type of baseplate connection used depends on the design considerations of each individual job. Selecting a baseplate connection is the engineer's responsibility. Drawing baseplate connection details is the drafter's responsibility.

DRAWING BASEPLATE CONNECTION DETAILS

When working on a job involving columns, the engineer gives the drafter instructions such as:

- What type of baseplate connection is desired
- Baseplate size and configuration specifications
- Any other pertinent information about the baseplate

With this information known, the drafter may begin preparing a baseplate connection detail.

Baseplate connection details are drawn according to the following procedures:

1. Examine the baseplate and column dimensions and the amount of space available for drawing the detail. After considering these factors, select an appropriate scale for drawing the connection detail. Commonly used scales are 3/4″ = 1′-0″, 1″ = 1′-0″, 1 1/2″ = 1′-0″. Do not draw connection details to a scale smaller than 3/4″ = 1′-0″.
2. Construct an elevation view of the baseplate and a portion of the column showing the baseplate, the anchor bolts, the footing, leveling nuts with washers, and tightening nuts with washers, Figures 13-1 and 13-2.
3. Cut a section horizontally through the column to show a sectional view looking down on top of the baseplate and column and draw the sectional view, Figures 13-1 and 13-2.
4. Completely label all components of the connection detail.

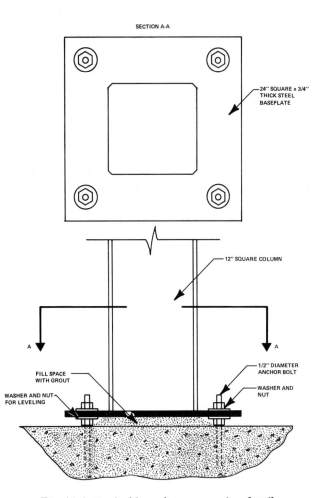

Fig. 13-1 Typical baseplate connection detail

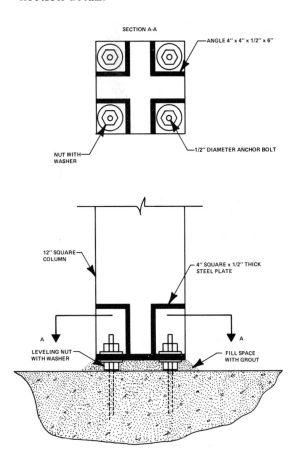

Fig. 13-2 Typical baseplate connection detail

BOLTED BEAM-TO-COLUMN CONNECTION DETAILS

In precast concrete construction, beams are connected to columns by bolting. Each place where a beam and column are attached requires a connection detail. Connection details are first identified and then called out on the roof and floor framing plans with letter designations, Figure 13-3.

A connection detail must be drawn for each separate letter designation called out on the framing plan(s). However, in the case of connections that are exactly opposite or are very similar, the number of details required can be reduced by adding abbreviations for the words *opposite hand* or *similar*. Opposite hand is abbreviated *Opp Hd*. This means that the connection being called out is an exact mirror image of another connection that has already been called out. The callout on the framing plan would read *Conn Det A (Opp Hd)*. Similar is abbreviated *Sim*. This means that the connection being called out is substantially the same as another connection that has already been called out. The abbreviation *Sim* may be used in any situation where the beams, columns, and connection are exactly alike, but other factors differ slightly, Figure 13-3.

There are six basic bolted beam-to-column connections used in precast concrete construction. In each case the connection procedures are the same. A long threaded rod is cast into the top of the column that protrudes a distance equal to the thickness of the beam it will support. The beam is fabricated with a round, 3" diameter blockout in the connection end(s) to fit over the threaded rod, and 4" x 4" x 4" blockout(s) to provide a com-

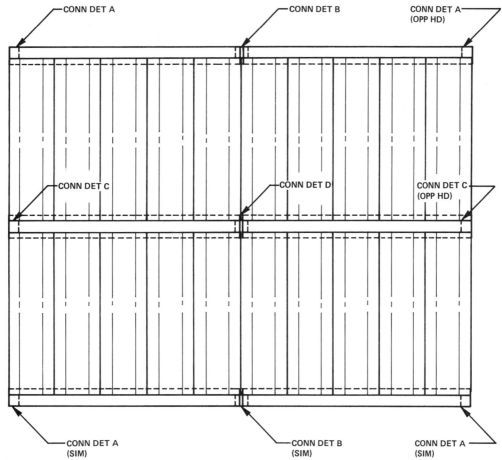

Fig. 13-3 Connection detail callouts

EXPLANATION OF CONNECTION DETAIL CALLOUTS
1. Connection Detail A — L-shaped beam bolted to a square column.
2. Connection Detail B — Two L-shaped beams bolted to the same square column.
3. Connection Detail C — Inverted-tee beam bolted to a square column.
4. Connection Detail D — Two inverted-tee beams bolted to the same square column.

partment for screwing a washer-nut combination onto the threaded rod, Figures 13-4 through 13-9.

To avoid damaging the bearing surfaces of the column or the beam, a 3/4" thick neoprene bearing pad is placed on top of the column before the beam is erected. This bearing pad cushions the bearing surface. The pad also lifts the beam just enough so that the top of the threaded rod is slightly below the top surface of the beam, Figures 13-4 through 13-9.

The six basic beam-to-column connections in precast concrete construction are:

- Single rectangular beam bearing fully on a column, Figure 13-4.
- Two rectangular beams bearing on the same column. In this case, a 1/2" to 1" joint is left between the abutting beams, Figure 13-5.
- Single L-shaped beam bearing fully on a column, Figure 13-6.
- Two L-shaped beams bearing on the same column. In this case, a 1/2" to 1" joint is left between the abutting beams, Figure 13-7.
- Single inverted-tee beam bearing fully on a column, Figure 13-8.
- Two inverted-tee beams bearing on the same column. In this case, a 1/2" to 1" joint is left between abutting beams, Figure 13-9.

DRAWING BOLTED BEAM-TO-COLUMN CONNECTION DETAILS

Regardless of the types of connections used on a given precast concrete job, the procedures for drawing connection details are the same.

1. Determine from the framing plan(s) how many connection details are required and the configuration of each.
2. Draw a plan view of each connection showing only the immediate area around the connection and cut a section horizontally through it, Figures 13-4 through 13-9.
3. Draw the sectional view showing an elevation of the column, beam(s), blockouts, threaded rod(s), washers/nuts, neoprene bearing pad(s), and any other information pertinent to the connection, Figures 13-4 through 13-9.

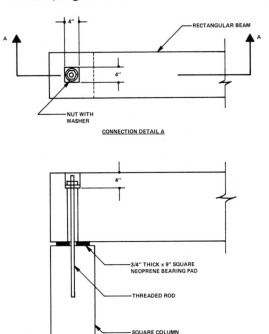

Fig. 13-4 Rectangular beam-to-column connection detail

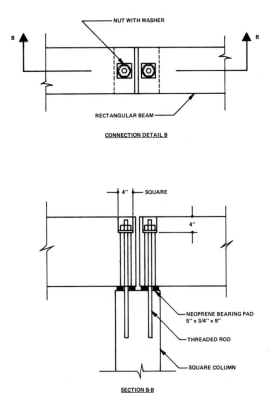

Fig. 13-5 Two rectangular beams-to-column connection details

Unit 13 Precast Concrete Connection Details 185

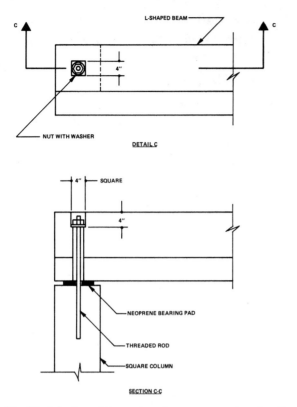

Fig. 13-6 L-shaped beam-to-column connection detail

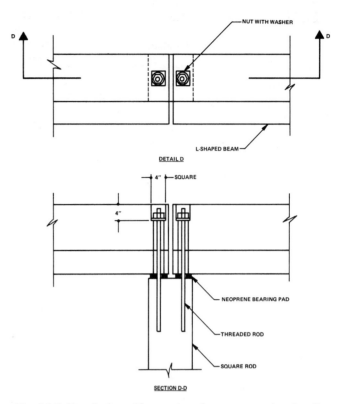

Fig. 13-7 Two L-shaped beams-to-column connection details

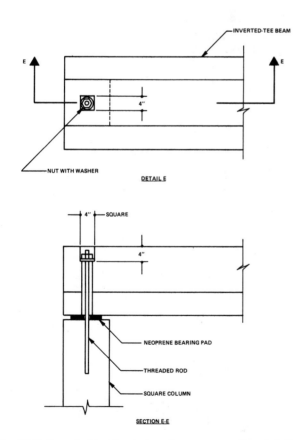

Fig. 13-8 Inverted-tee beam-to-column connection detail

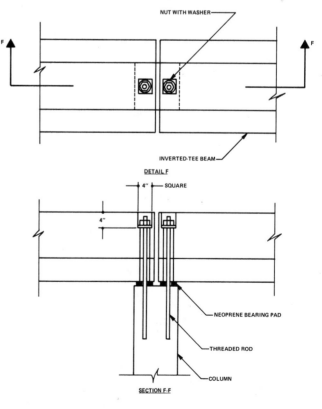

Fig. 13-9 Two inverted-tee beams-to-column connection details

WELDED CONNECTIONS

The most frequently used connection method in precast concrete construction is welding. The primary reason for the popularity of welded connections is its versatility in the shop and in the field. By casting weld plates, weld angles, and weld strips into precast concrete members, virtually any required connection can be made. The following are the most common types of welded connections in precast concrete construction.

- Flange connections in double-tee and single-tee members
- Base and top connections for wall panels
- Connections between beams and roof or floor members
- Stem connections for double tees and single tees

Refer to Figures 13-10 through 13-13 for examples of these types of welded connections.

Flange Weld Plate Connections

Roof and floor tee members must be joined together at the flanges so that several single members become fused into one solid roof or floor. This requires special welded connectors called *flange weld plates*. A flange weld plate consists of a rectangular steel plate welded to a continuous, bent reinforcing bar anchor, Figure 13-10.

The steel plate provides a smooth metal surface for making a field welded connection. The reinforcing bar anchor secures the weld plate firmly in the flange of the tee member so that once the structure is erected and the connection welded, forces acting on the structure do not dislodge the plate causing the connection to fail.

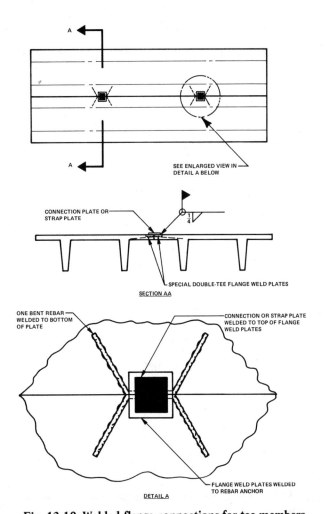

Fig. 13-10 Welded flange connections for tee members

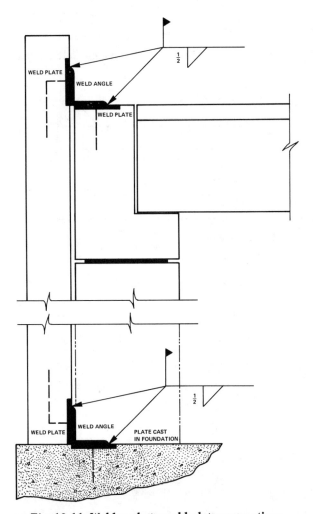

Fig. 13-11 Weld angle to weld plate connections

Abutting tee members in a roof or floor system are fabricated with aligning flange weld plates. Once the tee members are erected in the field, a *single, steel strap plate* (sometimes called *connection plate*) is welded to matching flange weld plates, Figure 13-10. By connecting all matching flange weld plates, the individual roof or floor members are fused into one continuous structural unit.

Weld Plate-Weld Angle Connections

An effective method of fastening precast members together or to other components of a structure involves weld plate connected by a weld angle. Weld plates are square or rectangular steel plates with reinforcing bar anchors to hold them securely in the concrete member. They are cast into precast concrete members at specific locations so as to match up, upon erection, with plates cast in other components of the structure. Matching weld plates are then connected by a weld angle, Figure 13-11.

Weld plate-weld angle connections are very versatile and can be adapted to almost any connection situation. A particular advantage of this type connection is that if a mistake during fabrication causes weld plates to be misaligned in the field, the error can often be corrected. The error is corrected by simply shifting the weld angle in one direction or the other.

Weld Plate-Strap Plate Connections

Similar to weld plate-weld angle connections are weld plate-strap plate connections. Again, weld plates are cast into precast concrete members. They are located so that they align with plates cast in other components of the structure after erection. Matching weld plates are connected by a single strap plate welded to both, Figure 13-12. This is also a versatile connection method that can be adapted for use in numerous connection situations. Like weld plate-weld angle connections, this method allows greater tolerances in casting the weld plate in the precast members.

Stem Plate-Weld Strip Connections

A connection method specifically designed for permanently fastening double-tee and single-tee stems to beams is the stem plate-weld strip

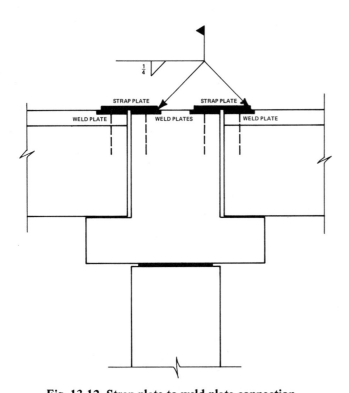

Fig. 13-12 Strap plate to weld plate connection

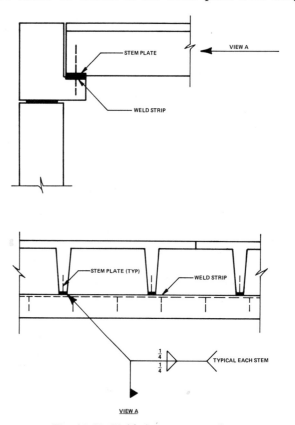

Fig. 13-13 Welded stem connections

connection. A stem plate is a steel plate with two or more reinforcing bar anchors that is cut to fit into the bottom of a double-tee or single-tee stem. A weld strip is one long, continuous weld plate cut to the length of the beam in which it will be cast, Figure 13-13.

Weld strips are frequently cast in beams that support tee members instead of weld plates to avoid the difficulty of matching the many weld plates and stem plates. This results in the utilization of more steel, but requires less labor time. Therefore, the cost factor is eliminated while the assurance of a connection for each tee stem is increased.

DRAWING WELDED CONNECTIONS

Welded connections are designed by engineers or experienced drafters working as designers. Before drawing welded connection details, the drafter should obtain from the engineer or designer such information as plate sizes, angle sizes, and weld information. In addition to drawing the connection details, it is the drafter's responsibility to locate and dimension all metal connectors on the appropriate framing plans.

Welded connection details differ from bolted connection details. Welded connections are drawn according to the following procedures:

1. Select an appropriate scale for drawing the details. The scale chosen should be at least 3/4" = 1'-0". The scale selected for drawing welded connection details depends on the number of details to be drawn, the space available in which to draw them, and individual company drafting procedures.
2. Obtain the engineer's or designer's specifications for each connection. These instructions are frequently given orally since most connections are repeated in use so often as to breed familiarity with them.
3. At the selected scale, draw only those views required to convey the connection configuration and design. A properly drawn welded connection detail will show the members being connected, all metal connectors involved, and appropriate weld symbols, Figures 13-10 through 13-13.

PRECAST CONCRETE HAUNCH CONNECTIONS

There are special situations in precast concrete construction that require special connection methods. The two most common are multistoried structures and structures in which precast concrete wall panels are designed to support roof members without involving columns or beams. The special connection method in both cases involves adding a haunch or haunches to the columns or wall panels.

A *haunch* is a bearing surface that is cast onto a column or wall panel as a secondary pour that supports beams or roof members after erection. Haunch connections on columns are usually bolted, while haunch connections on wall panels are usually welded. Figures 13-14 and 13-15 show isometric

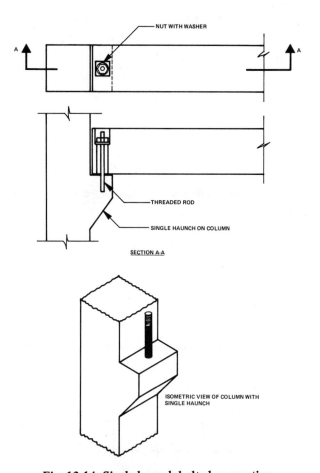

Fig. 13-14 Single-haunch bolted connection

Unit 13 Precast Concrete Connection Details

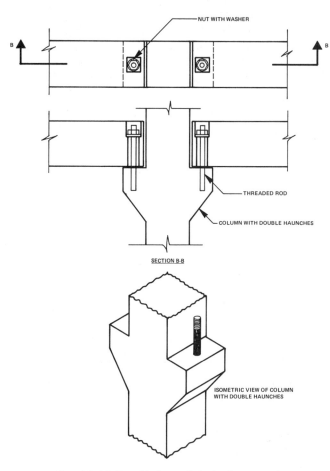

Fig. 13-15 Double-haunch bolted connection

illustrations of haunches on columns. Figures 13-16, 13-17, and 13-18 show isometric illustrations of haunches on wall panel members.

Haunched Beam-to-Column Connections

Columns in a multistories structure must support more than one system of beams. For example, a two-story structure requires a system of beams to carry the roof and another system of beams to carry the second floor. Beams that support the roof bear on top of the columns. However, beams that support the floor(s) run into the sides of columns, requiring a special connection. This special connection is accomplished by the addition of a haunch at every point where a floor beam intersects a column.

Haunch connections for beams to columns are bolted. This is accomplished by casting a threaded rod into the haunch, Figures 13-14 and 13-15. Bolted connections involving haunches are drawn very much like normal beam-to-column connections, with the exception that haunch configurations and dimensions must be supplied by an engineer or designer.

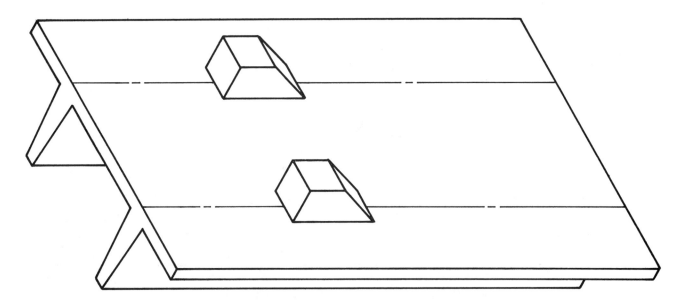

Fig. 13-16 Double-tee wall panel with two haunches

190 Section 3 Structural Precast Concrete Drafting

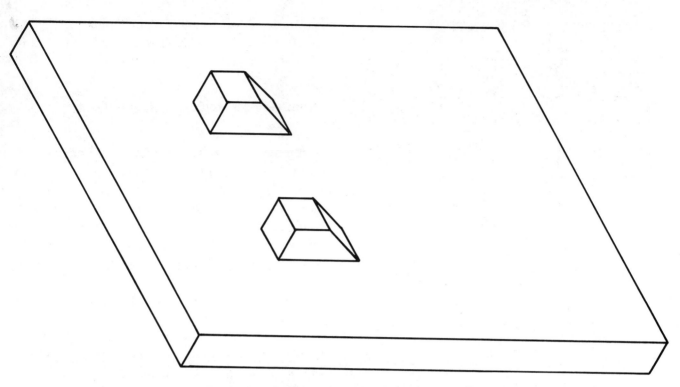

Fig. 13-17 Flat-slab wall panel with two haunches

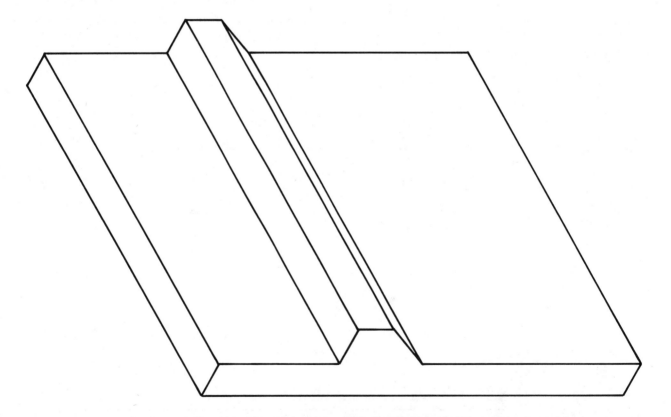

Fig. 13-18 Flat-slab wall panel with continuous haunch

Drawing Bolted Haunch Connections

Once an engineer or designer has supplied haunch configuration and dimensioning information, the procedures for drawing bolted haunch connection details are the following:

1. Determine from the framing plan(s) how many connection details are required and the configuration for each.
2. Draw a plan view of each connection showing only the immediate area around the connection and cut a section through it, Figures 13-14 and 13-15.
3. Draw the sectional view showing an elevation of the column, beam(s), haunch(es), blockouts, threaded rod(s), washers/nuts, neoprene bearing pad(s), and any other information pertinent to the connection, Figures 13-14 and 13-15.

SUMMARY

... Precast concrete members are connected at the jobsite by bolting, welding, or a combination of both.
... The most common connection situations in precast concrete construction are column baseplate connections, bolted beam-to-column connections, welded connections, and haunch connections.
... Columns are provided with a heavy steel baseplate cast in one end that is specially designed to fit over anchor bolts for connecting with washers and nuts.
... Baseplates for precast concrete columns are of two types: those that are the same width as the column in which they are cast (Figure 13-2) and those that extend beyond the edges of the column (Figure 13-1).
... Baseplate connection details contain the following information: the baseplate, anchor bolts, the footing, leveling nuts with washers, and tightening nuts with washers.
... Baseplate connection details are drawn to a scale of 3/4" = 1'-0" or larger.
... Connection details are called out by letter designation on the framing plans.
... The number of connection details that must be drawn is reduced by labeling connection situations that are exactly opposite *Opp Hd*, and situations that are very similar *Sim*.
... There are six basic beam-to-column connections in precast concrete construction: single rectangular beam bearing fully on a column, two rectangular beams bearing on the same column, single L-shaped beam bearing on a column, two L-shaped beams bearing on the same column, single inverted-tee beam bearing on a column, and two inverted-tee beams bearing on the same column.
... Bolted beam-to-column connection details contain a plan view of the connection and a sectional elevation showing: the column, beam(s), blockouts, threaded rod(s), washers and nuts, neoprene bearing pad(s), and any other information that is pertinent to the connection.
... The most frequently used connection method in precast concrete construction is welding because of its versatility in the shop and the field.
... The most common types of welded connections in precast concrete construction are: flange connections in tee members, base and top connections for wall panels, connections between beams and roof/floor members, and stem connections for tee members.
... Welded connection details show only those views required to convey the connection design. A properly drawn welded connection contains: the members being connected, all metal connectors involved, and appropriate weld symbols.
... Additional bearing surfaces may be added to columns or wall panels through the addition of haunches.
... Haunch connection details for bolted connections contain plan and sectional elevation views of the connection including: column, beam(s), haunch(es), blockouts, threaded rod(s), washers/nuts, neoprene bearing pad(s), and any other information pertinent to the connection.

REVIEW QUESTIONS

1. What are the two basic connecton methods used in precast concrete construction?
2. List the four most common connection situations in precast concrete construction.
3. Sketch an example of the two types of column baseplates.
4. List the information that should be contained in baseplate connection details.
5. List two abbreviations that may be added to connection details called out on the framing plans to reduce the number of details required.
6. List the six basic beam-to-column connections in precast concrete construction.
7. List the information that should be contained in bolted beam-to-column connection details.
8. What are the most common welded connections in precast concrete construction?
9. List the information that should be included in haunch connection details.
10. Select the most appropriate scale for use in preparing connection details in precast concrete drafting.
    a. 1/4" = 1'-0"
    b. 3/8" = 1'-0"
    c. 3/4" = 1'-0"
    d. 3/32" = 1'-0"

REINFORCEMENT ACTIVITIES

The title block as shown in Figure 1-21 and 1/2" borders are required on all activities.

Figure 13-19 contains an example of how connection details appear when drawn as part of a set of precast concrete connection details. Students should refer to this example when laying out the connection details required in the following drafting activities. Beam, column, and haunch configuration dimensions for the Unit 13 Reinforcement Activities are contained in Figure 13-20.

1. Construct connection detail "A" according to the following specifications: Rectangular beam bearing fully on a square column with a 1" diameter threaded rod and 3/4" neoprene bearing pad.

2. Construct connection detail "B" according to the following specifications: Two rectangular beams bearing on the same rectangular column with a 1" joint, 1" diameter threaded rod, and 3/4" neoprene bearing pad.

3. Construct connection detail "C" according to the following specifications: L-shaped beam bearing fully on a square column with a 1" diameter threaded rod and 3/4" neoprene bearing pad.

4. Construct connection detail "D" according to the following specifications: Two L-shaped beams bearing on the same rectangular column with a 1" joint, 1" diameter threaded rod, and 3/4" neoprene bearing pad.

5. Construct connection detail "E" according to the following specifications: Inverted-tee beam bearing fully on a square column with a 1" diameter threaded rod and 3/4" neoprene bearing pad.

6. Construct connection detail "F" according to the following specifications: Two inverted-tee beams bearing on the same rectangular column with a 1" joint, 1" diameter threaded rod, and 3/4" neoprene bearing pad.

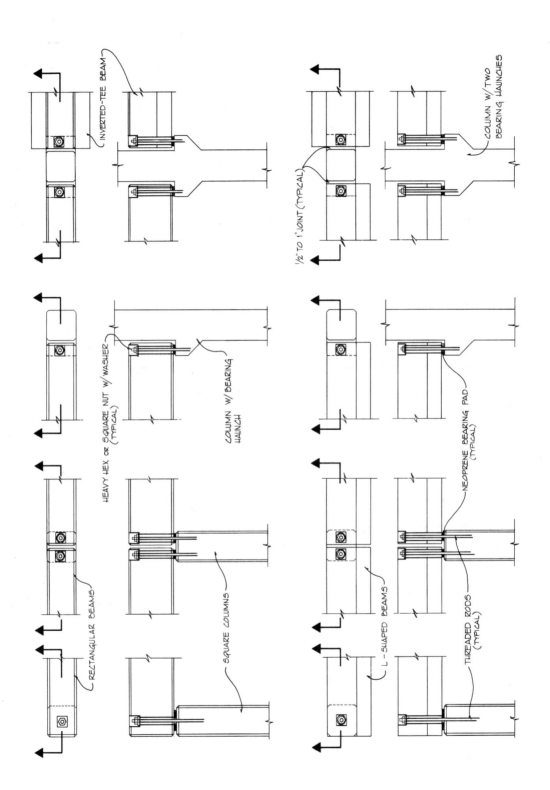

Fig. 13-19 Connection details as they appear in a set of structural drawings

194 Section 3 Structural Precast Concrete Drafting

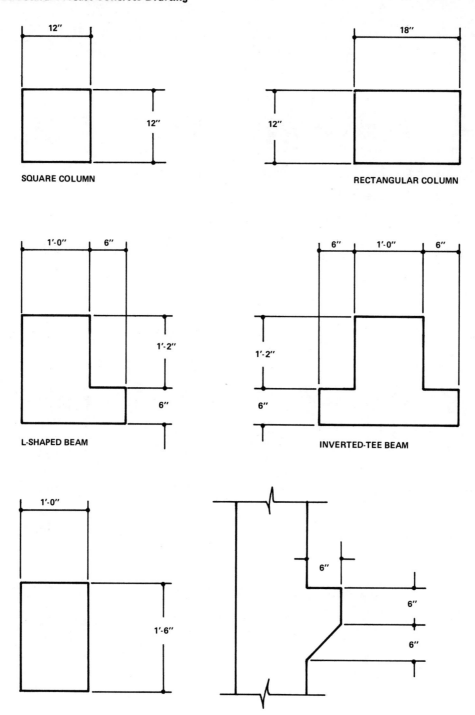

Fig. 13-20 Configuration dimensions for drafting activities

Unit 14
Precast Concrete Fabrication Details

OBJECTIVES

Upon completion of this unit, the student will be able to

- define shop drawings and fabrication details.
- explain how shop drawings fit into a set of precast concrete working drawings.
- explain how fabrication details fit into the shop drawings.
- construct fabrication details of precast concrete columns, beams, wall panels, floor/roof members, and metal connectors.

SHOP DRAWINGS DEFINED

In Unit 1, it was learned that a set of structural working drawings consists of engineering drawings and shop drawings. Engineering drawings include framing plans, sections, and connection details. These were covered in Units 11, 12, and 13.

Shop drawings are comprehensive, precisely detailed drawings prepared for use by shop workers in fabricating precast concrete products and the metal connectors they contain. Shop drawings consist of fabrication details and bills of materials.

The information required in preparing shop drawings is taken from the engineering drawings. This unit deals with the preparation of fabrication details for precast concrete products and metal connectors. Bills of materials are covered in the next unit.

FABRICATION DETAILS DEFINED

Fabrication details are orthographic drawings of precast concrete beams, columns, wall panels, roof/floor members, and metal connectors. They contain all of the information required by shop workers to fabricate all of the precast products and metal connectors used in a job. Each concrete product and each metal connector requires a separate fabrication detail.

Figures 14-1 through 14-4 illustrate examples of fabrication details for precast concrete products. Miscellaneous metal fabrication details are shown in Figure 14-5.

CONSTRUCTING FABRICATION DETAILS

To construct fabrication details, the drafter must have copies of the framing plans, sections, and connection details for the job. In addition, sketches showing the design calculations for all precast products and metal connectors must be provided by the engineer assigned to the job.

Figure 14-6 shows the engineering calculation sketches for the fabrication details in Figures 14-1 through 14-4. Examine the engineering calculation sketches closely. Compare them with the fabrication details to begin developing an understanding of their important relationship. Figure 14-7 shows the engineering calculation sketches for the metal connectors detailed in Figure 14-5.

Column Fabrication Details

Precast concrete column fabrication details contain the following information:

- At least two orthographic views of the column
- Section(s) showing reinforcing bar or strand patterns
- A baseplate configuration diagram
- A special view showing the location of threaded rod(s) in relation to the top of the column
- Complete dimensions

Refer to Figure 14-1 for examples of precast concrete column fabrication details.

Column fabrication details are drawn according to the following procedures:

1. Column length, width, and depth dimensions are determined from the framing plan and products schedule and at least two orthographic views of the column are drawn. It should be noted that it is common practice in precast concrete drafting to draw column lengths on fabrication details *not to scale*. This is because columns are often very long, making it difficult to fit them on the sheet. Widths and depths are drawn to scales of 3/4" = 1'-0" or larger.

2. Baseplate width, depth, and thickness are determined from the connection detail. The baseplate is drawn at the bottom end of the column and the baseplate configuration diagram is added, Figure 14-1.

3. Locations of threaded rods are determined from the appropriate beam-to-column connection detail and the rod is added to the orthographic views of the column. A special end view is then drawn to show how the threaded rod(s) fit into the top of the column. These special views are called out and drawn exactly as if they were sections, Figure 14-1.

4. Reinforcing bar or strand patterns for the column are determined from the engineering calculation sketches and represented in the orthographic views of the column by long, dashed lines. All reinforcing bars used in the column must be assigned numeric designations such as 301, 401, 501, 601, etc., and entered into the master legend. The master legend for Figures 14-1 through 14-4 is shown in Figure 14-5.

5. A section is cut on the column and drawn to show how reinforcing bars or strands are arranged and spaced within the column. See Section C-C, Figure 14-1. The column detail is completed by adding all necessary dimensions including: column width, depth, and length; reinforcing bar lengths; reinforcing bar spacing along the column length; and reinforcing bar or strand spacing in a section of the column, Figure 14-1.

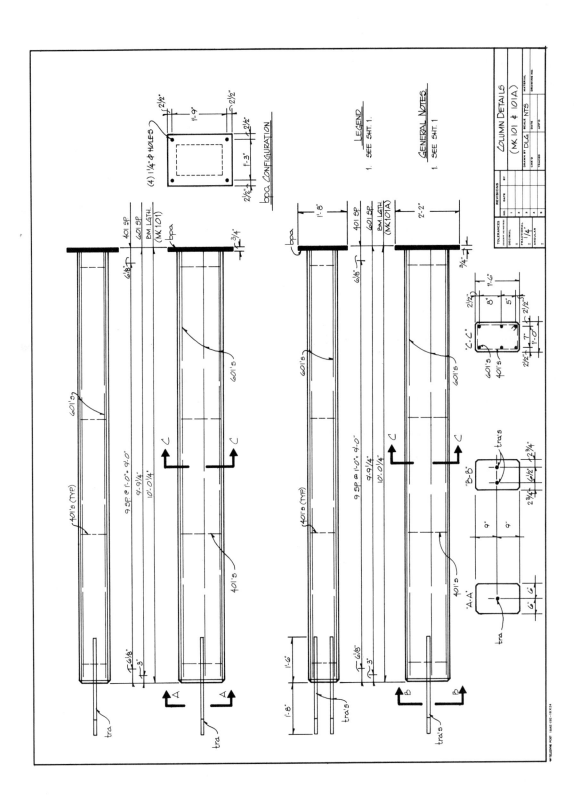

Fig. 14-1 Column fabrication details

Beam Fabrication Details

Precast concrete beam fabrication details contain the following information:

- At least two orthographic views of the beam
- Section(s) showing reinforcing bar or strand patterns
- Blockouts
- Metal connectors
- Complete dimensions

Refer to Figure 14-2 for examples of precast concrete beam fabrication details.

Beam fabrication details are drawn according to the following procedures:

1. Beam configuration and length dimensions are determined from the framing plan, products schedule, and connection details or sections. When this information has been obtained, at least two orthographic views of the beam are drawn. It should be noted that it is common practice in precast concrete drafting to draw beam lengths on fabrication details *not to scale*. This is because beams are often very long, making it difficult to fit them on the sheet. All other dimensions are drawn to scales of 3/4" = 1'-0" or larger.

2. Blockout and metal connector sizes and locations are determined from the framing plan and appropriate connection details and drawn on the orthographic views of the beam, Figure 14-2.

3. Reinforcing bar or strand patterns are determined from the engineering calculation sketches and represented in the orthographic views of the beam by long, dashed lines. All reinforcing bars must be assigned numeric designations such as 301, 401, 501, 601, etc., and entered into the master legend. All metal connectors must be assigned appropriate letter designations such as wpa, spa, cpa, waa, etc., and entered into the master legend. The master legend for Figures 14-1 through 14-4 is contained in Figure 14-5.

4. A section is cut on the beam and drawn to show configuration dimensions and how bars or strands are arranged and spaced within the beam. The beam detail is completed by adding all necessary dimensions including: beam length and configuration dimensions; reinforcing bar lengths; reinforcing bar spacing along the length of the beam; blockout sizes and locations; metal connectors sizes and locations; and reinforcing bar or strand spacing in the sectional view of the beam, Figure 14-2.

Unit 14 Precast Concrete Fabrication Details 199

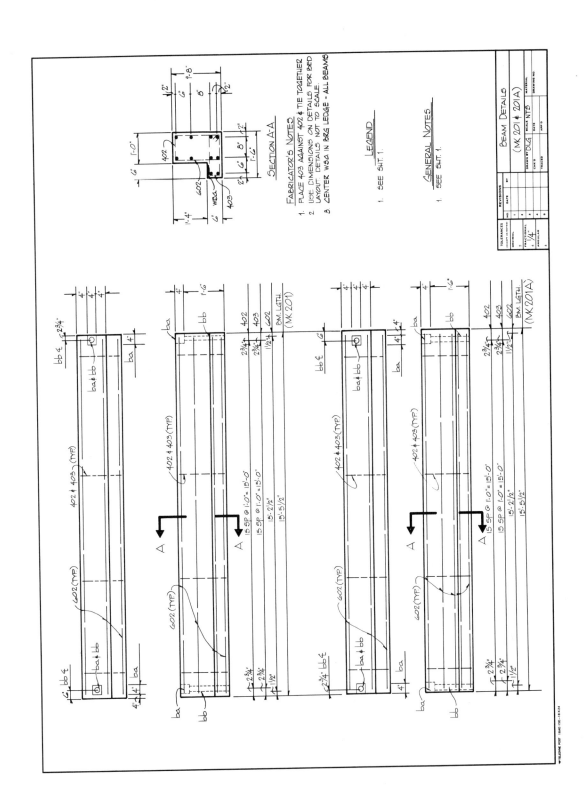

Fig. 14-2 Beam fabrication details

Floor/Roof Member Fabrication Details

Precast concrete roof and floor members may be double tees, flat slabs, or cored flat slabs. The most common floor and roof members are double tees. However, the procedures for detailing the members are the same regardless of their type. Refer to Figure 14-3 for an example of a precast concrete floor or roof member fabrication detail.

Floor/roof member fabrication details are drawn according to the following procedures:

1. Member configuration and length dimensions are determined from the framing plan, products schedule, or illustrations of standard sizes in the case of tee members. It should be noted that it is common practice in precast concrete drafting to draw floor/roof member lengths on fabrication details *not to scale*. This is because floor/roof members are often very long, making it difficult to fit them on the sheet. All other dimensions are drawn to scales of 3/4" = 1'-0" or larger.
2. Blockout and metal connector sizes and locations are determined from the framing plan and appropriate connection details and drawn on the orthographic views of the floor/roof member, Figure 14-3. Blockouts and metal connectors must be assigned letter designations and entered in the master legend. The master legend for Figures 14-1 through 14-4 is contained in Figure 14-5.
3. Precast concrete roof and floor members receive prestressing strands rather than reinforcing bars. Strand patterns are shown in a sectional view only. Since prestressing strands run the entire length of the precast product, they are not dimensioned in the orthographic views of the product. All stranding information required on floor/roof fabrication details is taken from the engineering calculation sketches.
4. The floor/roof member fabrication detail is completed by adding all necessary dimensions including: member length and configuration dimensions; reinforcing bar lengths and spacing (if used they must be assigned numeric designations and entered in the master legend); blockout sizes and locations; metal connector sizes and locations; strand patterns in the sectional view, Figure 14-3.

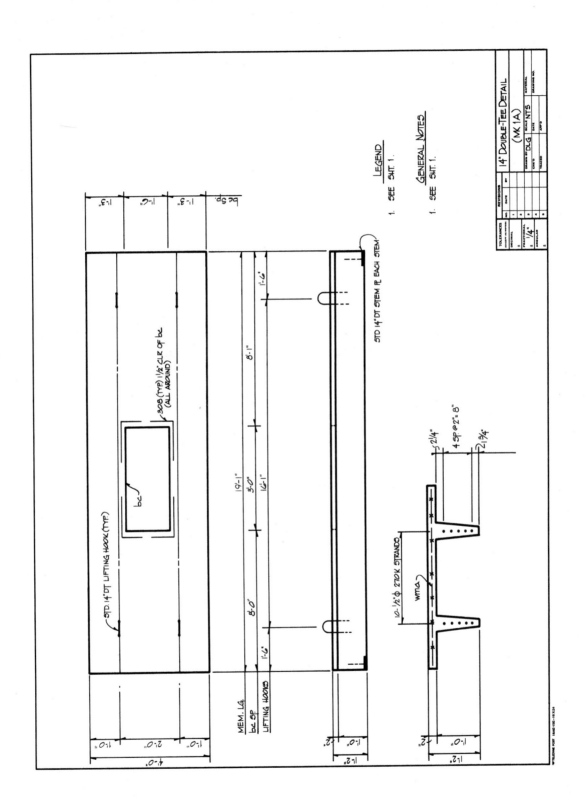

Fig. 14-3 Double-tee roof member fabrication details

Wall Panel Fabrication Details

Like precast concrete floor and roof members, wall panels may be double tees, flat slabs, or cored flat slabs. The most common wall panels are flat slabs. Regardless of the product type, the procedures for detailing the members are the same. Refer to Figure 14-4 for an example of a precast concrete wall panel fabrication detail.

Wall panel fabrication details are drawn according to the following procedures:

1. Wall panel length and configuration dimensions are determined from the framing plan and products schedule. It should be noted that it is common practice in precast concrete drafting to draw wall panel lengths on fabrication details *not to scale*. This is because wall panel lengths are often too long to fit comfortably on the sheet. All other dimensions are drawn to scales of 3/4" = 1'-0" or larger.

2. Blockout and metal connector sizes and locations are determined from the framing plan and appropriate connection details and drawn on the orthographic views of the wall panel. Later designations are assigned and entered in the master legend. The master legend for Figures 14-1 through 14-4 is contained in Figure 14-5.

3. Sections are cut at each point along the wall panel where the width of the panel varies and drawn to show the arrangement and spacing of reinforcing bars within the wall panel. All reinforcing bars of differing lengths and shapes must be assigned different numeric designations and entered in the master legend.

4. The wall panel fabrication detail is completed by adding all necessary dimensions including: wall panel length and configuration dimensions; reinforcing bar lengths and spacing; blockout sizes and locations; metal connector sizes and locations; and strand patterns in the sectional views (if stranding is used), Figure 14-4.

Unit 14 Precast Concrete Fabrication Details 203

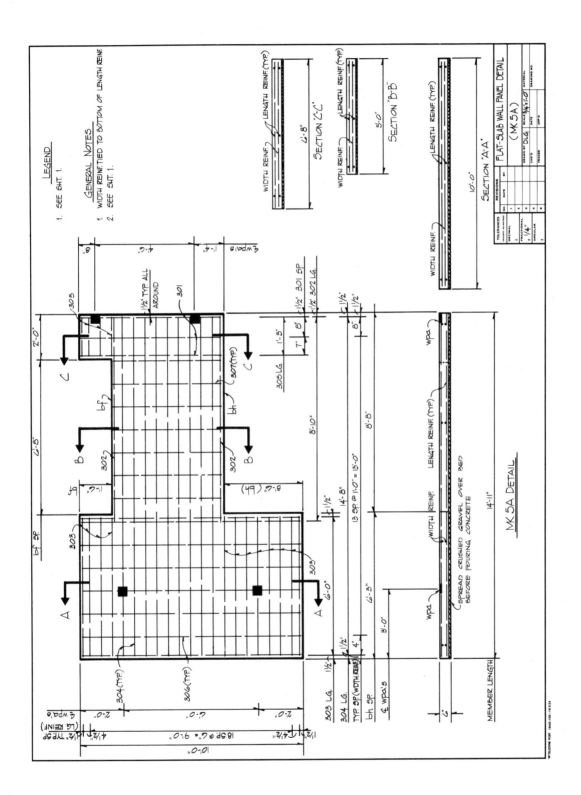

Fig. 14-4 Flat-slab wall panel fabrication detail

Metal Connector Fabrication Details

Metal connectors for a precast concrete job are designed by an engineer. The engineer may convey his or her design for the connectors to the drafter through sketches, Figure 14-7. The drafter's job is to draw orthographic fabrication details based on the engineer's design.

Metal connectors and miscellaneous materials such as neoprene bearing pads are detailed on the same sheet. This sheet then becomes the first sheet of the shop drawings. The metals sheet also contains the master legend for the job and is accompanied by a separate reinforcing bar schedule.

The *reinforcing bar schedule* is a form showing a number of bar diagrams with letters substituted for dimensions. In addition, the form has columns for listing the bar number, type, and bending dimensions, Figure 14-8. All reinforcing bars used on a job are tabulated on the reinforcing bar schedule. The reinforcing bar schedule for the members detailed in Figures 14-1 through 14-4 is shown in Figure 14-9.

Reinforcing bar schedule forms may be produced by the individual companies that use them or purchased from commercial producers. The format used in Figure 14-8 should be used in completing the Reinforcement Activities for this unit.

In the event that the engineering drawings are combined with the shop drawings to form a set of precast concrete working drawings, the metals sheet becomes the first sheet in the set and contains the master legend and the general notes for the job. Figure 14-5 shows an example of a metals sheet used as the first sheet in a set of precast concrete working drawings.

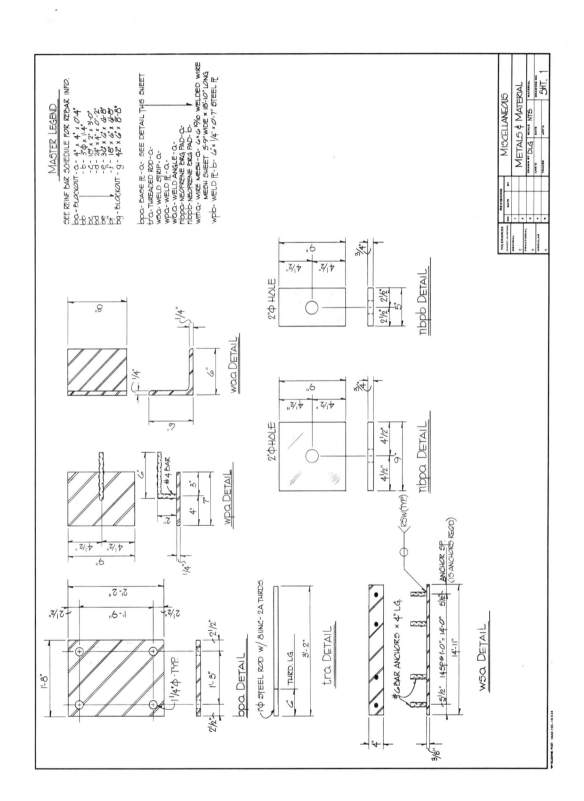

Fig. 14-5 Metal connectors fabrication details

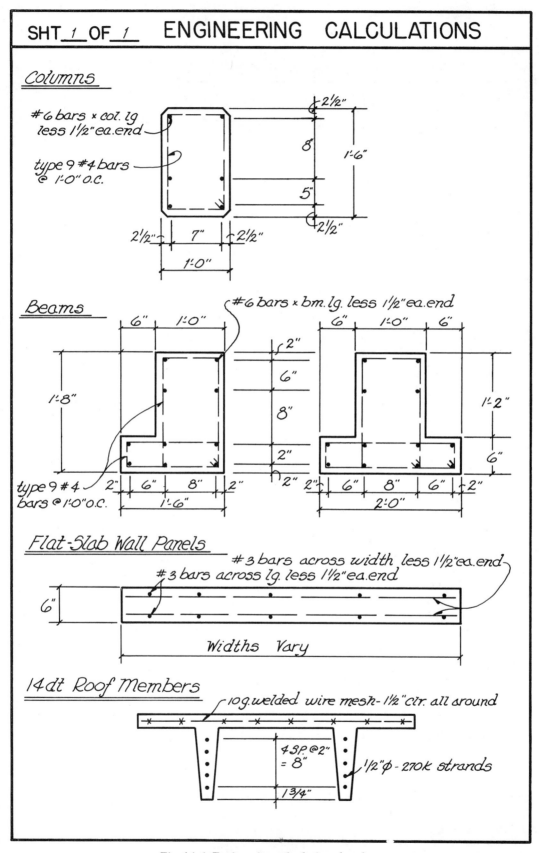

Fig. 14-6 Engineering calculation sketches

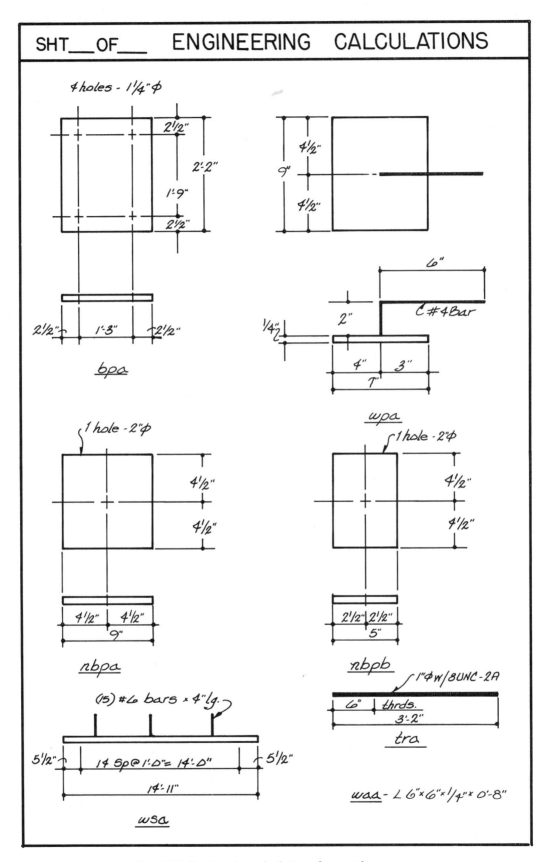

Fig. 14-7 Engineering calculations for metal connectors

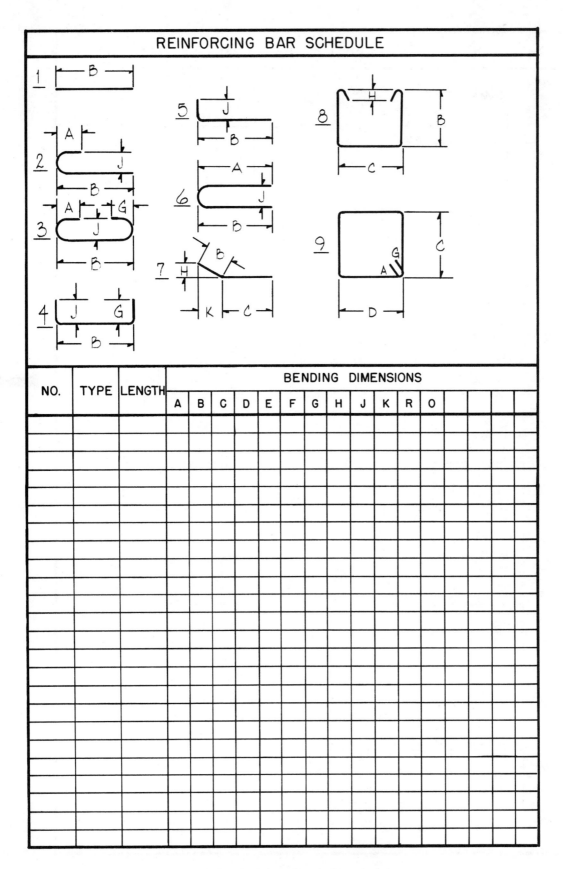

Fig. 14-8 Reinforcing bar schedule

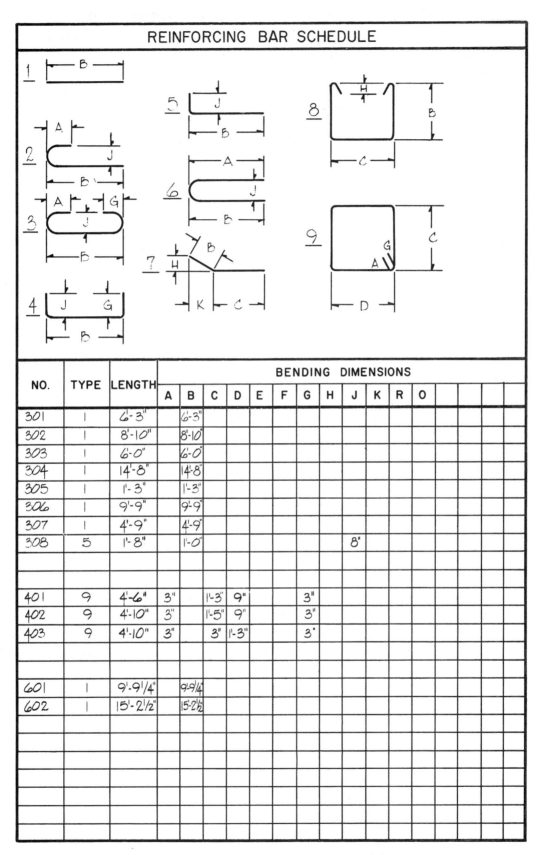

Fig. 14-9 Completed reinforcing bar schedule

SUMMARY

... Shop drawings are prepared for use by shop workers in fabricating precast concrete products and consist of fabrication details and bills of material.
... Information required on shop drawings is obtained from the engineering drawings and engineering calculation sketches.
... Fabrication details are orthographic drawings or precast concrete products or metal connectors that show all of the information needed by shop workers in fabrication.
... In order to construct the fabrication details for a job, the drafter must have the framing plan(s), sections, connection details, and engineering calculation sketches.
... Fabrication details are constructed according to specific procedures that include: drawing at least two orthographic views of the subject; locating, sizing, and labeling all blockouts, reinforcing bars, and metal connectors; and providing complete dimensions.
... Metal connectors for a precast concrete job are designed by an engineer. The design is conveyed to drafters by sketches.
... The metals sheet, accompanied by a reinforcing bar schedule, contains the master legend and is the first sheet in a set of shop drawings.
... When a set of engineering and shop drawings are combined into a set of working drawings, the metals sheet becomes the first sheet of the new set.

REVIEW QUESTIONS

1. Define *shop drawings*.
2. Where is the information that is required on shop drawings located?
3. Define *fabrication details*.
4. What must the drafter have in order to construct fabrication details?
5. Briefly, and in general terms, summarize the procedures for preparing precast concrete fabrication details.
6. Who designs the metal connectors in a precast concrete job?
7. Which sheet is first in a set of shop drawings? Working drawings?

REINFORCEMENT ACTIVITIES

The following drafting exercises are based on the engineering drawings provided in Figures 14-10 through 14-15 and the engineering calculation sketches in Figure 14-16. All activities should be completed on C-size paper with 1/2" borders and the title block as shown in Figure 1-21. Where the use of a scale is required, a scale appropriate for the task being performed should be selected.

Column Fabrication Details

1. Prepare a complete fabrication detail for MK 100 in Figure 14-10. Refer to Figure 14-1 for examples.*

2. Prepare a complete fabrication detail for MK 101 in Figure 14-10. Refer to Figure 14-1 for examples.*

   *The student may select the type and size of baseplate used in completing these activities.

Beam Fabrication Details

3. Prepare a complete fabrication detail for MK 200 in Figure 14-11. Refer to Figure 14-2 for examples.

4. Prepare a complete fabrication detail for MK 200A in Figure 14-11. Refer to Figure 14-2 for examples.

5. Prepare a complete fabrication detail for MK 201 in Figure 14-11. Refer to Figure 14-2 for examples.

6. Prepare a complete fabrication detail for MK 201A in Figure 14-11. Refer to Figure 14-2 for examples.

Wall Panel Fabrication Details

7. Prepare a complete fabrication detail for MK 10 in Figure 14-12. Refer to Figure 14-4 for an example.

8. Prepare a complete fabrication detail for MK 10A in Figure 14-12. Refer to Figure 14-4 for an example.

9. Prepare a complete fabrication detail for MK 10B in Figure 14-12. Refer to Figure 14-4 for an example.

10. Prepare a complete fabrication detail for MK 10C in Figure 14-12. Refer to Figure 14-4 for an example.

Roof Member Fabrication Details

11. Prepare a complete fabrication detail for MK 1 in Figure 14-13. Refer to Figure 14-3 for an example.

12. Prepare a complete fabrication detail for MK 1A in Figure 14-13. Refer to Figure 14-3 for an example.

Miscellaneous Metals Fabrication Details

13. Referring to Figures 14-10 through 14-15 and Figure 14-16, prepare the legend and general notes for a metals fabrication details sheet. Refer to Figure 14-5 for an example.

14. Referring to Figures 14-10 through 14-15 and Figure 14-16, prepare all required metal fabrication details. Refer to Figure 14-9 for an example.

15. Using the fabrication details completed in activities 1 through 12, prepare a complete bar reinforcing schedule. Refer to Figure 14-9 for an example.

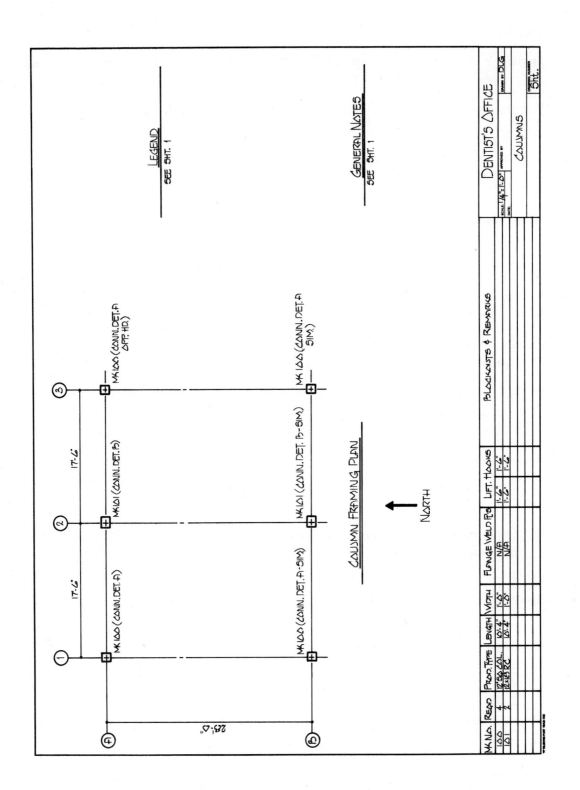

Fig. 14-10 Column framing plan for Reinforcement Activities

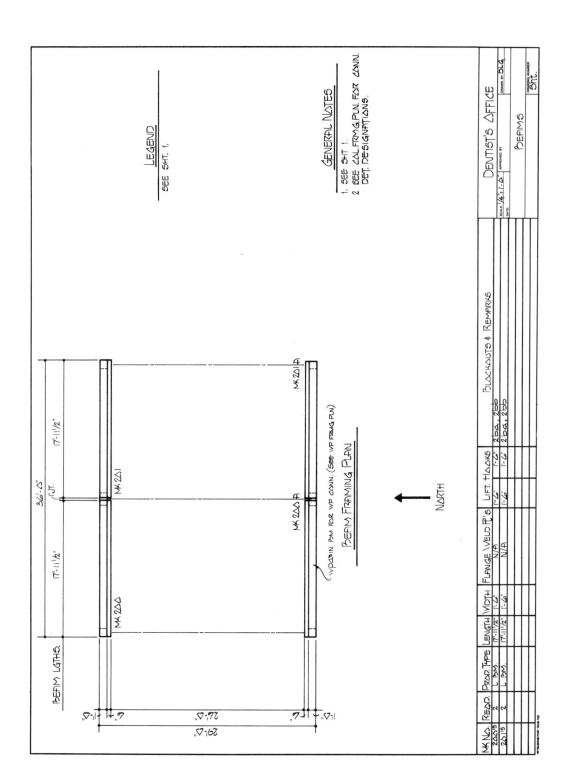

Fig. 14-11 Beam framing plan for Reinforcement Activities

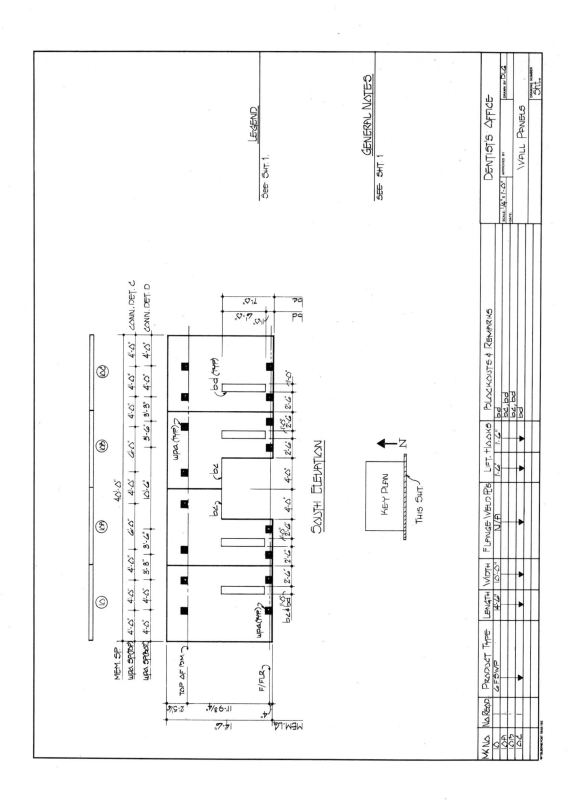

Fig. 14-12 Wall panel framing plan for Reinforcement Activities

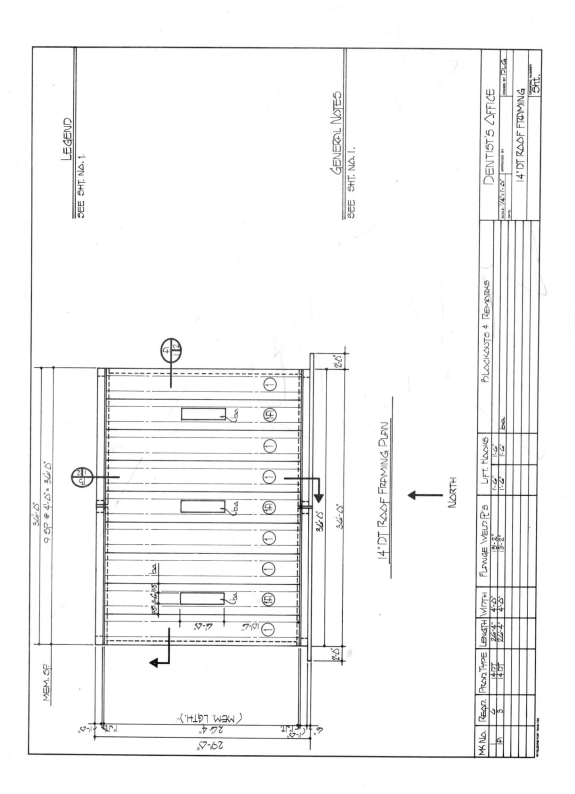

Fig. 14-13 Roof framing plan for Reinforcement Activities

216 Section 3 Structural Precast Concrete Drafting

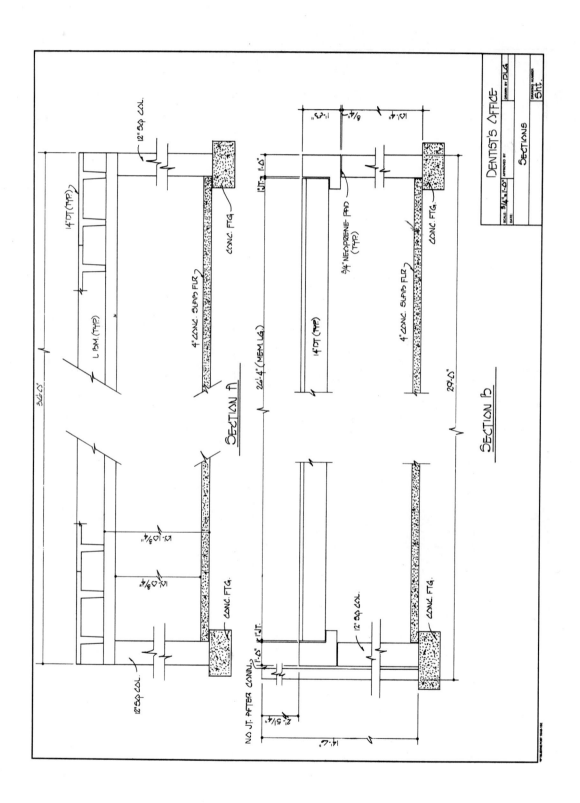

Fig. 14-14 Sections for Reinforcement Activities

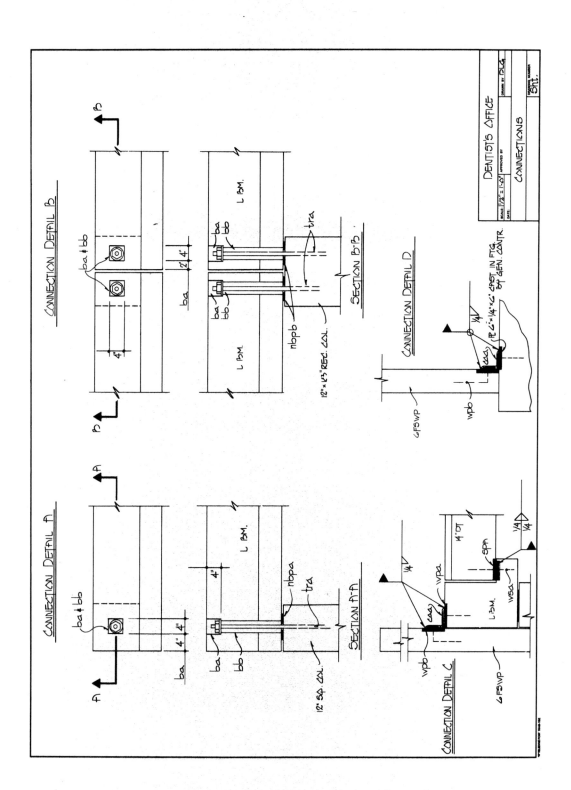

Fig. 14-15 Connection details for Reinforcement Activities

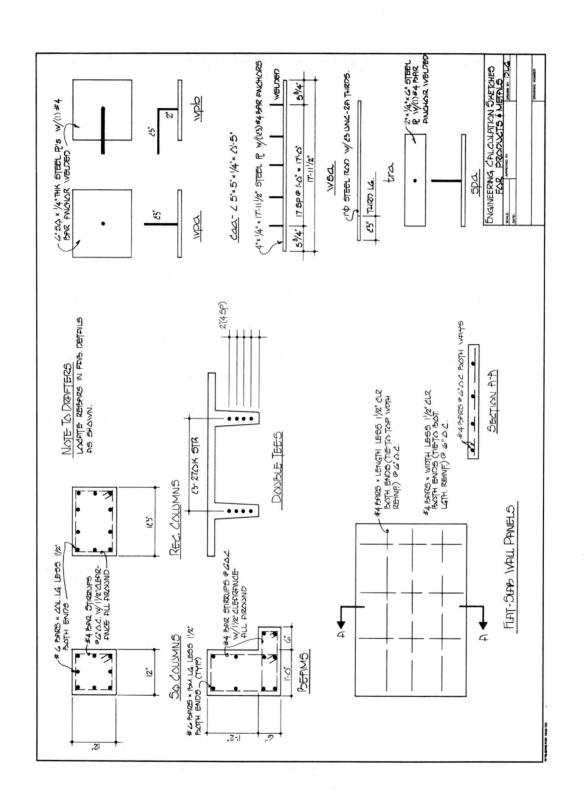

Fig. 14-16 Engineering calculation sketches for Reinforcement Activities

Unit 15
Precast Concrete Bills of Materials

OBJECTIVES

Upon completion of this unit, the student will be able to

- define bill of materials and explain its purpose.
- prepare bills of materials for columns, beams, floor/roof members, and wall panels.

DEFINITION AND PURPOSE

A *bill of materials* is a comprehensive listing of all materials that are used in the fabrication and erection of the precast concrete products for a given job. Bills of materials are used by shop workers. From the supply bin, the shop workers draw all of the material they need to manufacture a given member. The purchasing department also uses a bill of materials for ordering the materials that will be needed in the fabrication of the precast products for a job.

Several different procedures are used in preparing a bill of materials:

- One complete bill of materials covering all of the precast concrete members in a job
- Individual bills of materials for each individual precast concrete member in a job
- One bill of materials for each separate precast concrete product type used on a job — for example, one bill of materials for columns, one for beams, one for wall panels, and one for floor/roof members.

The last method mentioned is the most popular. This is because it can serve the same purposes as both of the other two and it is more efficient. In this method, one drafter may be preparing the bill of materials for the columns, while another prepares a bill for the beams. At the same time, another drafter prepares a bill for the wall panels, while still another prepares a bill for the floor/roof members. When each drafter finishes the bill of materials in his or her area of responsibility, the resultant bills may be stapled together and used as one complete bill for the entire job.

Most companies either prepare or purchase standard bill of materials forms. Forms vary according to the needs of the individual company. However, all bills list the following:

- Type of product
- The mark numbers for all members of that product
- All materials contained in the members
- How many of each type of material is contained in each member

These amounts are totaled by product type and then totaled again for the entire job. Figure 15-1 shows an example of a blank bill of materials form. Bills of materials prepared for columns, beams, wall panels, and roof/floor members are shown in Figures 15-2 through 15-5.

219

220 Section 3 Structural Precast Concrete Drafting

Fig. 15-1 Blank bill of materials form

Fig. 15-2 Completed bill of materials for columns

SHEET 1 OF 1   JOB NUMBER 3201   DATE 2-5-80
PREPARED BY TB
CHECKED BY DLG

BILL OF MATERIALS

Product Type: L-SHAPED BEAMS

MK NUMBERS	200	201	202	203	204	205	Total
w5a	1	–	–	–	–	–	1
w5b	–	1	–	–	–	–	1
w5c	–	–	1	–	–	–	1
w5d	–	–	–	1	–	–	1
w5e	–	–	–	–	1	–	1
w5f	–	–	–	–	–	1	1
wpc	4	–	3	2	–	5	14
wpd	–	5	–	–	6	–	11
wpe	2	2	2	2	2	2	12
301's	3	3	3	3	3	3	18
302's	3	3	3	3	3	3	18
402's	16	12	14	15	18	19	94
403's	16	12	14	15	18	19	94
601's	10	10	10	10	10	10	60

Fig. 15-3 Completed bill of materials for beams

SHEET 1 OF 1   JOB NUMBER 3201   DATE 2-5-80
PREPARED BY DM
CHECKED BY DLG

BILL OF MATERIALS

Product Type: 6" FLAT-SLAB WALL PANELS

MK NUMBERS	1	1A	1B	1C	2	3	3A	3B	4	5	6	6A	6B	Total
wpa	2	2	2	2	2	2	2	2	2	2	2	2	2	26
wpb	2	2	2	2	2	2	2	2	2	2	2	2	2	26
404's	3	3	3	5	4	4	4	–	–	–	–	–	–	29
405's	–	–	–	–	–	–	–	3	4	5	5	5	–	22
406's	10	10	10	10	10	10	10	10	10	10	10	10	10	260
407's	–	–	–	–	–	1	1	1	–	–	1	1	1	6
408's	–	–	–	–	–	1	1	1	–	–	1	1	1	6
409's	–	–	–	–	1	1	1	–	–	1	1	1	–	6
410's	12	12	12	12	12	12	12	12	12	12	12	12	12	156
701's	–	–	–	2	–	–	2	2	–	–	–	–	–	6

Fig. 15-4 Completed bill of materials for flat-slab wall panels

SHEET 1 OF 1 JOB NUMBER 3201 DATE 2-5-80

PREPARED BY RT

CHECKED BY DLG

BILL OF MATERIALS											
MATERIAL	PRODUCT TYPE										
	14" DOUBLE-TEE ROOF MEMBERS										Total
MK NUMBERS	10	10A	10B	10C	11	12	13	14	14A	15	
wpc	2	2	2	2	2	2	2	2	2	2	20
spa	4	4	4	4	4	4	4	4	4	4	40
tha	4	4	4	4	4	4	4	4	4	4	40
4I1's	–	2	2	2	–	–	–	–	–	–	6
4I2's	–	–	–	–	–	–	–	2	2	2	6
270K STR × L.F.	60	60	60	60	90	90	90	120	120	120	870

Fig. 15-5 Completed bill of materials for double-tee roof members

CONSTRUCTING BILLS OF MATERIALS

Bills of materials for precast concrete jobs are prepared from the fabrication details. A drafter assigned to prepare a bill of materials for the columns used in a job needs copies of the fabrication details of all of the columns. This is also true of the drafter assigned the bill of materials for any portion of a job. The procedures used in preparing the bill of materials for columns, beams, wall panels, and roof/floor members are the same.

1. Collect all of the fabrication details in the area for which the bill of materials is being prepared.
2. Examine the details and compile a list of every bar, plate, angle, bolt, washer, weld strip, rebar, bearing pad, etc., that is used in any of the members. These should be listed according to the abbreviations used on the fabrication details under the heading *material* on the bill of materials form, Figures 15-2 through 15-5.
3. In the blank on the bill of materials form marked *Type of Product,* the product type should be entered: columns, beams, roof/floor members, or wall panels, Figures 15-2 through 15-5.
4. A row of boxes on the bill of materials form is labeled *MK Numbers* and each mark number for the product type being billed is entered into a box. For the convenience of the people who will use the bill of materials, the numbers should be listed in order from smallest to largest, Figures 15-2 through 15-5.

5. The bill of materials is now ready to complete. The fabrication detail for each mark number listed on the bill is examined closely, the quantity of each individual material used is counted, the count for each material is recorded in the appropriate box, and the process is repeated for every mark number under the subject product type, Figures 15-2 through 15-5.

COUNTING MATERIAL QUANTITIES

To save drafting time and for ease of reading the prints, every piece of rebar or every metal connector in a group of rebars or connectors used in a member is not shown. Rather, a few samples of the rebar or connector may be shown with one being called out. This indicates there are more used than have been shown.

In counting material quantities, the drafter should be careful to count according to dimensional specifications and not according to the picture presented by the fabrication detail. For example, the detail for a precast member that is to be cast 20'-0" long and contain #4 bars running across its width at 6" On Center, would not show each individual piece of rebar. The drafter would have to read the #4 bar dimension line to determine the number of bars used, Figure 15-6.

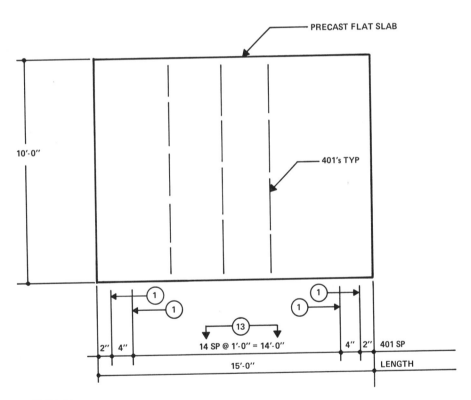

PROBLEM
DETERMINE THE NUMBER OF 401 BARS USED IN THE FLAT SLAB ILLUSTRATED ABOVE.

SOLUTION
1. LOCATE THE DIMENSION LINE THAT GIVES THE SPACING FOR 401 BARS. IN THE ILLUSTRATION ABOVE, IT IS DIRECTLY ABOVE THE LENGTH DIMENSION LINE.
2. COUNT THE 401 BARS ALONG THE DIMENSION LINE. EACH LINE EXTENDED FROM THE FLAT SLAB TO THE DIMENSION LINE REPRESENTS ONE 401 BAR. THE DIMENSION STATEMENT "14 SP @ 1'-0" = 14'-0" " CONVERTS TO 13 — 401 BARS, OR ALWAYS ONE LESS THAN THE NUMBER OF SPACES CALLED OUT. THIS METHOD OF DETERMINING BAR QUANTITIES FOR BILLS OF MATERIAL APPLIES FOR ALL PRECAST PRODUCTS.

Fig. 15-6 Counting bar quantities for the bill of materials

The same rule applies in counting quantities of metal connectors or any other material used in a precast product. Quantities must be determined from the dimensions for that material with the exception of materials whose location is obvious from the picture. In these cases, the quantity of the material shown is the quantity used. An example of how metal connector quantities are counted for the bill of materials is shown in Figure 15-7.

SUMMARY

... A bill of materials is a detailed listing of all materials used in the fabrication and erection of a job.

... The most common method used in the preparation of a bill of materials is to prepare one bill for each product type used in a job.

... Most companies prepare or purchase standard bill of materials forms.

... Standard forms vary from company to company, but all contain the type of product used, mark numbers for each product being billed, a quantity count of each material used in each product, and totals for each type of material used.

... Bills of materials are prepared from information contained in the fabrication details.

... Material quantities are determined from dimensional specifications rather than the picture presented by the fabrication details except in cases where materials are located from the picture rather than by dimensions.

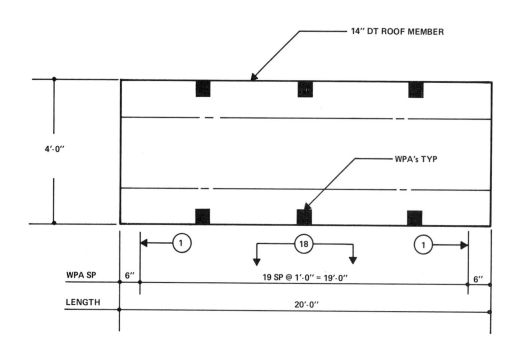

Fig. 15-7 Counting plate quantities for the bill of materials

REVIEW QUESTIONS

1. Define the term *bill of materials* as it is used in precast concrete drafting.
2. Explain the most common method used in preparing precast concrete bills of materials. Why is this method the most popular?
3. List four common entries that are usually found on any bill of materials form.
4. Where do drafters find the information needed in completing a bill of materials?
5. The dimensional specification *12 SP @ 1'-0"* converts to what quantity of material?

REINFORCEMENT ACTIVITIES

In order to complete the following reinforcement activities, the student should prepare four blank bill of materials forms. A format similar to that shown in Figure 15-1 is recommended. Due to the length of the Reinforcement Activities that follow, A-size paper may be used.

1. Prepare a complete bill of materials for the precast columns contained in Figures 15-8 and 15-9.

2. Prepare a complete bill of materials for the precast beams contained in Figures 15-10 and 15-11.

3. Prepare a complete bill of materials for the precast flat-slab wall panels contained in Figures 15-12 and 15-13.

4. Prepare a complete bill of materials for the precast concrete double-tee roof members contained in Figures 15-14 and 15-15.

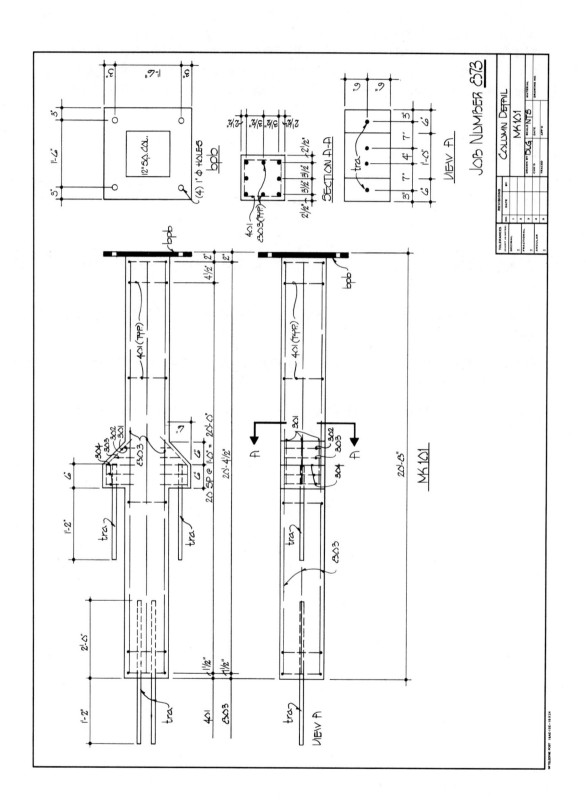

Fig. 15-8 Column details for Reinforcement Activities

Unit 15 Precast Concrete Bills of Materials 227

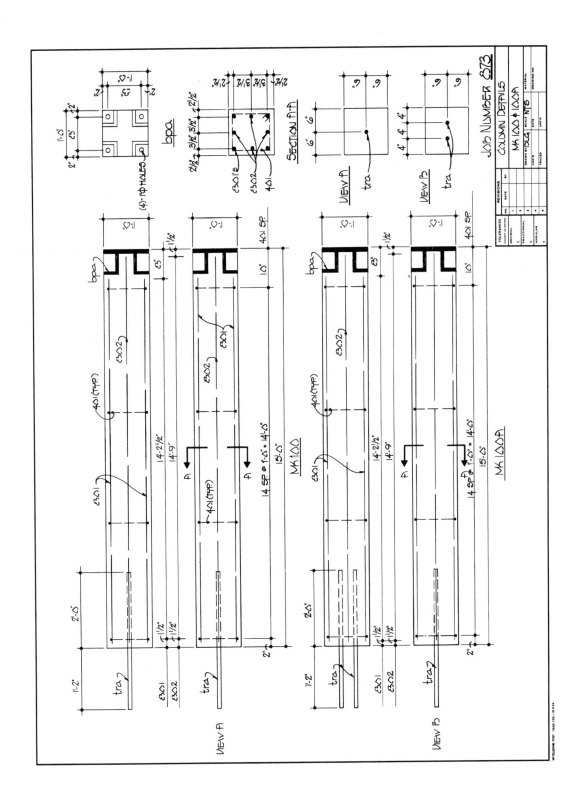

Fig. 15-9 Column detail for Reinforcement Activities

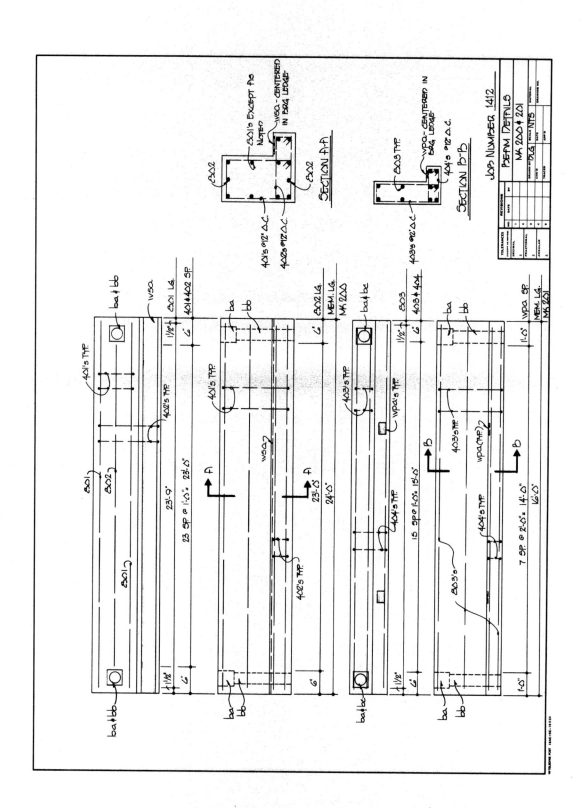

Fig. 15-10 Beam details for Reinforcement Activities

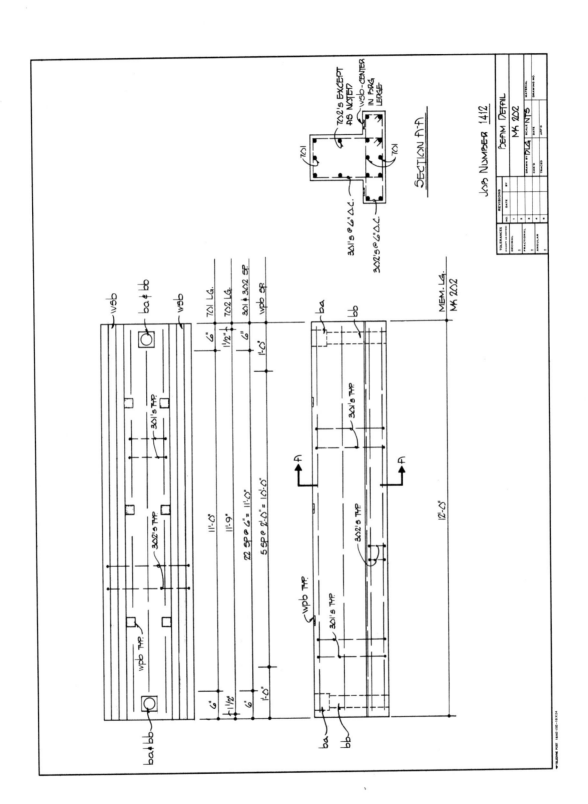

Fig. 15-11 Beam detail for Reinforcement Activities

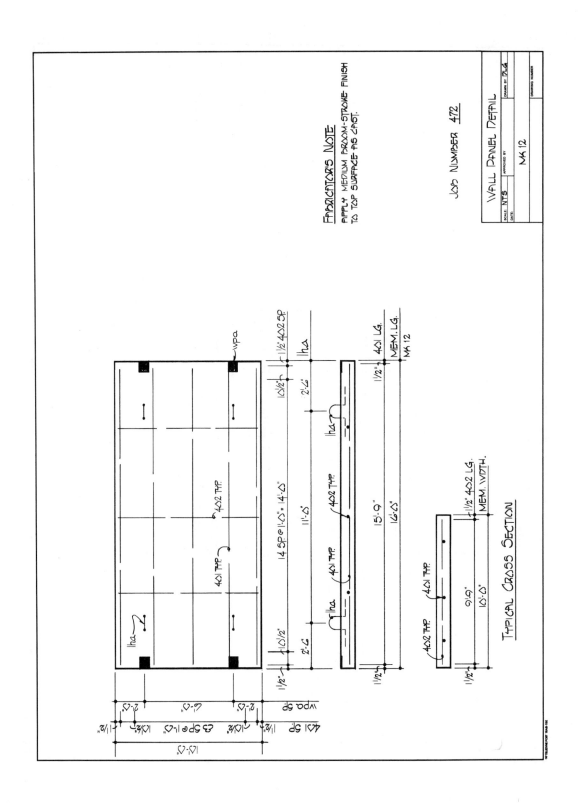

Fig. 15-12 Flat-slab wall panel detail for Reinforcement Activities

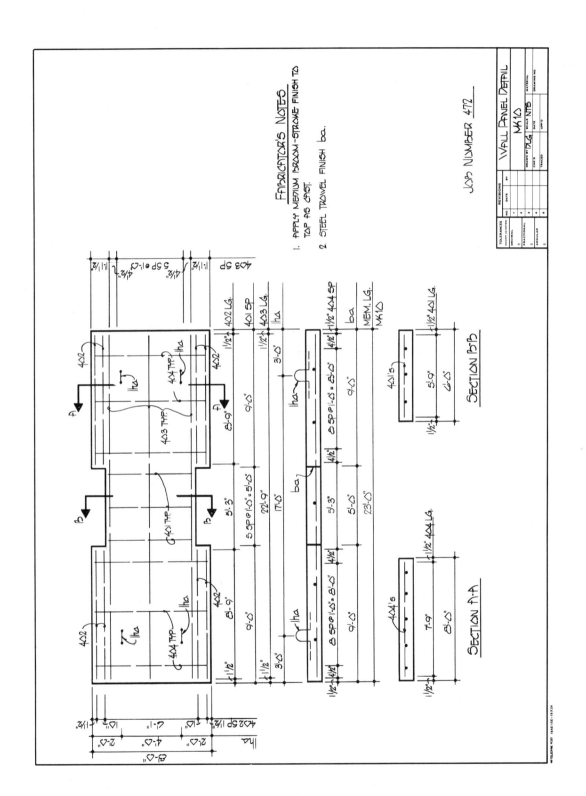

Fig. 15-13 Flat-slab wall panel detail for Reinforcement Activities

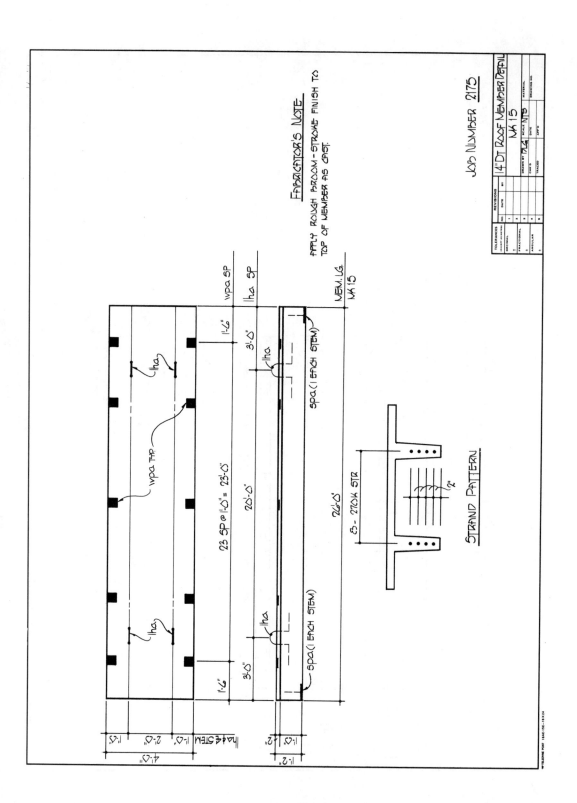

Fig. 15-14 Double-tee roof member detail for Reinforcement Activities

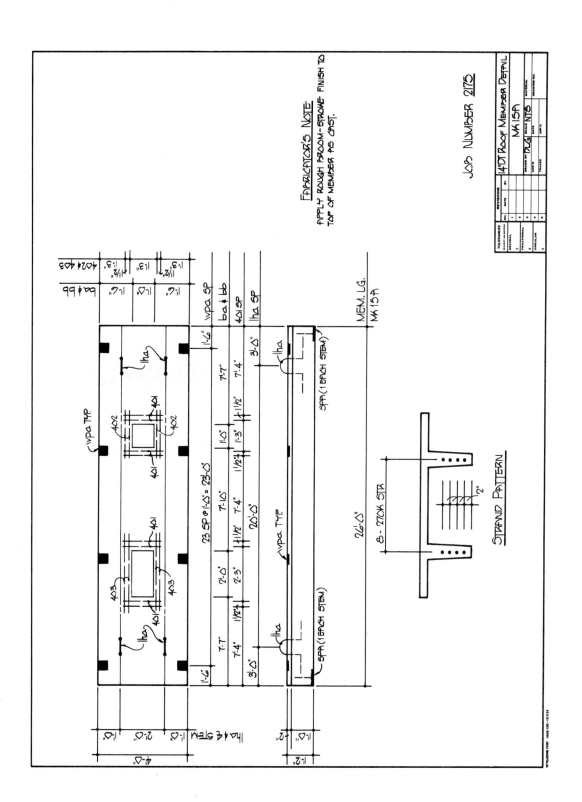

Fig. 15-15 Double-tee roof member detail for Reinforcement Activities

Section 4
Structural Poured-in-Place Concrete

Unit 16
Poured-in-Place Concrete Foundations

OBJECTIVES

Upon completion of this unit, the student will be able to

- define poured-in-place concrete.
- define engineering drawings and placing drawings as used in poured-in-place concrete drafting.
- identify common abbreviations and symbols used in poured-in-place concrete drafting.
- assign and interpret mark numbers for structural members.
- prepare engineering and placing drawings for poured-in-place concrete foundations.

POURED-IN-PLACE CONCRETE CONSTRUCTION

Poured-in-place concrete construction involves the following:

- Building forms in place at the jobsite
- Placing steel reinforcing bars into the forms
- Pouring concrete into the forms around the reinforcing bars

Once the concrete has been allowed to harden sufficiently, the forms are stripped away. This leaves an in-place, fully erected structural member.

Poured-in-place indicates that the concrete is poured at the jobsite into forms that are in place. This differs from precast concrete that is poured into forms at a manufacturing plant and then shipped to the jobsite and erected in pieces. Poured-in-place concrete, with the exception of special members such as tilt-up walls, is actually erected as it is poured.

POURED-IN-PLACE CONCRETE DRAWINGS

Poured-in-place concrete drawings are of two types: engineering drawings and placing drawings. *Engineering drawings* show the general layout of the structure, size and spacing of structural components (foundations, walls, beams, columns, etc.), and various notes containing information helpful to the reader in understanding the drawings.

Placing drawings are more detailed drawings. They show the actual size, shape, and locations of all reinforcing bars in a structural member and how the bars are to be placed into the forms. Placing drawings also include comprehensive schedules containing information on every piece of reinforcing bar used in the structural members in question.

SHEET LAYOUT AND SCALES

The format chosen for laying out engineering and placing drawings tends to vary from company to company. The American Concrete Institute

(ACI) recommends a standard sheet layout that is followed by many companies. However, factors such as the amount of information that must be placed on a sheet, individual company preferences, and sheet sizes used affect the layout of a sheet. Because of this, sheet layout formats are only loosely standardized. A format that represents generally what is used in industry for laying out engineering drawings is shown in Figure 16-1. A similar example for placing drawings is illustrated in Figure 16-2.

Engineering and placing drawings of poured-in-place concrete structures are drawn to scale. The scale selected for each drawing depends on several factors:

- Overall size of the structure being drawn
- Size of sheet the structure is being drawn on
- Complexity of the drawing
- Individual company preferences

There are some industry-wide accepted guidelines for drawing the various components (plan views, elevations, sections, and details) of engineering and placing drawings:

Plan Views	Elevations	Sections and Details
1/8" = 1'-0"	1/4" = 1'-0"	1/4" = 1'-0"
1/4" = 1'-0"	3/8" = 1'-0"	3/8" = 1'-0"
	1/2" = 1'-0"	1/2" = 1'-0"
		3/4" = 1'-0"
		1" = 1'-0"
		1 1/2" = 1'-0"

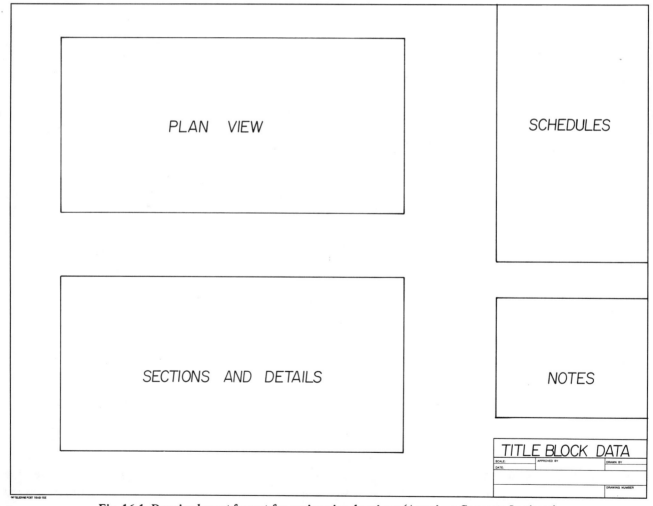

Fig. 16-1 Drawing layout format for engineering drawings (American Concrete Institute)

SYMBOLS AND ABBREVIATIONS

Industry-wide standardization of symbols, abbreviations, and practices would reduce drafting time and make drawings more easily understood. However, standardization of symbols and abbreviations on poured-in-place concrete drawings is not always achieved. Some companies have their own symbols and abbreviations that have been developed over the years and are usually explained in a legend.

In spite of individual company differences, there are a number of abbreviations and symbols that have achieved industry-wide recognition, if not standardization, Figure 16-3. A similar listing of common abbreviations is shown in Figure 16-4.

MARK NUMBERING SYSTEMS

In order to manage the design, drawing, and construction of a structure composed of separate but interrelated parts, a numbering system is necessary. *Component* or *mark numbers* are used on poured-in-place concrete drawings. A well-developed mark numbering system relates not only the component number, but where the component fits into the overall job. For example, the mark number 3B 4 refers to beam number four on the third floor.

Mark numbering systems, like other aspects of poured-in-place concrete drawings, vary from company to company. However, most systems use numbers to designate floor levels, letters to designate structural components, and numbers to

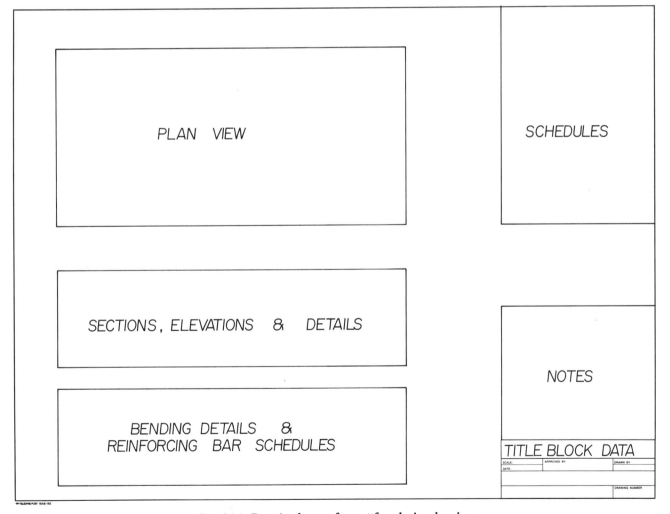

Fig. 16-2 Drawing layout format for placing drawings

SYMBOL	DESCRIPTION
#	SYMBOL FOR NUMBER. USED TO INDICATE THE WORD NUMBER WHEN DESCRIBING BARS. EXAMPLE: #3 BAR, #4 BAR, ETC.
∅	SYMBOL FOR DIAMETER. EXAMPLE: 1/2" ∅ BAR OR 10" ∅ COLUMN.
□	SYMBOL FOR SQUARE. EXAMPLE: 12" □ BEAM.
@	SYMBOL FOR AT. USED IN DIMENSIONING FOR CENTER-TO-CENTER SPACING. EXAMPLE: #3 BARS @ 7" ON CENTER.
←——→	INDICATES DIRECTION OF REINFORCING BARS. USED ON PLAN VIEWS.

Fig. 16-3 Common drafting symbols

identify specific components (i.e., beam number 1, 2, 3, 4, etc.). Letter designations for beams, columns, joists, and other structural components of poured-in-place concrete jobs have achieved a high degree of industry-wide recognition. Some letter designations commonly used on poured-in-place concrete drawings are:

- B — Beams
- C — Columns
- F — Footings or foundations
- G — Girders
- J — Joists
- L — Lintels
- S — Slabs
- T — Ties
- U — Stirrups (U is used to avoid conflicts with S as used for slabs)

Mark numbers are also assigned to reinforcing bars shown on poured-in-place concrete placing drawings. Various methods are used throughout the industry. Most methods use a combination of numbers and letters which designate the bar size, bar type, and specifically identify each individual length of bar. For example, the bar mark number

Bot	BOTTOM (AS IN BOTTOM OF MEMBER)
Bt	BENT (AS IN BENT BAR)
CT	COLUMN TIE
EF	EACH FACE
EW	EACH WAY (AS IN BARS SPACED @ 6" ON CENTER EACH WAY)
FF	FAR FACE
IF	INSIDE FACE
NF	NEAR FACE
OF	OUTSIDE FACE
Pl	PLAIN BAR
St	STRAIGHT
Stir	STIRRUP
Sp	SPIRAL
T	TOP (AS IN TOP OF MEMBER)

Fig. 16-4 Commonly used abbreviations

4A1 indicates a #4 bar, type A, and it is the first length of bar used in the job. A bar numbered 4A2 would also be a type A, number 4 bar, but it would have a different length than a 4A1.

Designations of bar types vary from company to company, but most firms arrive at some type of standard system. Straight bars may be designated type A, J-shaped bars type B, U-shaped bars type C, etc. Typical bent bar types found in industry with sample designations are shown in Figure 16-5.

Unit 16 Poured-in-Place Concrete Foundations 241

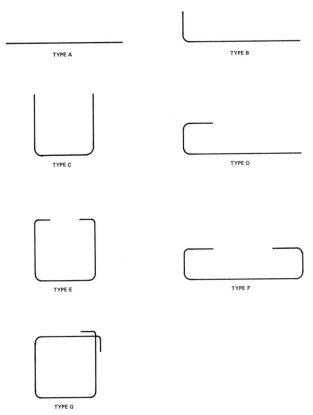

Fig. 16-5 Common bent bar types

Another common system of numbering bars uses only numbers. This method identifies the bar size and separates the bars according to lengths, but does not designate the bar type. For example: Bar 601 would be a number 6 bar of some specific length. Bar 602 would also be a number 6 bar, but of a different length. The bar bending schedule would be consulted to determine the bar type.

SCHEDULES

Schedules are an important part of engineering and placing drawings. Engineering drawings contain beam schedules, column schedules, joist schedules, etc., depending on the structural members being drawn, Figures 16-6 and 16-7. Placing drawings contain schedules similar to those on engineering drawings, Figure 16-8. Placing drawings also contain schedules showing the bending information for all reinforcing bars used in the job, Figure 16-9.

			BEAM SCHEDULE			
			REINFORCING BARS			
			LONGITUDINAL REINFORCING		STIRRUPS AND TIES	
MARK NUMBERS	BEAM WIDTH	BEAM DEPTH	BAR SIZE	REMARKS	BAR SIZE	BAR SPACING
3B1	12"	20"	4 - #6	Straight	16 - #3	8 @ 12"
3B2	12"	20"	4 - #6	Straight	16 - #3	4 @ 12", 4 @ 10"
3B3	10"	18"	4 - #6	Straight	12 - #3	6 @ 12"
3B4	10"	18"	4 - #6	Straight	12 - #3	3 @ 12", 3 @ 10"
3B5	8"	14"	4 - #5	Straight	8 - #3	8 @ 12"
3B6	8"	14"	4 - #5	Straight	9 - #3	4 @ 12", 5 @ 10"

Fig. 16-6 Sample beam schedule for engineering drawings

COLUMN SCHEDULE	B2, B3	C1, C4	A1, A4	COLUMN MARK NUMBERS
ROOF	C2, C3	A2, A3	B1, B4	
	18 x 24	18 x 18	16 x 16	COLUMN SIZE
12'-0"	4 - #8	4 - #8	4 - #8	VERTICAL BARS
FIRST FLOOR	#3 @ 12"	#3 @ 12"	#3 @ 12"	COLUMN TIES
	18 x 24	20 x 20	16 x 16	COLUMN SIZES
10'-4"	6 - #8	6 - #8	6 - #8	VERTICAL BARS
BASEMENT	#4 @ 12"	#4 @ 12"	#4 @ 12"	COLUMN TIES

Fig. 16-7 Sample column schedule

				BEAM SCHEDULE								
				LONGITUDINAL REINFORCEMENT						STIRRUPS		
				BOTTOM OF BEAM			TOP OF BEAM					
MARK NO.	NO. REQ'D.	WIDTH	DEPTH	NO. REQ'D.	BAR SIZE	LENGTH	NO. REQ'D.	BAR SIZE	LENGTH	NO. REQ'D.	BAR SIZE	STIRRUP SPACING
2B1	3	12"	16"	4	6	21'-2"	4	5	21'-2"	22	3G1	20 @ 12" 2 @ 6"
2B2	1	12"	16"	4	6	20'-6"	4	5	20'-6"	21	3G1	19 @ 12" 2 @ 6"
2B3	1	10"	14"	3	5	16'-5"	3	4	16'-5"	17	3G2	15 @ 12" 2 @ 6"
2B4	2	10"	14"	3	5	15'-6"	3	4	15'-6"	16	3G2	14 @ 12" 2 @ 6"
2B5	1	8"	12"	3	4	10'-4"	3	4	10'-4"	11	3G3	9 @ 12" 2 @ 4"
2B6	2	8"	12"	3	4	10'-2"	3	4	10'-2"	11	3G3	9 @ 12" 2 @ 3"

Fig. 16-8 Sample beam schedule for placing drawings

BAR BENDING DETAILS AND DIMENSIONS											
BAR MARK NUMBER	BAR SIZE	TYPE	TOTAL LENGTH	A	B	C	D	E	F	G	H
6C1	6	C	5'-2"	24"	14"						
6C2	6	C	4'-10"	22"	14"						
4F3	4	F	4'-0"			4"	10"	20"			
3C4	3	C	4'-4"	20"	12"						
5F5	5	F	4'-6"			3"	12"	24"			
6F6	6	F	4'-0"			3"	10"	22"			

Fig. 16-9 Sample bar bending schedule

POURED-IN-PLACE FOUNDATION DRAWINGS

The first drawing in a set of poured-in-place concrete plans is the foundation plan. A complete foundation plan will require both engineering and placing drawings. The engineering drawings are used in the design and planning process and are prepared first.

Preparing Engineering Drawings for Foundations

Like so many aspects of poured-in-place concrete drafting, the procedures followed in preparing engineering drawings are not completely standardized. However, the American Concrete Institute has established guidelines for preparing engineering drawings that can simplify the process if followed.

The following are some ACI recommendations for preparing engineering drawings of poured-in-place concrete structures:

- All information relating to the size and arrangement of concrete members and the size, positioning, and arrangement of reinforcing bars in the members must be provided. This may be accomplished through drawings, notes, or schedules.
- Time may be saved by providing typical details and then developing a schedule containing the necessary information for members not specifically detailed.
- All dimensions are not required, but a sufficient number should be provided to accommodate design and to indicate where bars are to be bent. More specific dimensions are left to the placing drawings.
- Details of construction joints or splice points are helpful and should be provided to the various trades who must build the structure.

An example of an engineering drawing of a foundation for a commercial structure is shown in Figure 16-10. Examine it closely and become familiar with its layout and contents.

Preparing Placing Drawings for Foundations

Placing drawings are prepared after the engineering drawings have been completed. A common practice is to trace the outlines of the various structural components (i.e., beams, columns, walls, etc.) from the engineering drawings and then add the necessary detail. Placing drawings are prepared from information contained on the engineering drawings. They are used for fabricating the structural members for a job and ordering the materials necessary to do a job.

Placing drawings must show:

- Size and shape of the structural members to be fabricated
- Size, shape, and location of all bars in the structural member(s)
- How the bars are to be placed in the forms

Figure 16-11 is an example of a placing drawing for a commerical foundation. Compare this drawing with Figure 16-10, making note of similarities and differences.

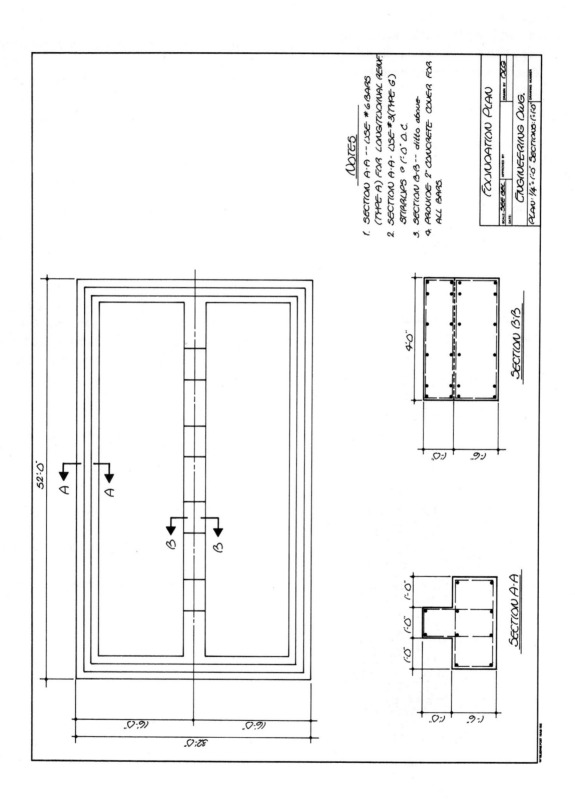

Fig. 16-10 Sample engineering drawing

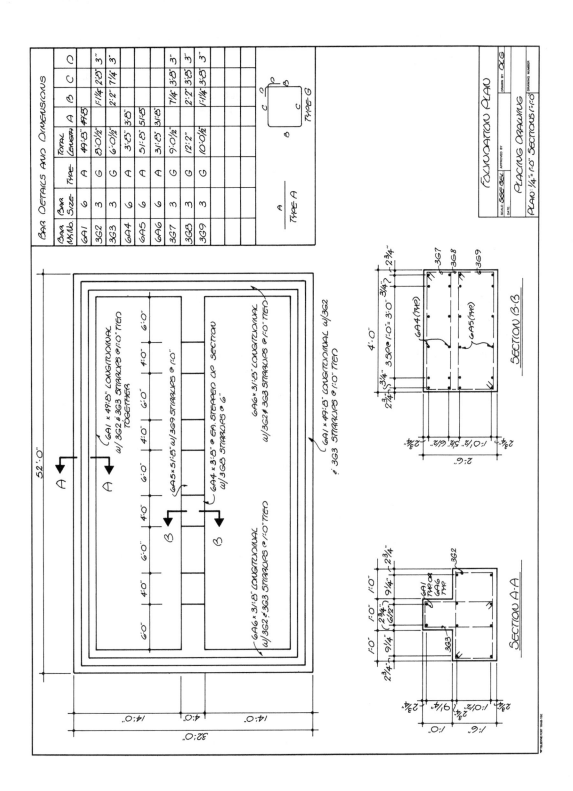

Fig. 16-11 Sample placing drawing

SUMMARY

... Poured-in-place concrete construction involves building forms in place at the jobsite, placing steel reinforcing bars into the forms, and pouring concrete into the forms around the reinforcing bars.

... Poured-in-place concrete differs from precast concrete in that it is poured into temporary forms at the jobsite while precast members are poured in a manufacturing plant and shipped to the jobsite.

... Poured-in-place concrete drawings are of two types: engineering drawings and placing drawings.

... Engineering drawings show the general layout of the structure, size and spacing of structural components, and various notes containing information helpful in understanding the drawings.

... Placing drawings are more detailed than engineering drawings. They show the actual size, shape, and locations of all reinforcing bars in each structural member as well as how the bars will fit into the forms.

... Sheet layouts for poured-in-place concrete drawings are only loosely standardized, but individual companies should adhere to a general pattern of sheet layout.

... Engineering and placing drawings are drawn to a scale selected according to the overall size of the structure being drawn, size of sheet the structure is being drawn on, complexity of the drawing, and individual company preferences.

... Structural drafters should use poured-in-place concrete symbols that are recognized industry wide when preparing drawings.

... Mark numbering systems are required in order to organize and manage the design, drawing, and construction of a large structure composed of separate, but interrelated parts. The mark number 3B4 refers to beam number four on the third floor.

... Schedules are an important part of engineering and placing drawings. Schedules for engineering drawings are member schedules such as beam, column, joist, etc., schedules. Schedules for placing drawings contain bending information for reinforcing bars.

REVIEW QUESTIONS

1. Explain how poured-in-place concrete construction differs from precast concrete construction.
2. Name the two types of poured-in-place concrete drawings.
3. Make a sketch to show the general layout of a poured-in-place concrete engineering drawing.
4. Make a sketch to show the general layout of a poured-in-place concrete placing drawing.

REINFORCEMENT ACTIVITIES

The title block as shown in Figure 1-21 and 1/2" borders are required for all activities.

1. Prepare four, C-size sheets of paper with title blocks and borders.

2. On one of the sheets prepared in Reinforcement Activity #1, redraw the foundation plan in Figure 16-12, excluding the asphalt pavement area, as an engineering drawing. Refer to Figure 16-10 for an example. Revision notes and symbols should also be excluded.

3. On the second sheet of paper from Reinforcement Activity #1 prepare a placing drawing for the engineering drawing prepared in Reinforcement Activity #2. Refer to Figure 16-11 for an example.

4. Redraw an engineering drawing of Figure 16-12, making the following changes:
    a. Change the cross-sectional dimension of Section A-A to 3'-0" and adjust the longitudinal reinforcement spacing accordingly.
    b. Add an identical row of reinforcing bars 2" down from the top of Section A-A and surround both rows with Type G stirrups at 1'-0" On Center.

5. Prepare a complete placing drawing for the engineering drawing developed in Reinforcement Activity #4.

248 Section 4 Structural Poured-in-Place Concrete

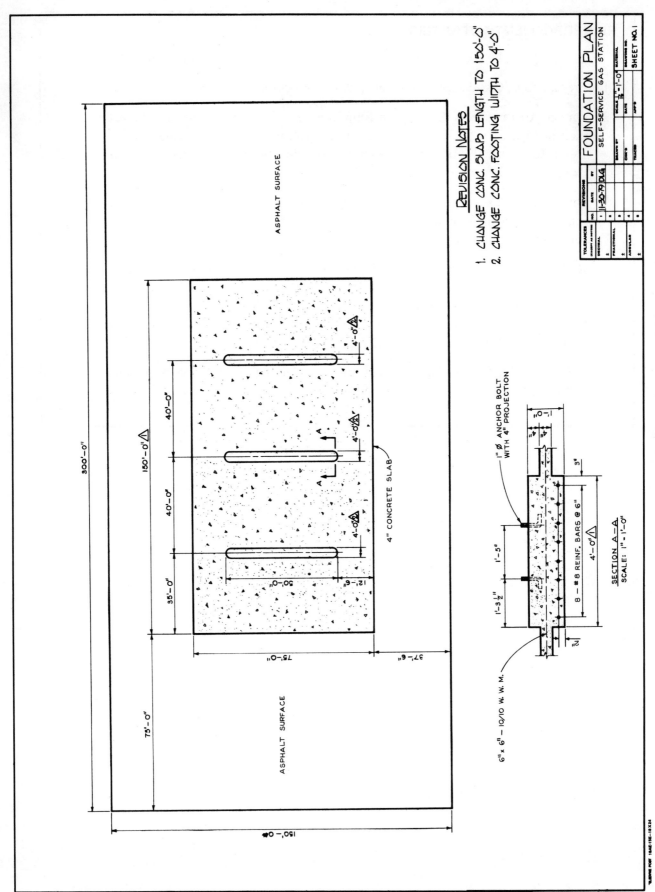

Fig. 16-12 Foundation plan for Reinforcement Activities

Unit 17
Poured-in-Place Concrete Walls and Columns

OBJECTIVES

Upon completion of this unit, the student will be able to

- list the four basic categories of poured-in-place concrete walls and give a definition of each.
- define and distinguish between the two basic types of poured-in-place concrete columns.
- prepare complete engineering drawings of poured-in-place concrete wall systems and drawings of poured-in-place concrete columns.
- prepare complete placing drawings of poured-in-place concrete wall components and drawings of poured-in-place concrete columns.

POURED-IN-PLACE CONCRETE WALLS AND COLUMNS

Two major components of most poured-in-place concrete structures are walls and columns. Walls fall into four basic categories: security walls, shear walls, tilt-up walls, and retaining walls. Columns fall into two categories: tied columns and spiral columns. The terms *tied* and *spiral* refer to the type of reinforcing used in the columns.

Security Walls

Security walls may be used for a number of different applications. As the name implies, their primary function is usually to provide some type of security. Common applications are to provide security from fire, break-ins, or noise. Bank vaults are very often composed of poured-in-place concrete security walls. Firewalls in apartments and other commercial structures are also a common application.

Shear Walls

Shear walls can be made of poured-in-place concrete. They are used to brace tall structures against lateral loading from wind and swaying motions. Elevator shafts and stairwells may be designed as shear walls, Figure 17-1. Poured-in-place concrete shear walls, such as those illustrated in Figure 17-1, provide support for the floors of the building as well as overall lateral support for the building.

Tilt-Up Walls

Exterior walls of a structure are commonly constructed of tilt-up wall components. A tilt-up wall may be precast or poured-in-place, as may all structural concrete walls listed in this unit. A poured-in-place tilt-up wall breaks the rule of poured-in-place concrete construction. This rule says that members are poured into forms that have been built in place at the jobsite, thereby speeding up pouring and erection concurrently.

250 Section 4 Structural Poured-in-Place Concrete

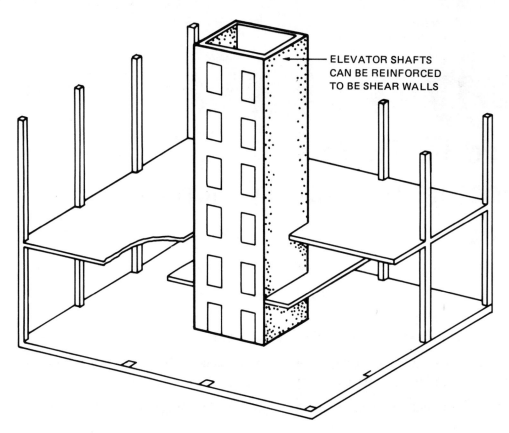

Fig. 17-1 Elevator shaft as shear walls (Charles D. Willis)

Tilt-up wall components are poured into forms that have been built at the jobsite, but they are not in place. The forms for tilt-up walls are built on the ground or on the floor of the building. Concrete is poured over the reinforcing bars in the forms and allowed to harden. The wall panels are then lifted into place by a hoist or crane, Figure 17-2.

Retaining Walls

A retaining wall is a freestanding, self-supporting wall designed to hold back earth, water, or other material. Basement walls hold back earth or water and must be reinforced to do so. However, since a structure rests on basement walls they are not freestanding and should not be classified as retaining walls. True retaining walls fall into two categories: gravity walls and cantilever walls, Figure 17-3.

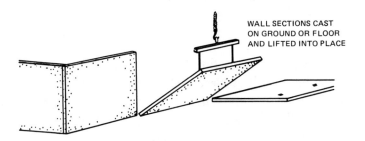

Fig. 17-2 Poured-in-place concrete tilt-up walls (Charles D. Willis)

Gravity walls are very heavy, poured-in-place concrete walls. Their use is usually restricted to applications requiring less than 6' of height. Applications requiring more height have a tendency to cause the wall to tip over. For these applications, the cantilever wall is more appropriate.

The *cantilever wall* is tapered from top to bottom with the thickest portion at the bottom

Unit 17 Poured-in-Place Concrete Walls and Columns 251

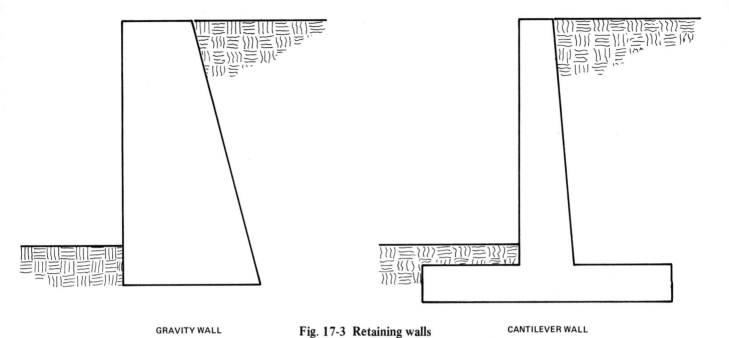

Fig. 17-3 Retaining walls

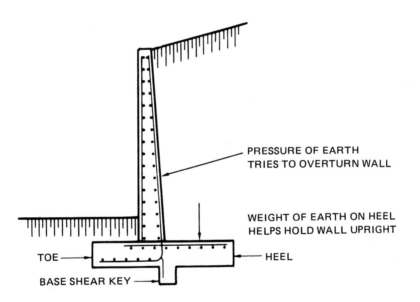

Fig. 17-4 Cantilever retaining wall in section (Charles D. Willis)

resting on a wide footing. The footing is poured in the ground and heavily reinforced. Its purpose is to prevent the forces of pressure from the material being held back from tipping the wall over. The vertical portion of the cantilever wall, known as the *stem,* is also heavily reinforced. A cantilever wall in section showing the steel reinforcing bars is shown in Figure 17-4.

Tied and Spiral Columns

Columns in poured-in-place concrete construction are classified as either tied or spiral according to the method of reinforcing used in them. *Tied* columns are square or rectangular in cross section, while *spiral* columns are round, Figure 17-5.

Tied columns are reinforced with long, straight reinforcing bars tied off with wire ties or

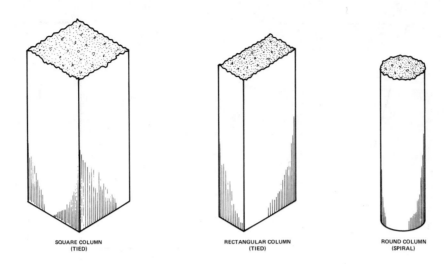

Fig. 17-5 Tied and spiral column configurations

smaller, bent reinforcing bars, Figure 17-6. Spiral columns are reinforced with long, straight reinforcing bars surrounded by spiral wire or reinforcing bars, Figure 17-6.

Poured-in-place columns are often used in multistory buildings. When this is the case, the columns must be spliced. Column splicing usually occurs at floor levels. Splicing a poured-in-place column involves extending the main reinforcing steel out of the bottom column so that it can be spliced to the main reinforcing of the column that will rest on top of it, Figure 17-7.

Most poured-in-place columns are square, round, or rectangular. However, on occasion, design and appearance may dictate the need for a different shape of column. When this is the case, several special column shapes are available, Figure 17-8.

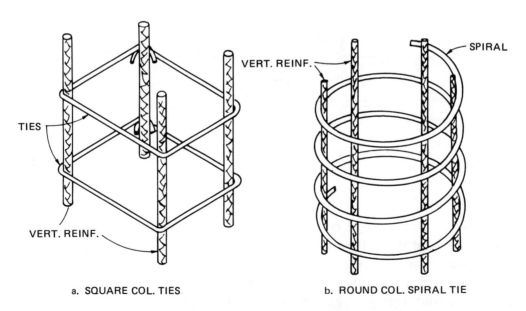

Fig. 17-6 Typical column reinforcing (Charles D. Willis)

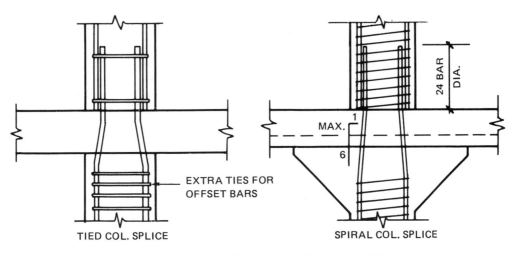

Fig. 17-7 Column splicing details (Charles D. Willis)

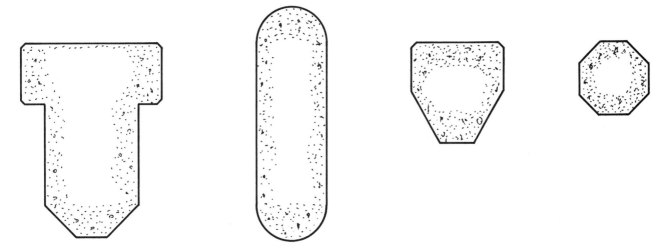

Fig. 17-8 Special column configurations

WALL AND COLUMN ENGINEERING DRAWINGS

Drafters prepare engineering drawings of column layouts and wall systems from sketches and specifications supplied by structural engineers or architects. Engineering drawings contain the following:

- A plan view of the walls or columns
- General dimensions
- An indication of the types and sizes of reinforcing bars to be used
- Sections and/or details
- Necessary notes
- Wall component or column schedules

Figures 17-9 through 17-11 illustrate wall and column engineering drawings.

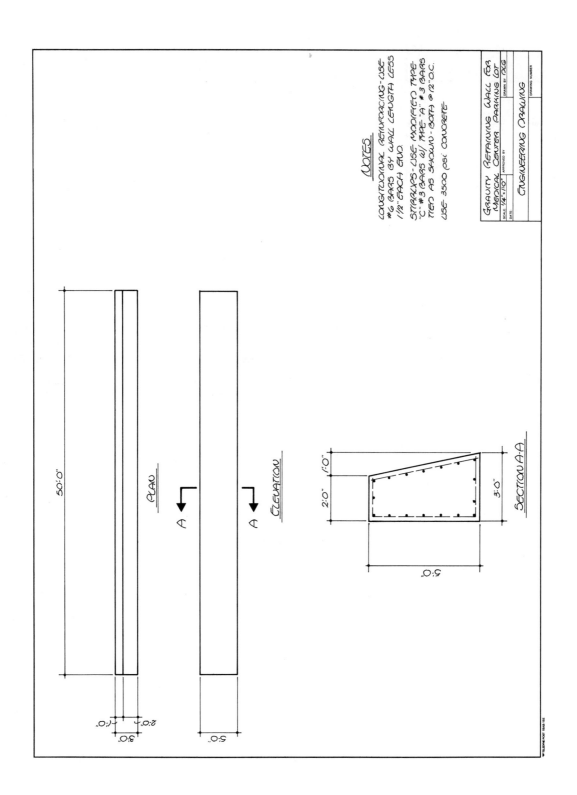

Fig. 17-9 Gravity retaining wall engineering drawing

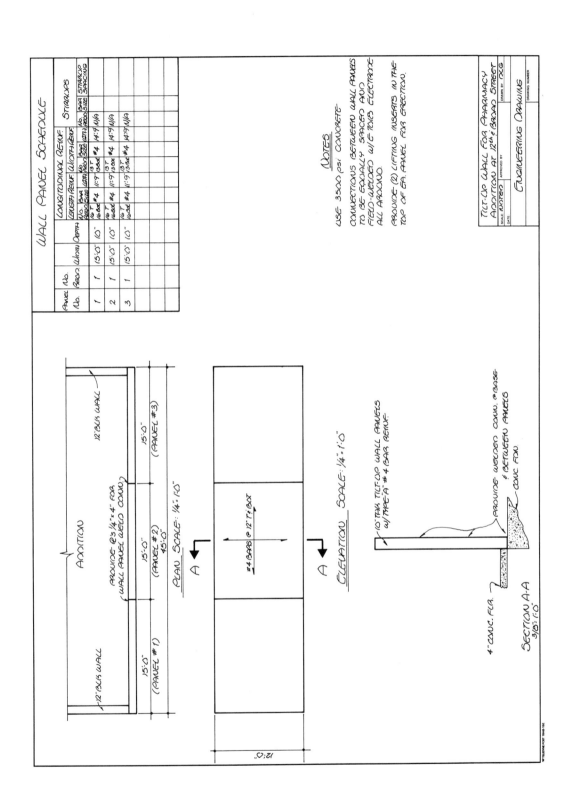

Fig. 17-10 Tilt-up wall engineering drawing

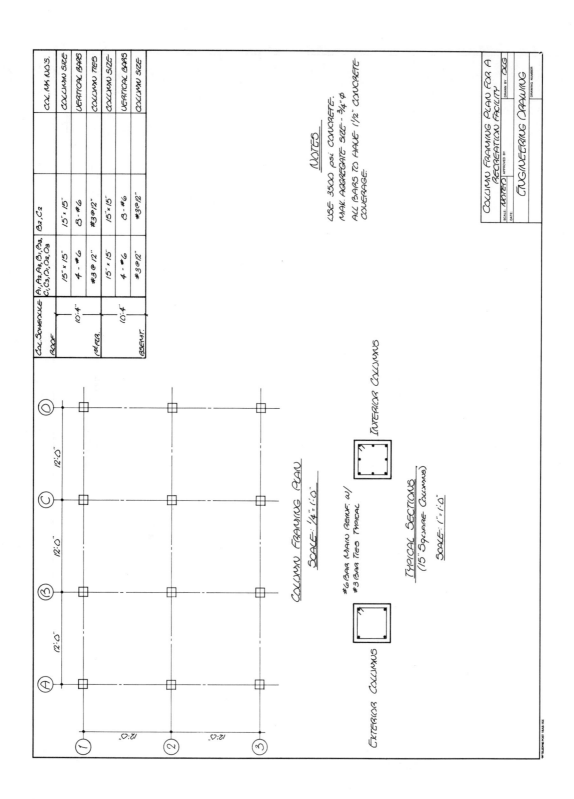

Fig. 17-11 Column engineering drawing

WALL AND COLUMN PLACING DRAWINGS

Drafters prepare placing drawings from engineering drawings and sketches and specifications provided by engineers. Placing drawings contain:

- Plan views
- Sections and/or details
- Schedules
- Bending details
- Notes

The placing drawings for a job are usually drawn to the same scale as the engineering drawings so that the outlines of structural components can be traced to save drafting time. Poured-in-place concrete wall and column placing drawings are shown in Figures 17-12 and 17-13.

SUMMARY

... Walls are a major component of most poured-in-place concrete structures. Walls fall into four categories: security walls, shear walls, retaining walls, and tilt-up walls.
... Columns are another major component of most poured-in-place concrete structures and fall into two categories: tied columns and spiral columns.
... Security walls are used to provide security from fire, break-ins, or noise.
...Shear walls are used to brace tall structures against lateral loading from wind or other forces.
... Exterior walls of a structure are commonly tilt-up walls that are poured in temporary forms on the ground or the floor of the building and erected with a hoist or crane.
... Retaining walls are freestanding, self-supporting walls designed to hold back earth, water, or other material. There are two types: gravity walls and cantilever walls.
... Gravity walls are very heavy, poured-in-place concrete walls usually restricted to applications requiring less than 6′ of height.
... Cantilever walls are poured on a wide footing that prevents them from tipping over. Consequently, they can be used for applications requiring more than 6′ of height.
... Tied columns are square or rectangular in cross section and are reinforced with stirrup ties wrapped around long, straight longitudinal reinforcing bars.
... Spiral columns are usually round and are reinforced with spiral wire wrapped around long, straight, longitudinal reinforcing bars.

258 Section 4 Structural Poured-in-Place Concrete

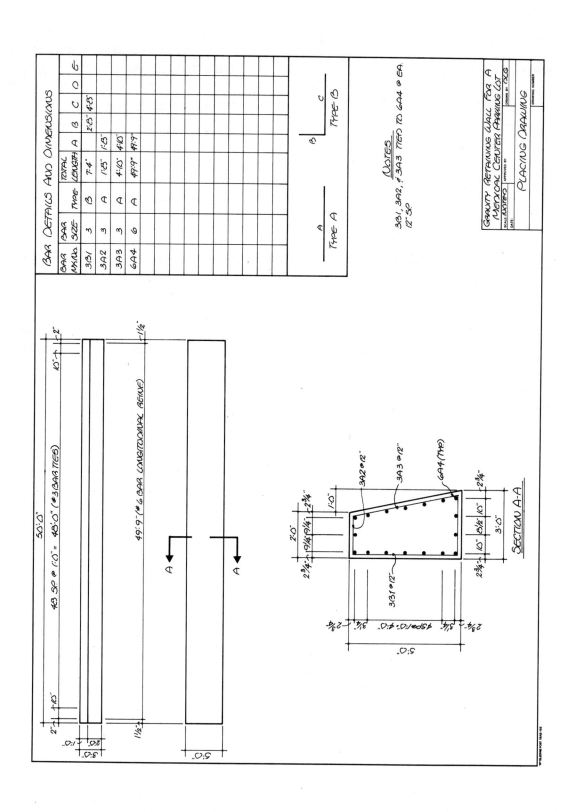

Fig. 17-12 Retaining wall placing drawing

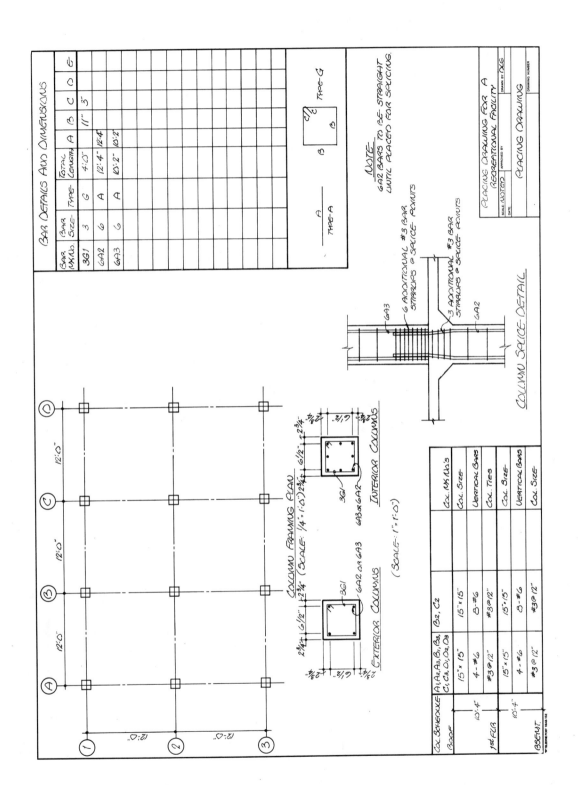

Fig. 17-13 Column placing drawing

REVIEW QUESTIONS

1. List the four basic categories of walls.
2. Sketch a simple illustration of each type of wall listed in question 1.
3. Make a sketch that illustrates the different reinforcing methods for tied and spiral columns.
4. List three uses of security walls.
5. What is the primary function of shear walls?
6. What is a tilt-up wall?
7. What are the two types of retaining walls?
8. Sketch an example of a cross section of each of the two types of retaining walls listed in question 7.

REINFORCEMENT ACTIVITIES

The title block as shown in Figure 1-21 and 1/2" borders are required for all activities.

1. Figure 17-14 shows a foundation plan for a self-service gas station. The concrete slab area is 75'-0" x 150'-0" in the center of a larger area. Draw an engineering drawing for a gravity retaining wall for the left side of the concrete slab area. The wall should be 4'-0" tall, 3'-0" wide at the top, and 4'-0" wide at the bottom. The plan should be similar to the example in Figure 17-9.

2. Prepare a placing drawing for the retaining wall drawn in Reinforcement Activity #1. The plan should be similar to the example in Figure 17-12.

3. Prepare an engineering drawing of a cantilever retaining wall for the right side of the concrete slab in Figure 17-14. Refer to Figure 17-4 for a sample cross-sectional illustration of a similar wall.

4. Prepare a placing drawing for the retaining wall drawn in Reinforcement Activity #3. Refer to Figure 17-12 for a similar plan that can be used as an example.

5. Figure 17-14 contains three, 50'-0" long concrete foundations equally spaced within the concrete slab area. Using the dimensions available, place three, 12" square columns (equally spaced) on each of the foundations. Using this as a starting point, prepare an engineering drawing for the columns. Each column is to be 15'-0" high, reinforced with 4 - #5 bars longitudinally, and #3 bar stirrups @ 12" On Center. Refer to Figure 17-11 for an example.

6. Prepare a placing drawing for the column plan drawn in Reinforcement Activity #5. Refer to Figure 17-13 for an example.

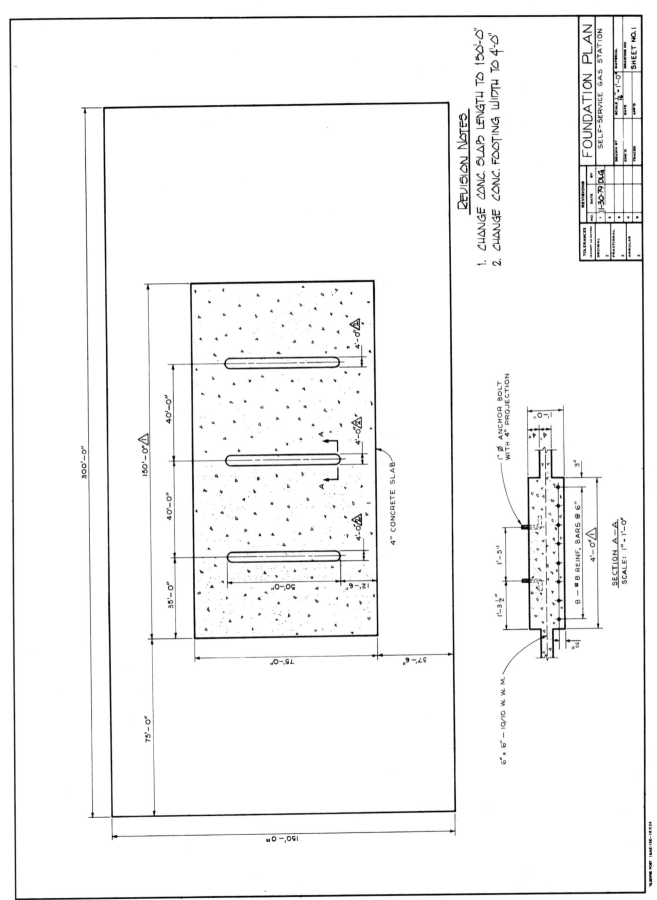

Fig. 17-14 Poured-in-place concrete drawing for Reinforcement Activity

Unit 18
Poured-in-Place Concrete Floor Systems

OBJECTIVES

Upon completion of this unit, the student will be able to

- distinguish between ground-supported and suspended floor systems.
- define and recognize one-way solid slab-and-beam floor systems; one-way ribbed or joist-slab floor systems; two-way solid slab-and-beam floor systems; two-way flat-plate floor systems; and waffle-slab floor systems.
- Prepare complete engineering drawings of poured-in-place concrete floor systems.
- Prepare complete placing drawings of poured-in-place concrete floor systems.

POURED-IN-PLACE CONCRETE FLOOR SYSTEMS

There are several different types of poured-in-place concrete floor systems used in the heavy construction industry. One-way solid slab and beam, one-way ribbed or joist, two-way solid slab and beam, two-way flat plate, and waffle-slab floor systems are the most common. Floor systems in poured-in-place concrete construction are either ground-supported systems or suspended systems, Figure 18-1.

ONE-WAY SOLID SLAB AND BEAM

One of the most commonly used floor systems in poured-in-place concrete is the one-way solid slab-and-beam system. This type of system consists of a 4" to 6" thick slab of concrete supported by parallel beams or walls. It is used primarily in short-span situations of less than 12'.

As in poured-in-place columns, the one-way slab-and-beam system derives its classification from the method of reinforcement used. The *one-way* in the name refers to the fact that the main slab reinforcing runs in only one direction, from beam to beam. A poured-in-place concrete one-way solid slab-and-beam floor system is shown in Figure 18-2.

ONE-WAY RIBBED OR JOIST SLAB

One disadvantage of the one-way solid slab-and-beam floor system is that it can be very heavy, particularly when used in situations involving spans of over 12". The weight problem can be solved by using a modified version of the one-way solid slab-and-beam system called the *one-way ribbed* or *joist slab*. This system combines a thinner slab with concurrently poured joists or ribs to create a floor system very similar to a precast concrete double-tee system.

The joist or ribs are usually spaced 24" to 35" On Center and are reinforced with two or more reinforcing bars. The slab between and above the joists is reinforced in the same manner as the one-way solid slab-and-beam system. Figure 18-3 shows an example of a one-way ribbed or joist system.

Unit 18 Poured-in-Place Concrete Floor Systems 263

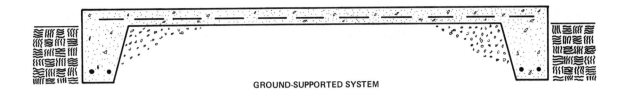

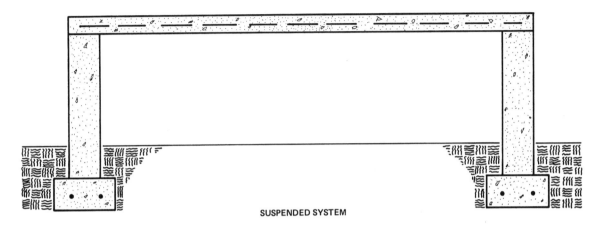

Fig. 18-1 Two classifications of poured-in-place concrete floor systems

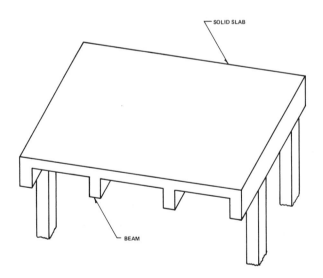

Fig. 18-2 One-way solid slab-and-beam floor system

Fig. 18-3 One-way ribbed or joist-slab system

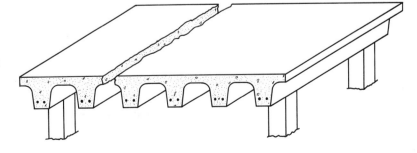

Lateral stability may be increased by adding bridging at intervals between the joists or ribs, Figure 18-4.

TWO-WAY SOLID SLAB AND BEAM

In construction situations that call for square slabs or rectangular slabs that are almost square, the two-way solid slab-and-beam floor system is used. As the name implies, the slab has beams running in two directions and is reinforced in two directions. This configuration allows for greater floor area with fewer beams and no girders. This type of floor system is slightly more difficult to construct. However, it is relatively economical in that it usually requires less building materials. A two-way solid slab-and-beam floor system is shown in Figure 18-5.

TWO-WAY FLAT-PLATE FLOOR SYSTEMS

The three floor systems just discussed are actually variations of the two-way solid-slab system. The flat-plate system is the most simple. It consists of a floor slab of uniform thickness supported by columns only. There are no beams or girders, Figure 18-6. The absence of beams and girders and the flat floor make this type of system ideal in applications where headroom is a problem.

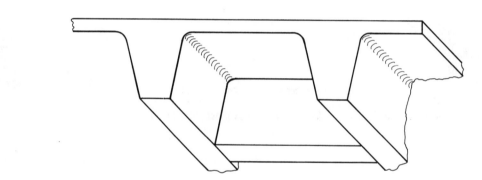

Fig. 18-4 Bridging between joists

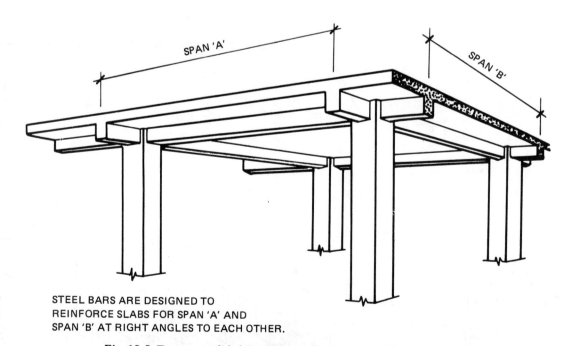

STEEL BARS ARE DESIGNED TO REINFORCE SLABS FOR SPAN 'A' AND SPAN 'B' AT RIGHT ANGLES TO EACH OTHER.

Fig. 18-5 Two-way solid slab-and-beam floor system (Charles D. Willis)

The *two-way* designation in the title indicates that the reinforcement is placed in both directions. Common applications of this type of floor system include hotels, motels, apartment buildings, and condominiums.

WAFFLE SLAB

In construction applications that are to be heavily loaded, extra thick slabs are required. However, the thicker the slab, the heavier it is, the more expensive it is, and often the less attractive it is. When a thick slab is required, these disadvantages may be overcome through the use of a waffle slab.

The *waffle-slab floor system*, Figure 18-7, is a thick slab with a series of geometric recesses formed into it. These recesses are formed with removable fillers placed in the forms before pouring. They cut down on the weight and cost of the slab without substantially decreasing its structural capabilities. These recesses also add a pleasing appearance to the slab.

POURED-IN-PLACE CONCRETE FLOOR SYSTEM DRAWINGS

Poured-in-place concrete floor systems require both engineering and placing drawings. The engineering drawings consist of a plan view of the floor, sections, slab schedules, and notes, Figure 18-8. Placing drawings contain a plan view of the floor, sections showing precise placement of the reinforcing bars, slab schedule, and a bar details schedule, Figure 18-9.

Slabs shown in plan view must be crossed to indicate the breadth of their length and width, Figures 18-8 and 18-9. They are also given mark number designations. For example, the slab designation 1S2 would be interpreted slab number 2 on the first floor. Slab designation 3S4 would be interpreted as slab number 4 on the third floor.

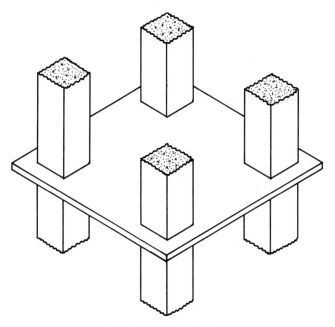

Fig. 18-6 Two-way flat-plate system

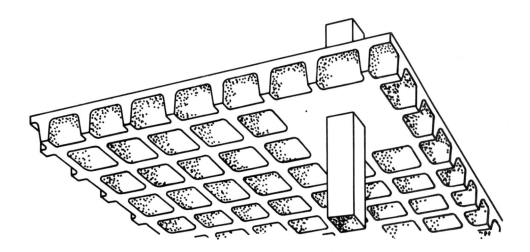

Fig. 18-7 Waffle-slab floor system

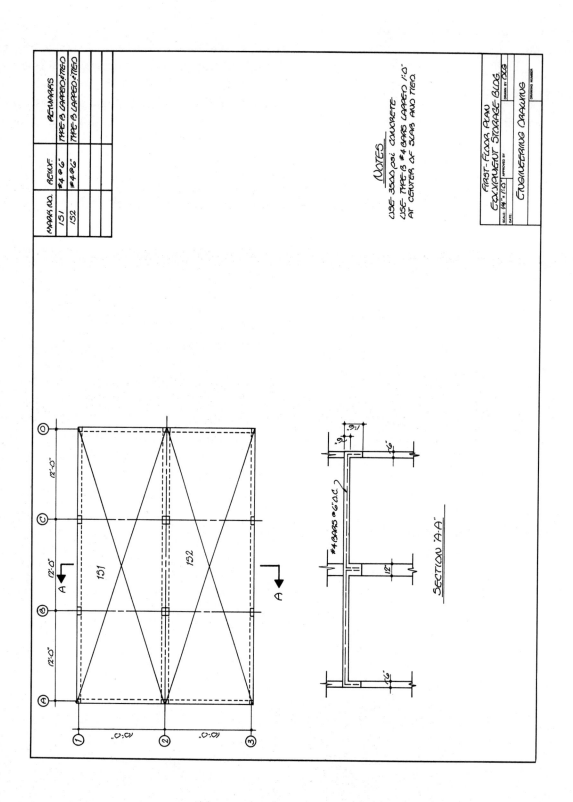

Fig. 18-8 Engineering drawing for poured-in-place concrete floor system

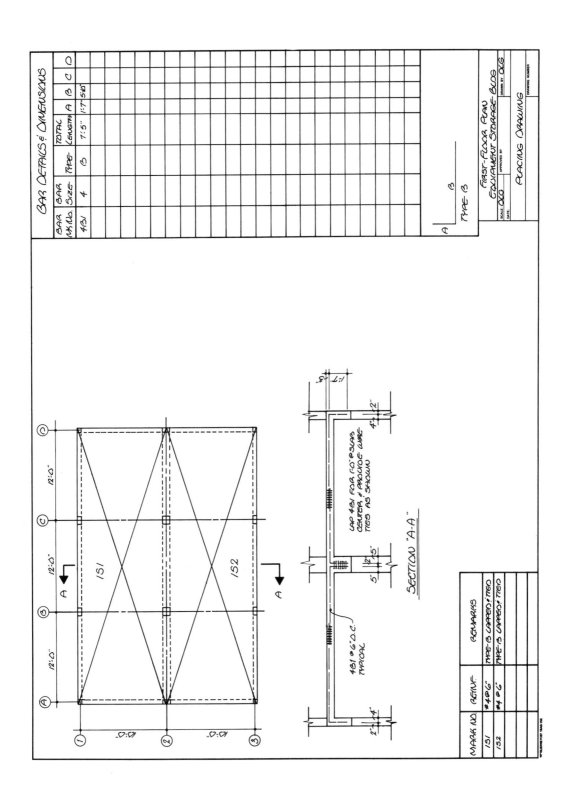

Fig. 18-9 Placing drawing for a poured-in-place concrete floor system

SUMMARY

... The most common types of poured-in-place concrete floors are: one-way solid slab and beam, one-way ribbed or joist, two-way solid slab and beam.
... Floor systems in poured-in-place concrete construction are either ground-supported or suspended systems.
... The one-way solid-slab system consists of a slab of uniform thickness supported by parallel beams or walls.
... The one-way ribbed or joist-slab system combines a thin slab with concurrently poured joists to create a floor system that is similar to the precast concrete double-tee system.
... Lateral stability in one-way ribbed or joist floor systems may be increased by adding bridging at intervals between the joists.
... The two-way solid slab-and-beam floor system is used in applications requiring square or rectangular slabs.
... The two-way solid-slab system is more difficult to construct than the one-way systems, but it is economical because it cuts down on the number of beams and requires no girders.
... The two-way flat-plate floor system is one of the simplest poured-in-place concrete floor systems to construct and is ideal in applications where headroom is a problem.
... Common applications of the two-way flat-plate system are hotels, motels, apartment buildings, and condominiums.
... The waffle slab may be used in applications requiring a thick slab.
... The waffle slab is reduced in cost and weight by a series of recesses formed into it with removable fillers. These necessary reductions do not substantially decrease its structural capabilities.

REVIEW QUESTIONS

1. List the most common types of poured-in-place concrete slabs.
2. Make a sketch that illustrates suspended floor systems and ground-supported systems.
3. Make a sketch to illustrate the one-way solid-slab system.
4. Make a sketch to illustrate the one-way ribbed or joist system.
5. How can the lateral stability of ribs or joists be increased?
6. Make a sketch to illustrate the two-way solid slab-and-beam system.
7. Make a sketch to illustrate the two flat-plate system(s).
8. List three common applications of the two-way flat-plate system.
9. Make a sketch to illustrate the waffle-slab system.
10. How are the recesses in a waffle slab formed?

REINFORCEMENT ACTIVITIES

The title block as shown in Figure 1-21 and 1/2" borders are required for all activities.

1. Prepare two, C-size sheets of paper with complete title block and borders to be used for engineering drawings. Prepare two, C-size sheets of paper with complete title block and borders to be used for placing drawings.

2. Redraw the engineering drawing in Figure 18-8 with the following changes:
   a. Increase the 12'-0" dimensions to 15'-0".
   b. Increase the 10'-0" dimensions to 12'-0".
   c. Make all columns 12" square and use #5 bars.

3. Prepare a complete placing drawing for the engineering drawing completed in Reinforcement Activity #2.

4. Convert the engineering drawing of the one-way solid slab-and-beam system to a two-way solid slab-and-beam system by adding 12" wide beams between columns B1 and B2, B2 and B3, C1 and C2, and C2 and C3. Provide type B #4 bars at 6" On Center in both directions. Use the same dimensions used in Figure 18-8.

5. Prepare a complete placing drawing for the engineering drawing completed in Reinforcement Activity #4.

Unit 19
Poured-in-Place Stairs and Ramps

OBJECTIVES

Upon completion of this unit, the student will be able to

- sketch examples of the various types of stairs.
- perform stair design computations.
- develop engineering and placing drawings of stairs and ramps.
- explain the purpose of a ramp.

TYPES OF STAIRS

Stairs are a set of steps of various designs that provide easy access between the floors of a building. Four types of stairs are very common in heavy construction: straight stairs, L-shaped stairs, double L stairs, and U-shaped stairs. Straight stairs are the most common and the simplest to construct. As the name implies, they run straight from one elevation to the next with no turns, Figure 19-1.

L-shaped stairs have one 90-degree turn between elevations. The two sides of the L are connected by a landing, Figure 19-2. Double L stairs are the same as L-shaped stairs except they make two turns and have two landings, Figure 19-3. U-shaped stairs traverse part of the distance between floors, usually half, in one direction and traverse the remaining distance in the opposite direction, Figure 19-4.

STAIR DESIGN

The structural drafter may be called upon to perform stair design computations from time to time. In order to do so, one must be familiar with the critical dimensions required in designing stairs.

The first is the total rise. The *total rise* is the distance from one floor to the next or, in other words, the height that the stairs must span. The next important term is the *total run* or horizontal distance the stairs cover. These two terms are often given, or known before the computations are performed. The two critical dimensions that must be computed are the *riser height* and the *tread width,* Figure 19-5.

When performing stair design computations, the structural drafter must work within certain parameters of proper design. The following are rules of thumb that insure proper design of stairs:

- There is always one less tread than risers.
- All risers are the same height and all treads are the same width.
- The sum of one riser height and one tread width should equal between 17" and 18".

Unit 19 Poured-in-Place Stairs and Ramps 271

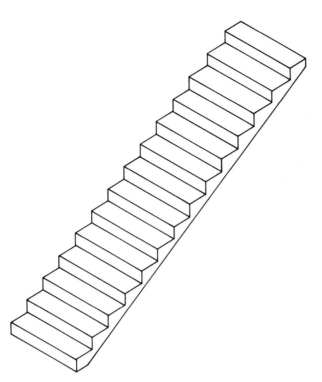

Fig. 19-1 Straight stairs

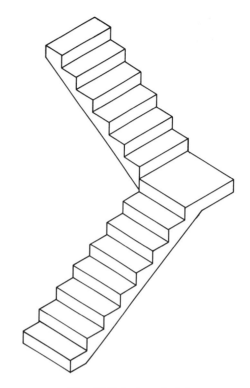

Fig. 19-2 L-shaped stairs

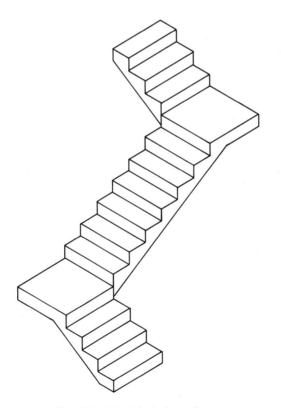

Fig. 19-3 Double L-shaped stairs

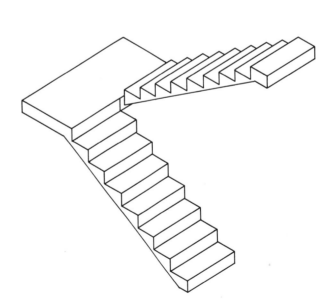

Fig. 19-4 U-shaped stairs

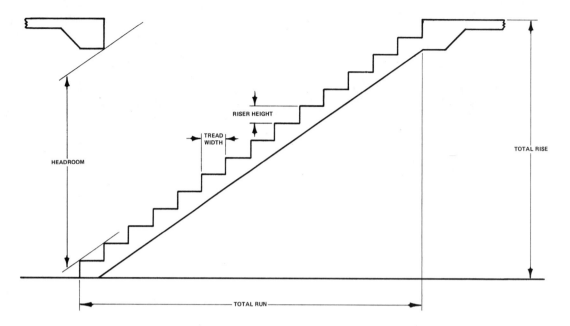

Fig. 19-5 Critical stair dimensions

STAIR DESIGN COMPUTATIONS

Stair design computations are not difficult. However, they should proceed in a certain predetermined order to insure correct results:

1. Determine the total height that the stairs will have to span and convert the distance to inches.
2. Divide the total height by 7" (the ideal riser height). This number should be rounded according to standard mathematical practices. For example, 14.57" would be rounded to 15"; 14.47" would be rounded to 14".
3. The final number from step 2 is the number of risers that this set of stairs will require. This rounded number is then divided into the total height dimension (stated in inches). The answer determined by this division is the exact height of the risers.
4. Since there is always one less tread than risers, the tread width can be easily computed. The number of risers less one is divided into the total run dimension stated in inches. The result is the exact width of each tread. Figure 19-6 shows how the result of stair computations are indicated on a structural drawing.

POURED-IN-PLACE CONCRETE RAMPS

Ramps are slightly sloped passages that may be used by handicapped persons or persons who for any number of reasons are unable to use stairs. They serve the same purpose as stairs in that they provide access from one elevation in a building to another.

Public awareness of the problems incurred by handicapped people in transporting themselves in and out of public buildings has rapidly increased the number of ramps being added to areas that have historically been provided for with stairs alone. Figure 19-7 shows how a ramp is shown in section on a structural drawing.

Unit 19 Poured-in-Place Stairs and Ramps 273

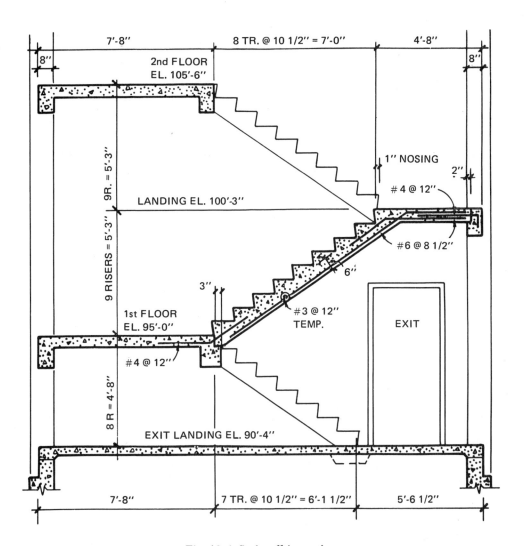

Fig. 19-6 Stairwell in section

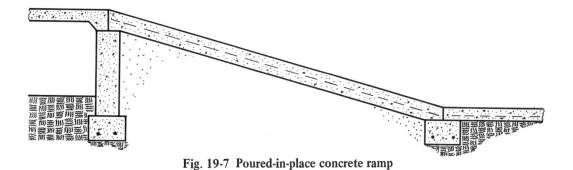

Fig. 19-7 Poured-in-place concrete ramp

SUMMARY

... There are four types of stairs commonly used in heavy construction: straight stairs, L-shaped stairs, double L stairs, and U-shaped stairs.
... The structural drafter must be aware of certain critical dimensions in stair design: total rise, total run, tread width, and riser height.
... In good stair design there is always one less tread than risers.
... All risers must be the same height and treads must be the same width.
... The sum of one riser height and one tread width should equal between 17" and 18".
... Ramps are slightly sloped passages that serve the same purpose for handicapped people that stairs serve for the nonhandicapped.

REVIEW QUESTIONS

1. Sketch and label an example of each of the four common types of stairs.
2. Sketch an example of a set of straight stairs and label the dimensions that are critical.
3. When designing stairs, what is the proper ratio of risers to treads?
4. What is the rule of thumb governing riser height and tread width?
5. The sum of any riser and any tread should equal _____.
6. What is the purpose of a ramp?

REINFORCEMENT ACTIVITIES

The title block as shown in Figure 1-21 and 1/2" borders are required for all activities.

1. Using the following specifications, design a set of stairs for a building:
   Total rise = 9'-0"   Total run = 12'-4"
   Using Figure 19-5 as an example, draw a side elevation of your stairs at a scale of 1" = 1'-0". Completely label the drawing and indicate the number of treads and risers.

2. Repeat the instructions in Reinforcement Activity #1 for the following specifications:
   Total rise = 9'-2"   Total run = 12'-6"

3. Repeat the instructions in Reinforcement Activity #1 for the following specifications:
   Total rise = 9'-11 1/4"   Total run = 13'-3"

4. Repeat the instructions in Reinforcement Activity #1 for the following specifications:
   Total rise = 10'-4 1/2"   Total run = 13'-8 1/2"

5. Redraw Figure 19-6 at a scale of 3/4" = 1'-0" after adding 6" to each rise and run shown and recomputing all numbers.

Section 5
Structural Wood Drafting

Unit 20
Structural Wood Floor Systems

OBJECTIVES

Upon completion of this unit, the student will be able to

- sketch an example of the following types of floor systems: joist and girder systems, plywood systems, and trussed floor systems.
- select floor joists from span data tables according to design requirements.
- prepare framing plans, sections, and details for structural wood floor systems.

JOIST AND GIRDER FLOOR SYSTEMS

Floor systems of commercial buildings are of two types: on grade and off grade. *On-grade floor systems* are made of poured-in-place concrete. *Off-grade floor systems* are usually made of wooden joists, girders, and planks. Off-grade floors are often used in commerical construction. This is because of the maintenance advantages they offer, such as easy access to plumbing and ductwork for heating and cooling.

Framing plans for off-grade floors may be drawn in either one of two ways:

- Joists may be shown on the floor plan with an arrow symbol indicating the direction of span and a note indicating the joist size and spacing, Figure 20-1.
- Or, a floor joist framing plan showing the joists and girders in plan may be used, Figure 20-2.

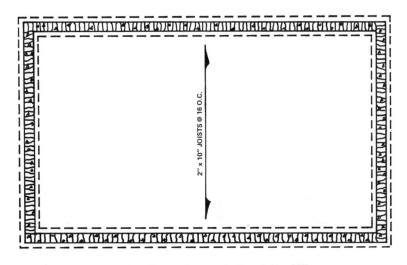

Fig. 20-1 Joists shown on foundation plan

277

278 Section 5 Structural Wood Drafting

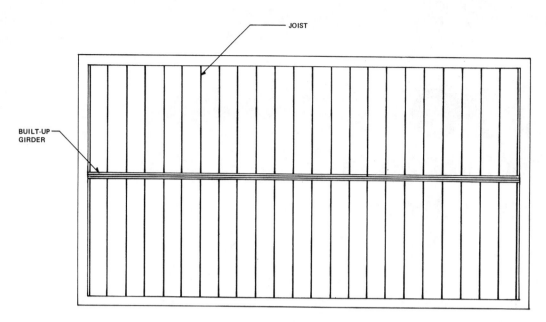

Fig. 20-2 Floor joist framing plan

Girders in wooden floor systems are of two types: built-up wooden girders and steel girders. In either case, joists frame into or over girders and are supported by them. Unless span and loading conditions are too great, built-up wooden girders are used because they are easier to work with and more economical than steel girders.

There are several construction methods used for connecting joists to wooden girders. Four of the most common are illustrated in Figure 20-3.

In certain heavy-loading or long-span situations, it may become necessary to use steel girders. Wooden joists may be framed on top of steel girders by first attaching a wooden nailer to the girder and then the joists to it, Figure 20-4. However, since framing over the top of a girder decreases the amount of crawl space under the floor, a more popular method is to frame into the side of the girder, Figure 20-5.

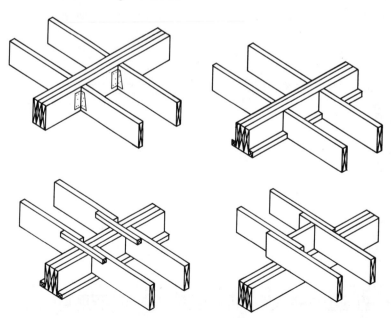

Fig. 20-3 Joist-to-girder connections (Goodheart-Wilcox Company, Inc.)

Unit 20 Structural Wood Floor Systems 279

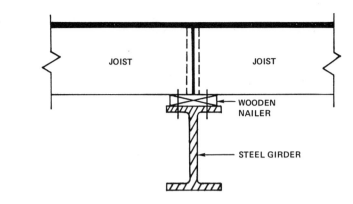

Fig. 20-4 Wood joist bearing on steel girder

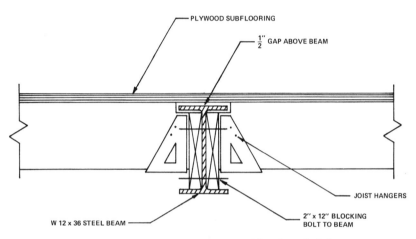

Fig. 20-5 Wood joist bearing into side of steel girders

PLYWOOD FLOOR SYSTEMS

A wooden floor system that is gaining in popularity is the plywood system. Tongue-and-groove plywood panels 1 1/8" thick have been developed for use simultaneously with wooden girders spaced at 4'-0" On Center to form a sound, economical floor, Figure 20-6. Since joists are not required, material costs are decreased. Because the large 4' x 8' plywood panels are quickly and easily installed, labor costs are also decreased, Figure 20-7.

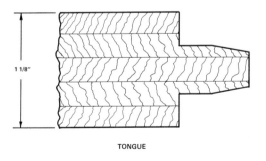
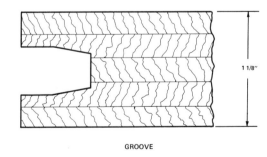

Fig. 20-6 Tongue-and-groove plywood

280 Section 5 Structural Wood Drafting

BRIDGING

Bridging is a term used to describe the construction method whereby joists in a wooden floor system are stiffened to prevent misalignment, twisting, or warping. Bridging also helps distribute the structural load on the floor over a larger floor area.

There are two basic types of bridging: solid bridging and cross bridging. Solid bridging consists of solid cuts of lumber placed between the joists and secured by nailing, Figure 20-8. Cross bridging may be made of wood, but in recent years, it is more commonly accomplished with two diagonal metal braces attached at intervals between the joists, Figure 20-9.

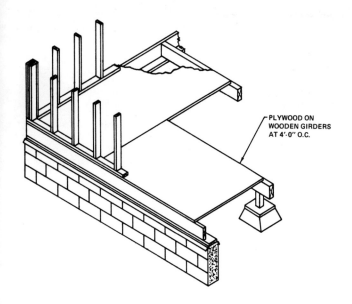

Fig. 20-7 Plywood floor system
(American Technical Society)

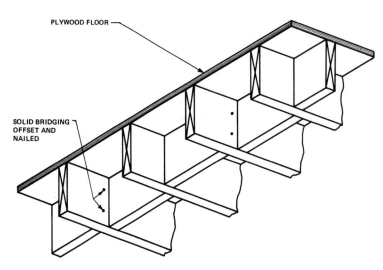

Fig. 20-8 Solid bridging between joists

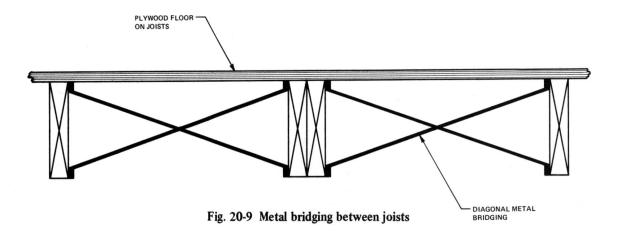

Fig. 20-9 Metal bridging between joists

FLOOR TRUSS SYSTEMS

A floor system that is sometimes used instead of the joist and girder system is the *floor truss system*. Floor trusses are made of wooden *chords* (top and bottom members) with wooden or metal diagonal cross braces in between. The braces that are used vary in length and depth, depending on the span and load requirements of the individual job, Figures 20-10 and 20-11. Another type of floor truss is actually a wooden reproduction of a steel beam. It is made of plywood flanges separated by a deep plywood web, Figure 20-12.

Floor trusses have several advantages over joists and girders. Since they span from one wall to the next wall, they require no intermediate supports. In addition, their ease of installation makes them more economical in some situations. The spaces between braces form convenient avenues for passing plumbing and ductwork through the floor. Trusses are usually spaced at 24" On Center and are shown on a floor truss framing plan.

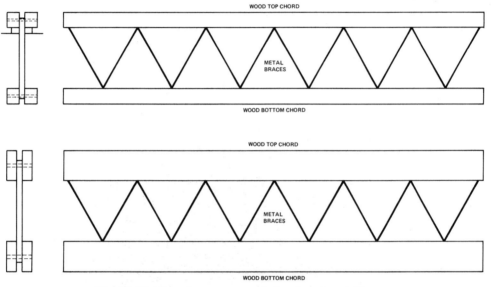

Fig. 20-10 Floor trusses — wood chords and metal braces

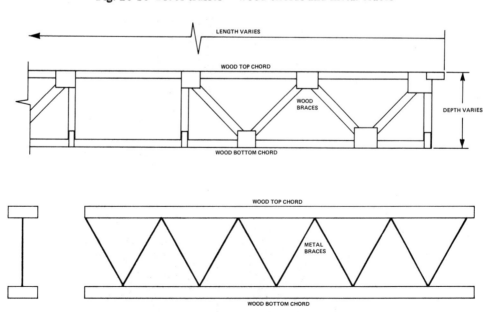

Fig. 20-11 Floor trusses — all wood and wood chords with metal braces

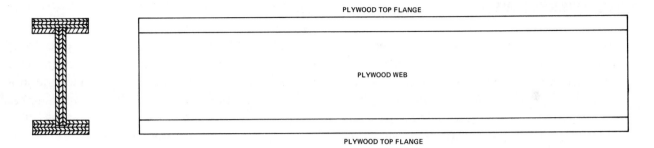

Fig. 20-12 Plywood floor truss

FLOOR JOIST SELECTION

Joists in wooden floor systems are made of lumber sized nominally as: 2 x 6, 2 x 8, 2 x 10, and 2 x 12 (refer to Appendix A). The structural drafter involved in the development of plans for a wooden floor system must be able to select the properly sized joist to meet the design requirements of the job. To assist the drafter in this task, the National Forest Products Association provides span data tables for floor joists. Refer to Appendix B in the back of this text.

These tables provide span data for joists made of all of the commonly used types and grades of wood. Lumber grades, joist sizes, joist spacings, and span capabilities for each size and spacing are also listed.

In order to select a particular joist from the table, the drafter must follow several steps:

1. Determine from the framing plan how far the joist has to span.
2. Determine what type and grade of wood is specified.
3. Go to the appropriate column in the span table and locate a joist and spacing that meets the span requirements.

Appendix B contains all appropriate span tables as well as directions explaining how to use them.

SUMMARY

... The most common structural wood floor systems are: joist and girder systems, plywood systems, and trussed floor systems.
... Joists are shown on the plans in one of two ways: by an arrow symbol on the foundation plan or by drawing a complete joist framing plan.
... Girders in wooden floor systems are of two types: built-up wooden girders and steel girders.
... Plywood floor systems combine tongue-and-groove plywood panels 1 1/8" thick and wooden girders spaced at 4'-0" On Center to make a sound, economical floor system.
... Bridging is used between joists to prevent misalignment, twisting, or warping.
... Bridging may be wood or metal and falls into two basic categories: solid bridging and cross bridging.
... Floor truss systems are sometimes used instead of joists and girders.
... Floor truss systems have several advantages over joist and girder systems: longer spanning capabilities, easier installation, and convenient for installation of plumbing and ductwork.
... Structural drafters should be able to select joist sizes and spacings from span data tables.
... Wooden joists are nominally sized as: 2 x 6, 2 x 8, 2 x 10, and 2 x 12.

REVIEW QUESTIONS

1. List the three most common types of structural wood floor systems.
2. Sketch an example of how joists may be indicated on a foundation plan.
3. List the two types of girders used in wooden floor systems.
4. Sketch an example of each of the four most common construction methods for joining joists and girders.
5. Sketch an example of how joists are attached to the side of a steel girder.
6. What is bridging? What are the two types? Sketch an example of each.
7. Sketch an example of four different types of floor trusses.
8. List the four different sizes of wooden floor joists.

REINFORCEMENT ACTIVITIES

The title block as shown in Figure 1-21 and 1/2" borders are required for all activities. The following activities should be completed on C-size paper.

1. Figure 20-13A contains the outline of a floor system that is to make use of joists to span from wall to wall. Convert this Figure to a foundation plan such as the one shown in Figure 20-1, using an 8" foundation wall centered on a 16" wide footing. Select the proper joist size and spacing from the span tables in Appendix A and place the proper indications on the foundation plan.

2. Figure 20-13B contains the outline of a floor system that is to make use of joists to span from wall to wall. Following the same instructions set forth in Reinforcement Activity #1, draw the foundation plan and place the proper indications on it.

3. Figure 20-13C contains the outline of a floor system that is to make use of floor joists to span from wall to wall. Select the proper joist size and spacing from the span tables in Appendix A. Draw a joist framing plan similar to the one shown in Figure 20-2.

4. Figure 20-13D contains the outline of a floor system that is to be split into two bays by 6 x 10 built-up wooden girders. Appropriate joist size and spacing is to be chosen. The joists are to span into the side of the girders and be connected with metal connectors. Select the joists from the span data tables in Appendix A and draw a floor joist framing plan.

5. Figure 20-13E contains the outline of a floor system that is to be split into three equal bays by 6 x 10 built-up girders. Appropriate joist size and spacing must be chosen. The joists are to span on top of the girders and be braced in between with solid bridging. Select the joists from the span tables in Appendix A and draw a floor joist framing plan.

6. Cut a section through the girder in Reinforcement Activity #4 to show how the joists are connected to the girder. Place this connection detail on the same sheet as the framing plan.

7. Cut a section through one of the girders in Reinforcement Activity #5 to show how the joists are connected to the girder. Place this connection detail on the same sheet as the framing plan.

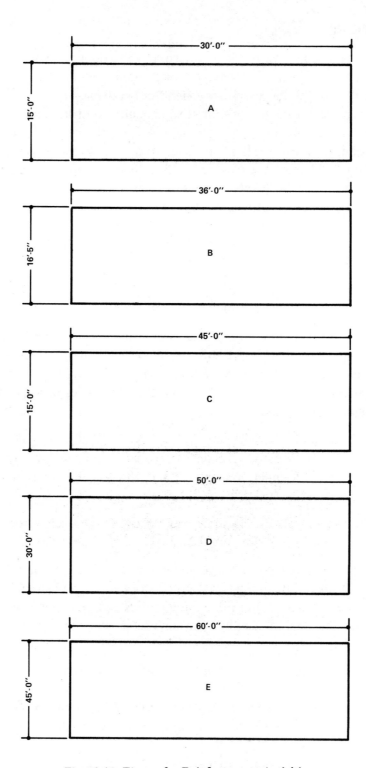

Fig. 20-13 Figures for Reinforcement Activities

Unit 21
Structural Wood Walls

OBJECTIVES

Upon completion of this unit, the student will be able to
- define and illustrate platform framing, balloon framing, and *MOD* 24 framing.
- define and illustrate bracing.
- prepare structural drawings of wall details and sections.

STRUCTURAL WOOD WALLS

One of the major structural components of a wooden structure is the wall. Walls are either load bearing or non-load bearing. *Load-bearing walls* are those that actually carry the load imposed by the floor(s) and roof. Exterior walls of a structure are almost always load bearing.

Non-load-bearing walls carry no load. They are simply built up from the floor to the ceiling and serve the purpose of partitioning space into rooms, closets, etc. Only load-bearing walls are considered structural.

Historically, structural wood walls were of two types: platform framed walls and balloon framed walls. A recent addition to these two framing methods is the MOD 24 wall.

PLATFORM FRAMING

Platform framing is the most common framing method used for building structural wood walls. Also called *western framing*, it begins with a 2 x 6 sill that is bolted to the foundation wall. Floor joists of the proper size and spacing (Unit 20) are nailed to the sill. Headers are attached to the ends of the joists and the subfloor to the tops of the joists. A 2 x 4 soleplate is then attached to the tops of the joists through the subfloor. Wall studs are nailed to the soleplate and capped with a double top plate composed of two 2 x 4's. A typical wall section that illustrates this method of wall framing is shown in Figure 21-1. Figure 21-2 shows an elevation of a platform wall showing the various components.

BALLOON FRAMING

Balloon framing is an older method than platform framing. However, it is seldom used any longer as the primary framing method in structural wood walls. Like platform framing, balloon framing begins with a 2 x 6 sill anchored to the top of the foundation wall. However, from this point on, the two types of framing begin to differ.

In balloon framing, the floor joists rest on the sill with headers or braces placed between the joists. The wall studs also rest on the sill and are placed against the floor joists. Second floors are accomplished by bearing the floor joist on ribbon boards that are nailed into slots cut in the wall studs. Again, braces are placed between the floor joists. These braces serve both as a source of lateral stability and as a fire stop.

286 Section 5 Structural Wood Drafting

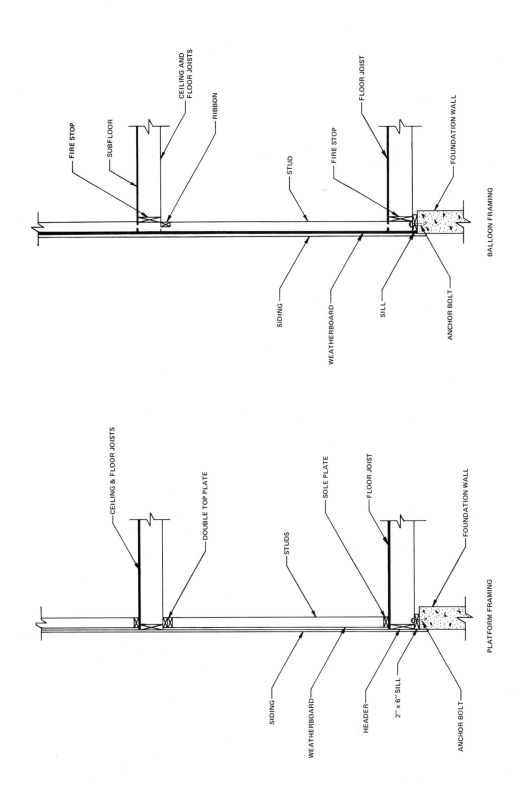

Fig. 21-1 Typical wall sections for platform and balloon framing

There are two important disadvantages to balloon framing that have decreased its popularity. The first is that the 18′ to 20′ long studs required are not readily available. Those that can be found are very expensive due to the additional length. A more practical problem with balloon framing is that the carpenters have no subfloor to stand on when building the stud wall. This method of wall framing is illustrated in Figure 21-1. An elevation of a balloon wall showing the various components is shown in Figure 21-3.

MOD 24 FRAMING

The ever-increasing price of labor and materials prompted the advent of a new framing method, MOD 24. The name comes from the 24″ modular spacing of wall studs. Floor and roof joists are also

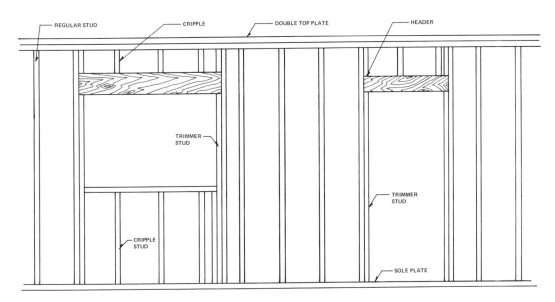

Fig. 21-2 Elevation of platform wall

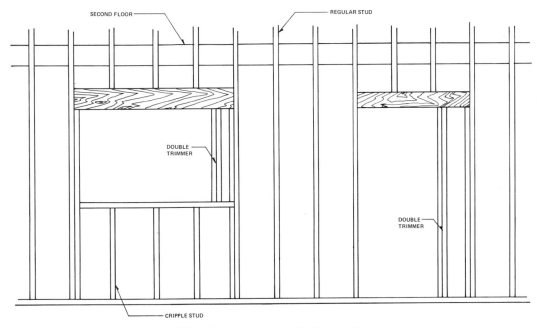

Fig. 21-3 Elevation of balloon wall

spaced on 24" centers. The most structurally sound MOD 24 framing arrangement has floor joists, wall studs, and roof joists aligned.

In addition to using less wood and requiring less labor time, MOD 24 framing produces less waste. This is because most building materials such as lumber, sheathing, plywood, etc., are produced in 8' lengths or widths. Since 8' can be equally divided into 24" modules, less waste results when a piece of lumber must be cut to fit.

Structural designers and architects with an interest in reducing cost design a structure so that it is completely modular in both length and width. For example, rather than designing a building that would be 43'-5" long x 25'-7" wide, the cost-minded designer would make the building 44'-0" x 26'-0". The latter dimensions are modules of 24". Therefore, the client would get a larger building for the same or a lesser price because labor time and material waste is reduced. An example of MOD 24 framing is shown in Figure 21-4.

BRACING

Wooden walls by themselves are not structurally sound enough to withstand the many forces acting on them after a structure has been erected. To provide additional support against bending, buckling, and wracking, wooden walls must be braced. Several methods are used. One of the most common methods has been to form a brace that extends diagonally from the top of the soleplate to the bottom of the double top plate, Figure 21-5. In this method, 2 x 4's are cut to fit between the studs at angles so that the result is one continuous brace from soleplate to top plate.

A similar bracing method is known as the *let-in brace*. In this method, a 1 x 6 brace is placed diagonally across the face of the studs in grooves that have been cut to allow the face of the brace and the face of the studs to be flush, Figure 21-6. There are advantages to this type of bracing. The primary advantages are that the brace, being flush

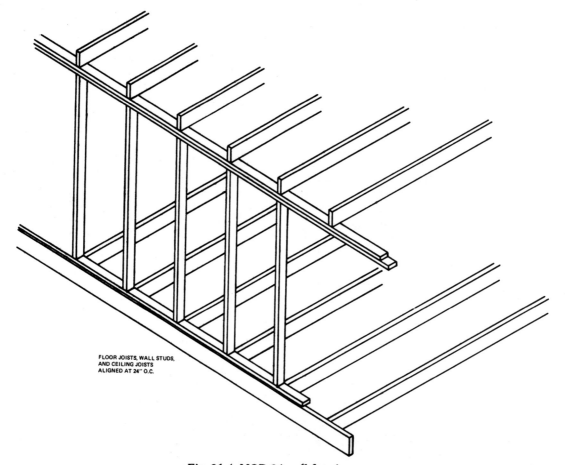

Fig. 21-4 MOD 24 wall framing

with the face of the studs rather than placed between them, does not interfere with the installation of insulation or electrical wiring.

Another effective method of bracing wooden walls is to attach a sheet of plywood to the exterior of the wall at each corner, Figure 21-7. This is a very effective, structurally sound bracing method. It has the same advantages as the let-in bracing method and to an even greater degree. However, since the cost of plywood is prohibitive, only the corners receive a plywood brace. The remainder of the wall is braced with sheathing or weatherboard.

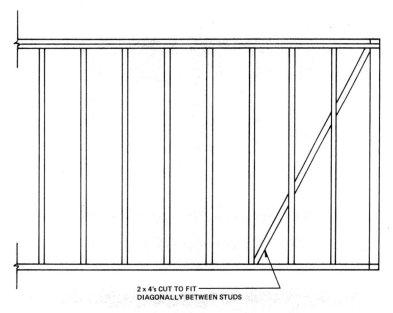

Fig. 21-5 Diagonally cut wall braces

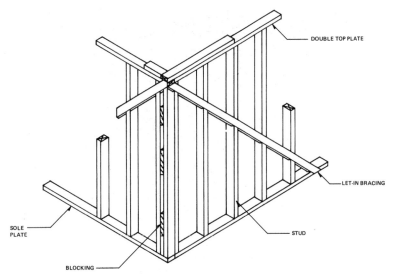

Fig. 21-6 Let-in wall bracing

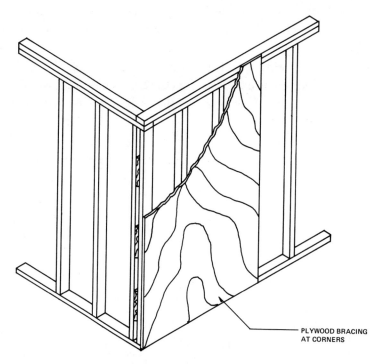

Fig. 21-7 Plywood wall bracing

SUMMARY

... Walls in a structure are either load bearing or non-load bearing.
... Load-bearing walls carry the weight of the floors in multistory buildings and the weight of the roof in multistory and single-story buildings.
... Non-load-bearing walls carry no weight (except their own) and are built up from the floor to the roof.
... Exterior walls are usually load bearing. Interior walls may be load bearing or non-load bearing, depending on the design circumstances of the individual job.
... Platform or western framing is the most common method of framing wooden walls.
... Balloon framing, though it is an older method than platform framing, is seldom used any longer due to the increased cost of the 18′ to 20′ studs required and practical building difficulties inherent in this method.
... MOD 24 framing came about in an attempt to cut material and labor costs.
... The most structurally sound MOD 24 framing arrangement has the joists, studs, and roof joists aligned.
... Bracing is used in wooden walls to prevent bending, buckling, or wracking.
... The three bracing methods commonly used in wooden wall construction are diagonal bracing between studs with specially cut 2 x 4's, let-in bracing with 1 x 6's, and plywood corner bracing.

REVIEW QUESTIONS

1. Which of the following are the two types of structural wood walls?
   a. Partition walls
   b. Load-bearing walls
   c. Interior walls
   d. Non-load-bearing walls
2. Which of the following types of walls carry the weight of the floors and the roof in a building?
   a. Platform walls
   b. Balloon walls
   c. Load-bearing walls
   d. MOD 24 walls
3. Which type of wall is more likely to be load bearing, an interior wall or an exterior wall?
4. Make a sketch that illustrates the difference between platform and balloon framing.
5. Make a sketch that illustrates MOD 24 framing.
6. Make a sketch that illustrates the diagonal-cut 2 x 4 method of wall bracing.
7. Make a sketch that illustrates the let-in method of wall bracing.
8. Make a sketch that illustrates the plywood method of wall bracing.

REINFORCEMENT ACTIVITIES

The title block as shown in Figure 1-21 and 1/2" borders are required for all activities. The Reinforcement Activities for this unit are based on the wall section shown in Figure 21-8.

1. Begin to understand how to draw typical wall sections for structural wood walls by redrawing Figure 21-8 to a scale of 3/4" = 1'-0".

2. Redraw Figure 21-8 to a scale of 3/4" = 1'-0" with the following changes:
   a. First-floor joists 2 x 10 @ 16" On Center
   b. Second-floor joists 2 x 10 @ 16" On Center
   c. Ceiling joists 2 x 10 @ 16" On Center

3. Redraw Figure 21-8 to a scale of 3/4" = 1'-0" with the following changes:
   a. All joists 2 x 8 @ 12" On Center
   b. All studs 2 x 4 @ 16" On Center

4. Using the same size and spacing for all structural components of the wall, convert Figure 21-8 to balloon framing.

5. Convert Figure 21-8 to balloon spacing and redraw it at a scale of 3/4" = 1'-0" with the following changes:
   a. All joists 2 x 10 @ 16" On Center
   b. All studs 2 x 4

292 Section 5 Structural Wood Drafting

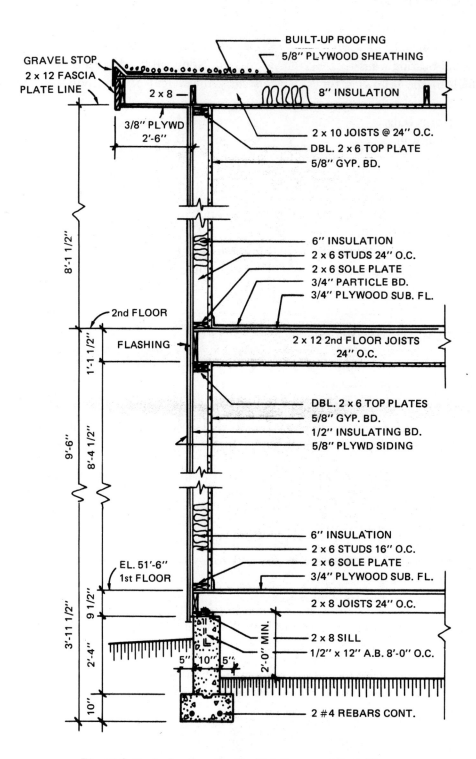

Fig. 21-8 Typical wall section for Reinforcement Activities

Unit 22
Structural Wood Roofs

OBJECTIVES

Upon completion of this unit, the student will be able to

- identify the most common classifications of roof configurations and draw roof configuration diagrams.
- calculate the slope and pitch of a roof.
- draw eave and ridge details.
- select ceiling joist sizes and spaces from span tables.
- identify the various types of roof trusses and select roof trusses from span tables according to design requirements.

COMMON ROOF CLASSIFICATIONS

Structural wood roofs are divided into two broad classifications: flat roofs and sloped roofs. *Flat roofs* encompass those that are actually flat as well as those that have a very gradual slope. *Sloped roofs* range from medium slopes to very steep slopes. Another important distinction between these two classifications of roofs is that flat roofs are usually built up, while sloped roofs are shingled.

Sloped roofs are the most popular in structural wood construction. This is because they provide for water runoff, snow runoff, additional attic or storage space, and an appearance that is pleasing to the eye. Some of the most popular sloped wooden roofs are the gable, hip, gambrel, dutch hip, mansard, and A frame, Figure 22-1.

ROOF SLOPE AND PITCH

The two most important terms when studying wooden roofs are slope and pitch. The *slope* of a roof is the amount of rise measured in inches that occurs for every 12" of run. The term *run* means a distance equal to half of the total span, Figure 22-2. A roof slope that rises at a rate of 3" per foot of run has a slope of 3:12. Slopes ranging from 3:12 to 5:12 are considered medium slopes. Slopes that are 6:12 or greater are considered steep slopes.

The *pitch* of a roof is a figure arrived at by dividing the total span into the total rise, Figure 22-2. A chart showing several of the most commonly used roof pitches and their corresponding slopes is shown in Figure 22-3. Roof slopes are indicated on structural drawings with a slope symbol that lists the units of rise for every 12" of run, Figure 22-2.

294 Section 5 Structural Wood Drafting

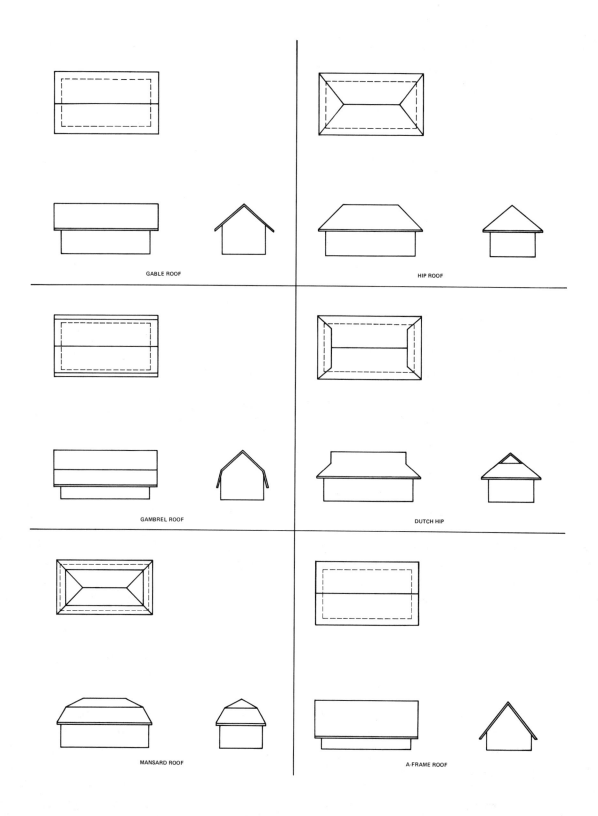

Fig. 22-1 Common roof types

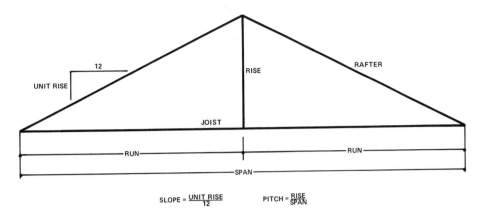

Fig. 22-2 Roof slope and pitch

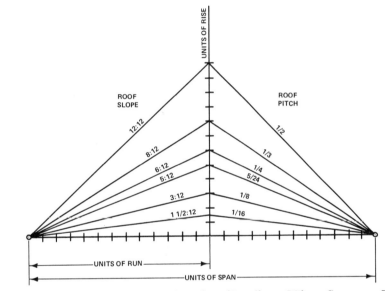

Fig. 22-3 Common roof slopes and pitches (Goodheart-Wilcox Company, Inc.)

EAVE AND RIDGE DETAILS

Most wooden roofs are designed to overhang the exterior walls a certain specified distance. This overhang area is called the *eave* or *cornice*. Structural drafters draw sectional views cut through the eaves of a building to guide the contractor in constructing the roof. The following are the four most common eave or cornice configurations used in wood construction (refer to Figure 22-4):

- Wide-box eave configuration
- Open eave configuration
- Wide-box eave configuration without lookouts
- Narrow-box eave configuration

Eave configuration details are drawn to show how the ceiling joists or roof trusses connect to the bearing wall, how far the rafters or trusses overhang the wall, and the construction of the eave or cornice, Figure 22-4.

A *ridge detail*, which is a section cut through the highest point on the roof, must also be drawn. Ridge details show the ridge board in section and how the rafters connect to it, Figure 22-5.

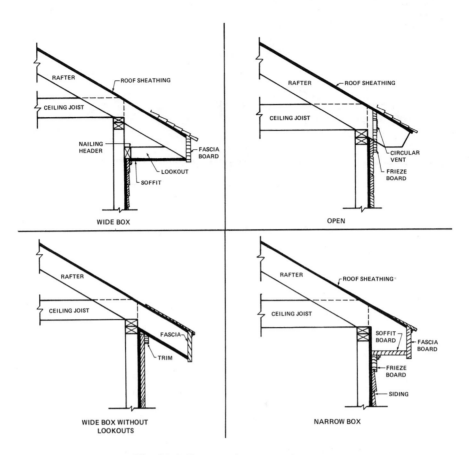

Fig. 22-4 Eave configuration details

WOODEN ROOF FRAMING

Wooden roofs are framed by conventional methods or with roof trusses. *Conventional* or *stick framing* is accomplished by spanning from wall to wall with ceiling joists and from double top plate to ridge board with rafters. When stick framing is to be used in constructing a roof, the structural drafter must select ceiling joists that will meet the design and span requirements. These joists are selected from a ceiling joist span table (see Appendix B) in the same manner as floor joists are selected.

Occasionally, the span in a building is so long that stick building the roof is not feasible. In these cases, roof trusses may be used. A *roof truss* is a prefabricated structural member composed of a joist member, rafter members, and various types of support members between the joist and rafters. Because of the additional support members, a roof truss spans further than a conventional joist-rafter combination.

There are numerous different types of trusses with several different configurations for each type. The four most common types of trusses are the common, scissors, mono pitch, and the flat, Figures 22-6 through 22-9. Trusses will span great distances, but they too have limitations. Trusses must be carefully selected to meet design requirements. Structural drafters select roof trusses from span tables in much the same manner as they select joists. The span tables for common, scissors, mono pitch, and flat roof trusses are shown in Figure 22-10.

Unit 22 Structural Wood Roofs 297

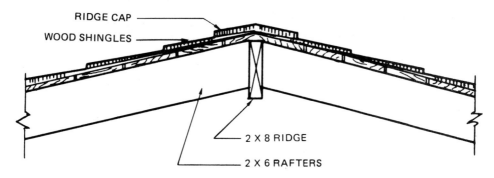

Fig. 22-5 Ridge detail (Charles D. Willis)

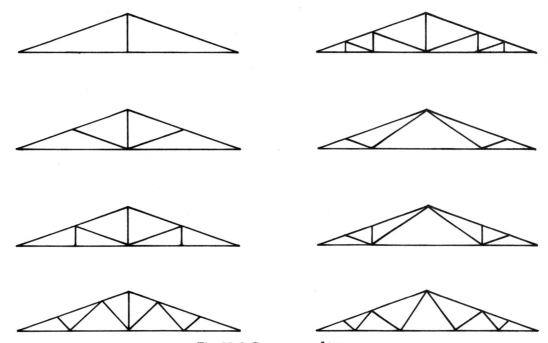

Fig. 22-6 Common roof trusses

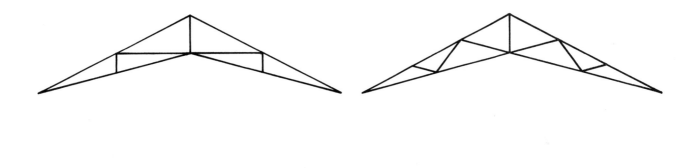

Fig. 22-7 Scissors roof trusses

Section 5 Structural Wood Drafting

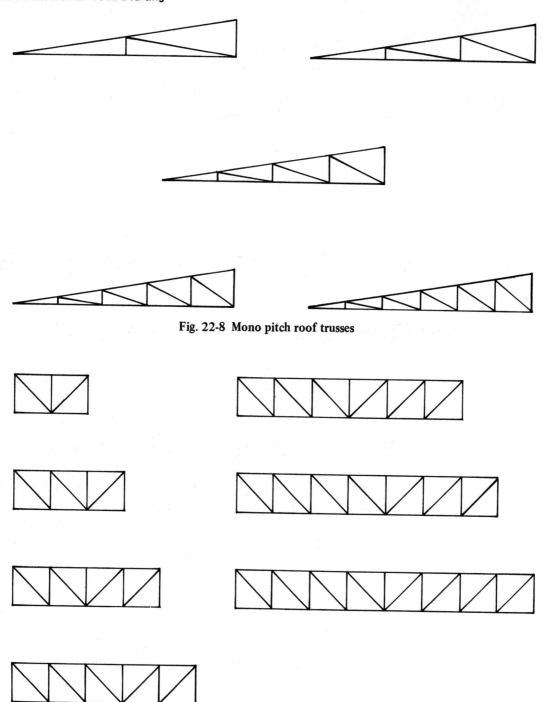

Fig. 22-8 Mono pitch roof trusses

Fig. 22-9 Flat roof trusses

		CHORD SIZE			CHORD SIZE			CHORD SIZE		
	PITCH	2 x 4 Top 2 x 4 Bot.	2 x 6 Top 2 x 4 Bot.	2 x 6 Top 2 x 6 Bot.	2 x 4 Top 2 x 4 Bot.	2 x 6 Top 2 x 4 Bot.	2 x 6 Top 2 x 6 Bot.	2 x 4 Top 2 x 4 Bot.	2 x 6 Top 2 x 4 Bot.	2 x 6 Top 2 x 6 Bot.
COMMON	2/12	22'	23'	34'	25'	25'	39'	28'	28'	44'
	3/12	29'	31'	44'	33'	34'	50'	37'	38'	56'
	4/12	33'	39'	49'	37'	42'	55'	41'	46'	62'
	5/12	35'	45'	53'	39'	48'	59'	44'	52'	66'
	6/12	37'	51'	55'	41'	53'	62'	44'	57'	68'
MONO PITCH	PITCH									
	2/12	22'	23'	34'	25'	25'	38'	28'	28'	44'
	3/12	30'	31'	45'	33'	35'	51'	38'	39'	57'
	4/12	33'	39'	50'	37'	42'	56'	42'	46'	63'
	5/12	35'	45'	53'	40'	48'	60'	44'	53'	67'
	6/12	37'	51'	56'	41'	53'	62'	45'	57'	68'
SCISSORS	6/2*	32'	38'	48'	36'	42'	54'	40'	46'	61'
	6/3	28'	30'	42'	31'	34'	48'	35'	38'	54'
	6/4	21'	22'	32'	24'	24'	36'	27'	27'	42'
	*6/12 = Top chord pitch, 2/12 = bottom chord pitch									
FLAT	16"	23'	—	—	24'	—	—	26'	—	—
	18"	24'	—	—	26'	—	—	28'	—	—
	20"	26'	27'	—	28'	28'	—	30'	30'	—
	24"	29'	30'	34'	31'	31'	37'	33'	33'	39'
	28"	31'	32'	38'	33'	34'	40'	36'	36'	43'
	30"	32'	33'	39'	34'	35'	42'	37'	37'	45'
	32"	33'	34'	41'	35'	36'	43'	38'	38'	47'
	36"	35'	36'	43'	37'	38'	46'	40'	40'	50'
	42"	37'	38'	47'	39'	41'	50'	43'	44'	54'
	48"	39'	41'	49'	41'	43'	53'	45'	47'	57'
	60"	42'	46'	54'	45'	49'	59'	49'	52'	64'
	72"	44'	50'	58'	48'	53'	63'	52'	57'	68'

Fig. 22-10 Roof truss span table (Alpine Engineered Products Inc.)

ROOF CONFIGURATION DIAGRAMS

Structural drafters involved in a job using a wooden roof are required to draw roof configuration diagrams. These are plan views of the completed roof showing the configuration of the roof. Roof diagrams may be drawn to scale or simply drawn out in the proper proportions but not to scale.

Roof configuration diagrams are extremely helpful to the carpenter who must build the roof. The diagrams give a completed picture of what the carpenter is trying to construct. Figure 22-11 is an example of a roof configuration diagram for a hip roof showing the proper proportions for each roof component. Examples of roof configuration diagrams for several other common types of sloped roofs are shown in Figure 22-12.

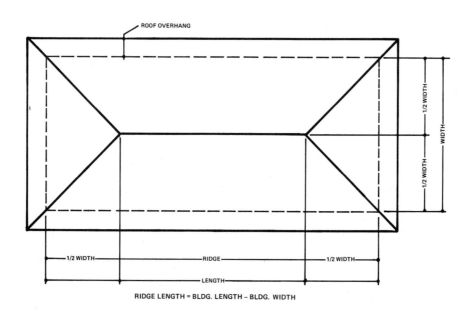

Fig. 22-11 Hip roof configuration diagram

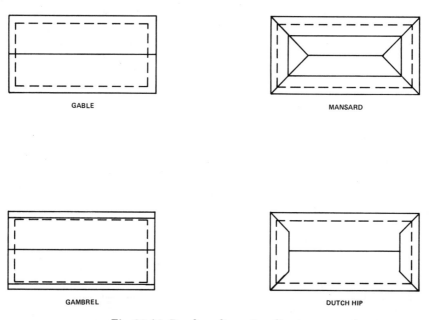

Fig. 22-12 Roof configuration diagrams

SUMMARY

... Structural wood roofs are divided into two broad classifications: flat roofs and sloped roofs.
... Flat roofs are those that range from actually flat to a very gradual slope.
... Sloped roofs range from medium slopes to very steep slopes.
... Flat roofs are built up, while sloped roofs are usually shingled.
... The most popular sloped roofs are the gable, hip, gambrel, dutch hip, mansard, and A frame.
... The slope of a roof is the amount of rise measured in inches that occurs for every 12" of run.
... The pitch of a roof is arrived at by dividing the rise by the span.
... The four most common eave configurations are the wide box, open eave, wide box without lookouts, and narrow box.
... Conventional or stick framing involves joists bearing on the double top plate and rafters spanning from double top plate to ridge board.
... The four most common types of roof trusses are the common, scissors, mono pitch, and flat.
... Roof configuration diagrams provide a completed picture to guide the carpenter in constructing a wooden roof.

REVIEW QUESTIONS

1. What are the two broad classifications of wooden roofs?
2. List the six most popular types of sloped roofs.
3. Sketch a plan view of each of the six roof types listed in question 2.
4. Sketch an example that shows how a 3:12 roof slope would be indicated on a drawing.
5. Calculate the roof pitches, given the following data:
   Roof A — 3' rise with 24' span
   Roof B — 6' rise with 24' span
6. Sketch an example of the four most common eave configurations.
7. List the four most common types of roof trusses.
8. Sketch an example of each of the four types of roof trusses listed in question 7.
9. Sketch an example of a roof configuration diagram for a gable roof.
10. Sketch an example of a roof configuration diagram for a dutch hip roof.

REINFORCEMENT ACTIVITIES

The title block as shown in Figure 1-21 and 1/2" borders are required for all activities.

1. Prepare a C-size sheet of paper. In the upper left-hand corner, draw a single-line representation of a plan view of a building that is 24'-0" long x 12'-0" wide. Use a 1/4" = 1'-0" scale.
   a. Select the proper ceiling joist size and spacing from the span tables in Figure 22-6 and enter them on the plan view.
   b. Assume a gable roof with 2 x 4 rafters, a 2' overhang, and a wide-box eave configuration. At a scale of 1" = 1'-0", draw an eave configuration detail.
   c. Use a 2 x 6 ridge board and at a scale of 1" = 1'-0", draw a ridge detail.
   d. Place a roof configuration diagram drawn not to scale immediately over the title block.

2. Repeat the instructions in Reinforcement Activity #1 for a building that is 36'-0" long x 15'-0" wide, has a 1' overhang, narrow-box eave configuration, and a hip roof.

3. Repeat the instructions in Reinforcement Activity #1 for a building that is 36'-0" long x 18'-0" wide, has a 1' overhang, open eave configuration, and a dutch hip roof.

4. Repeat the instructions in Reinforcement Activity #1 for a building that is 50'-0" long x 25'-0" wide, has a 2' overhang, wide-box eave configuration without lookouts, gable roof, and common roof trusses.

5. Repeat the instructions in Reinforcement Activity #1 for a building that is 48'-0" long x 24'-0" wide, has a 1' overhang, open eave configuration, a gable roof, and scissors roof trusses.

Unit 23
Structural Wood Posts, Beams, Girders, and Arches

OBJECTIVES

Upon completion of this unit, the student will be able to

- sketch examples of post-and-beam construction details, structural wood laminated arches, and various types of laminated beams and girders.

- prepare post, beam, girder, and arch drawings including: framing plans, sections, and connection details.

POST-AND-BEAM CONSTRUCTION

Post-and-beam wood construction is a popular method used for constructing commercial and large residential buildings. It involves three structural members: posts, beams, and planks, Figure 23-1.

There are several advantages to post-and-beam construction. In addition to its aesthetic qualities, post-and-beam construction allows for large glass areas in walls and wide overhangs. This is due to the longer spanning capabilities of the beams as compared to conventional wooden joists.

There are several types of beams used in post-and-beam construction. The most popular have been developed because of the lack of availability of solid beams cut from large trees. These include laminated beams, steel reinforced beams, and box beams, Figure 23-2.

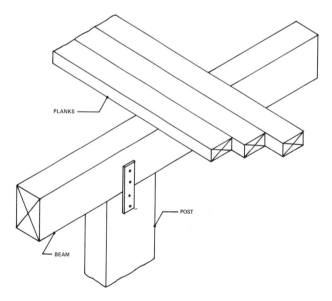

Fig. 23-1 Three components of post-and-beam construction (Goodheart-Wilcox Company, Inc.)

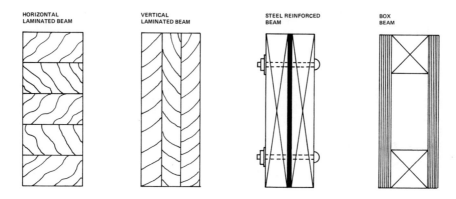

Fig. 23-2 Wooden beams (Goodheart-Wilcox Company, Inc.)

The most problematic area in post-and-beam construction involves connecting the posts to the footings and the beams to the posts. Several special metal connectors have been developed to solve the problem. Some of the most commonly used connectors and connection situations are illustrated in Figure 23-3.

LAMINATED ARCHES

Another popular construction method for building large commercial buildings of wood involves laminated arches. The term *laminated* means that the arch is composed of several layers of wooden planks glued together under intense pressure to form one larger, continuous structural member.

Laminated arches are a popular construction material for buildings requiring large, open areas such as churches, gymnasiums, hangars, etc. They are especially popular where the structural members used to span and support the building are to be exposed and must have a pleasing appearance.

There are several types of laminated arches used. Three of the most popular are the three-hinged arch, tudor arch, and A-frame arch, Figure 23-4. An example of one, half section of an arch with dimensions relating it to the completed building is shown in Figure 23-5. Notice that the roof slope can be varied to suit the needs of the individual job.

Laminated arches are attached to the footings in a manner similar to that used for connecting posts in post-and-beam construction. Figure 23-6 illustrates two commonly used connection methods.

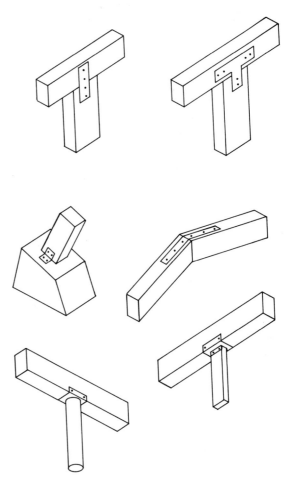

Fig. 23-3 Metal fasteners for post-and-beam connections (Goodheart-Wilcox Company, Inc.)

LAMINATED BEAMS AND GIRDERS

Beams and girders may also be laminated in much the same way as arches. Figure 23-2 illustrates the two types of laminated beams and girders,

Unit 23 Structural Wood Posts, Beams, Girders, and Arches 305

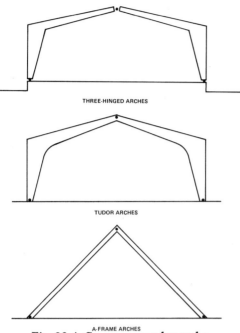

Fig. 23-4 Common wooden arches

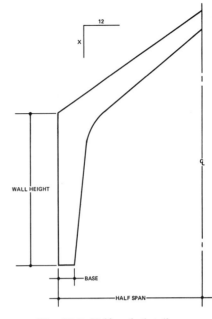

Fig. 23-5 Half-arch detail

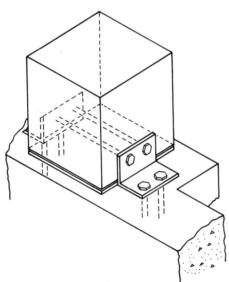

Fig. 23-6 Base connections — posts and arches

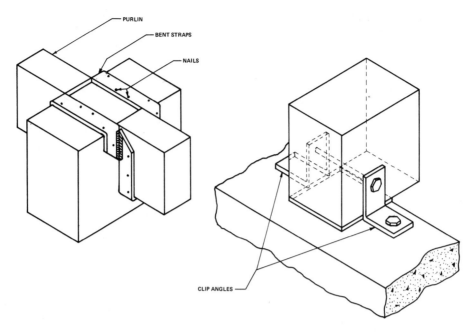

Fig. 23-7 Laminated beam connections

horizontal laminated and vertical laminated. Connecting laminated beams poses the same problems encountered with posts, beams, and arches.

Special metal connectors, similar to those shown in Figure 23-3, have been developed for connecting laminated beams and girders to other structural components and for connecting other structural members to them. Two connection methods commonly used for connecting laminated beams and girders are shown in Figure 23-7.

POST, BEAM, GIRDER, AND ARCH DRAWINGS

Structural drawings for buildings involving posts, beams, girders, and arches made of wood are very similar to steel and precast concrete drawings. The primary components of the drawings are the following:

- A plan view showing overall dimensions and spacings
- Sections showing height information and how the structural members fit together
- Connection details showing how the structural members are to be connected during erection

Figure 23-8 shows the drawing for a post, beam, and plank walkway to be constructed between two existing buildings. Note that as in all other types of structural drawings the connection details are drawn to a larger scale to emphasize detail. A structural drawing for a church using laminated tudor arches is shown in Figure 23-9.

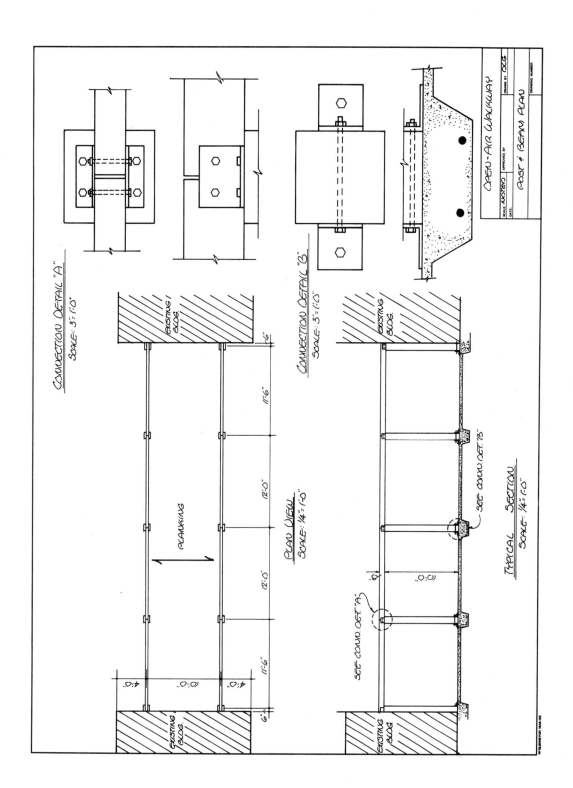

Fig. 23-8 Post-and-beam drawing

Section 5 Structural Wood Drafting

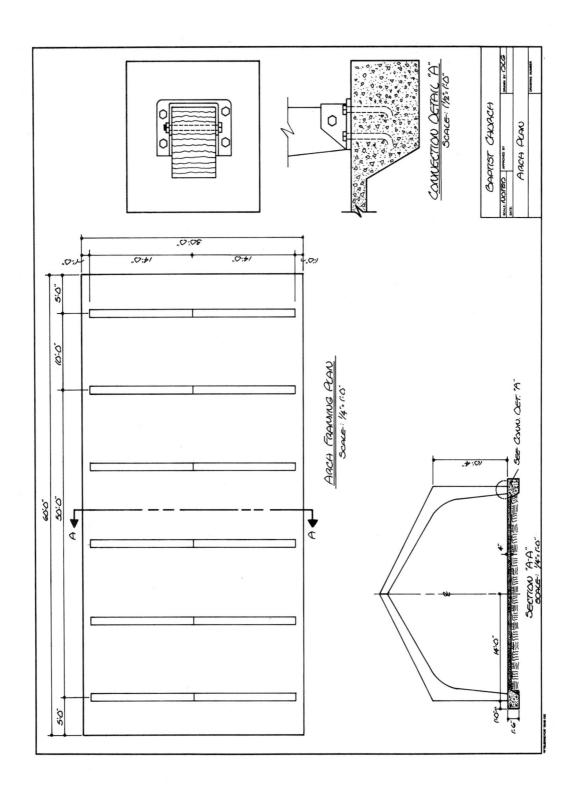

Fig. 23-9 Arch drawings

SUMMARY

... Post-and-beam construction involves three structural components: posts, beams, and planks.
... Post-and-beam construction is advantageous in that it allows for large glass-wall areas and greater overhangs.
... In order to circumvent the lack of solid wood structural beams, several types of built-up wooden beams are used: laminated beams, steel reinforced beams, and box beams.
... The most problematic area in post, beam, girder, and arch construction is connections. These problems have been overcome through the development of special metal connectors for virtually every connection situation that has been identified.
... Laminated arches are a popular structural building component in the construction of churches, gymnasiums, hangars, etc.
... Three of the most popular types of laminated arches are the: three-hinged arch, tudor arch, and the A-frame arch.
... Laminated beams and girders are either horizontal or vertical laminated.
... Structural drawings for post, beam, girder, and arch construction contain plan view(s), section(s), and connection details.

REVIEW QUESTIONS

1. Make a sketch and label it to illustrate the three basic components of post-and-beam construction.
2. List two advantages of post-and-beam construction over conventional wall and joist construction.
3. Make a sketch to illustrate four beam configurations that are used instead of solid wood construction.
4. How have the problems incurred in connecting wooden beams, girders, posts, and arches been solved?
5. List the three most common types of laminated arches.
6. Sketch an example of each of the three types of arches listed in question 5.
7. Sketch an example that illustrates the difference between a vertical and a horizontal laminated beam.
8. List the primary components of structural wood drawings.

REINFORCEMENT ACTIVITIES

The title block as shown in Figure 1-21 and 1/2" borders are required for all activities.

1. Redraw the isometric connection detail shown on the left in Figure 23-6 so that it is an actual connection detail showing a plan view with a section cut through it to provide a complete elevation view. The bottom of the arch resting on the footing should be 12" square, the angles 4" x 4" x 1/4" x 6", and the anchor bolts 3/4" in diameter. Use a scale of 1" = 1'-0".

2. Repeat the instructions in Reinforcement Activity #1 for the isometric connection detail on the right in Figure 23-6. The U-shaped strap is 1/4" thick, 2" wide, and 1'-2" long.

3. Repeat the instructions in Reinforcement Activity #1 for the isometric connection detail on the right in Figure 23-7. The beam is 4" wide x 12" deep and rests on a 12" wide concrete wall. The angle is 4" x 4" x 1/4" x 2" and the anchor bolts are 3/4" in diameter.

4. Redraw Figure 23-9 making the following changes: Use three-hinged arches and connect the arch to the footing in a manner similar to the isometric detail on the right in Figure 23-6.

5. Redraw Figure 23-9 making the following changes: Use A-frame arches and connect the arches to the footing in a manner similar to the isometric detail on the left in Figure 23-6. The arches are 6" wide x 12" deep and are 24'-0" high at the crest.

Section 6
Employment in Structural Drafting

Unit 24
Finding a Job in Structural Drafting

OBJECTIVES

Upon completion of this unit, the student will be able to

- list the primary employers of structural drafters.
- list the most productive job-finding aids for the structural drafter.
- demonstrate how to properly use job-finding aids for identifying potential employers.

EMPLOYMENT IN STRUCTURAL DRAFTING

Structural drafting is one of the most important occupations supporting the heavy construction industry. The structural plans for every apartment building, food store, medical center, educational facility, shopping mall, department store, and numerous other commercial or industrial buildings were prepared by structural drafters.

The heavy construction industry is one of the nation's largest industries, requiring a sizable number of drafters for support. The old adage "it must be drawn before it can be built" is particularly true in heavy construction. This makes structural drafting an occupation with high employment potential.

Students completing this textbook are prepared to enter the world of work and do the job of a structural drafter. However, being able to do the job is only the first step. Students must also learn how to find a job, get a job, and keep a job. This unit deals with finding a job in structural drafting.

PRIMARY EMPLOYERS OF STRUCTURAL DRAFTERS

The first step in finding a job as a structural drafter is learning where to look. Learning where to look for potential employers involves two stages. First, it must be determined who employs structural drafters, and second, how to locate these employers.

In any city, in any state in the United States, the primary employers of structural drafters are the following:

- Structural consulting engineering firms
- Structural steel fabrication companies
- Precast concrete manufacturing companies

There are also a number of secondary employers who occasionally hire drafters with a strong structural background: architects, large construction companies, and the federal civil service. Each of these different types of employers requires drafters with a specific set of skills and should be considered separately.

STRUCTURAL CONSULTING ENGINEERING FIRMS

Structural consulting engineers work with architects on heavy construction projects. They work on a contract basis and are responsible for preparing the structural drawings that must accompany the architectural plans. A complete set of architectural plans for a commercial or industrial building includes the following:

- The architectural drawings numbered A1, A2, A3, etc.
- The structural drawings numbered S1, S2, S3, etc.
- Drawings prepared by electrical and mechanical consulting firms

The structural drawings that go into the complete drawing package are prepared by structural drafters. Some structural consulting firms do specialize in a particular type of design and drawing such as steel, precast concrete, poured-in-place concrete, or wood. However, most firms are involved in all of these fields. Therefore, structural engineering consulting firms usually want to hire those drafters with a broad background encompassing all structural fields.

Consulting engineering firms are engaged only in design and the preparation of plans and do not fabricate products. Therefore, drafters employed in this setting can expect to deal primarily with engineering and layout drawings.

STRUCTURAL STEEL FABRICATION COMPANIES

The primary source of employment for persons wishing to specialize in structural steel drafting is the structural steel fabrication company. Structural steel fabrication companies employ drafters to prepare shop drawings, advance bills, and shop bills so that the steel members for a job may be ordered, designed, fabricated, and erected.

PRECAST CONCRETE COMPANIES

The primary source of employment for persons wishing to specialize in precast concrete drafting are the precast concrete manufacturing companies. Precast concrete companies employ structural drafters to prepare shop drawings and bills of materials so that the precast members for a job may be designed, cast, shipped, and erected.

SECONDARY EMPLOYERS OF STRUCTURAL DRAFTERS

In addition to the primary employers of structural drafters listed in the previous section, there are also secondary employers that should be familiar to the job-seeking structural drafter. Some of these secondary employers are architects, large construction companies, and the federal civil service.

Architects that specialize in the design of large commercial and industrial buildings often maintain a certified structural engineer on their staff. This engineer requires structural drafters for support.

In addition to architects, a number of construction companies around the country are large enough and do a volume of work sufficient enough to warrant maintaining a full-time staff of structural engineers and structural drafters. The federal civil service also employs a small amount of structural drafters. These drafters are involved in projects such as repair and renovations to military and government installations and facilities. The Army Corps of Engineers is a primary employer of federally employed structural drafters.

LOCATING POTENTIAL EMPLOYERS

Now that the potential employers of structural drafters are known, it must be learned how to locate them. There is a systematic approach that can be used to locate potential employers of structural drafters. Every city in every state in the United States has built-in, job-location aids for aspiring structural drafters.

The most productive job-finding aids available in any city are the following:

- The yellow pages in the telephone directory
- The want-ads in the Sunday newspaper
- The state employment office
- The Manufacturer's Directory of the local chamber of commerce

THE YELLOW PAGES

The yellow pages of the local telephone directory provide an alphabetized listing of potential employers of structural drafters. The yellow pages give an address and telephone number for each employer of structural drafters. Used properly, they can be a valuable tool for the job seeker. Architects, construction companies, structural steel fabricators, and precast concrete manufacturers all appear alphabetically in the yellow pages, Figure 24-1.

THE SUNDAY WANT-ADS

Most employers in search of qualified structural drafters solicit in the Sunday edition of the paper if they plan to use the want-ads. This is because Sunday newspapers generally have a wider distribution, thereby reaching more readers. Want-ads generally are printed in one of two formats: random printing and organized printing alphabetically in occupational categories, Figure 24-2.

In the first category, the job seeker must look carefully through all ads until those scattered throughout pertaining to structural drafting are located. In an organized format, the job seeker need only turn to the occupational category marked *Professional-Technical*. Under this category, potential jobs are found under *D* for *drafters* or *S* for *structural drafters*.

STATE EMPLOYMENT OFFICE

Most cities have a branch office of the state employment office. New jobs are posted periodically, usually once each week, on a predetermined day.

The job seeker wishing to investigate the possibility of a job in structural drafting is given a special code number and directed to a microfilm or microfiche viewer. By calling up this code number on the viewer, the job seeker is shown a listing of jobs available in his or her area of interest throughout the state.

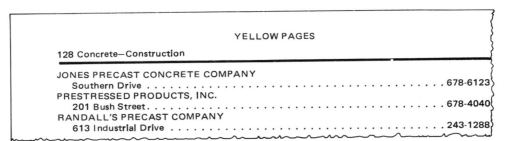

Fig. 24-1 Sample excerpt from the yellow pages

Fig. 24-2 Sample excerpt from the Sunday want-ads

A word of caution is in order for those aspiring structural drafters who have completed their training but have no job experience. Most employers that advertise through the state employment office insist that applicants have a certain amount of experience. Unless you have the experience, the state employment office is unable to send you out on an interview. This is a common problem incurred in using the state employment office. However, on occasion, jobs are available through the state employment office for individuals with no experience but the proper training.

LOCAL CHAMBER OF COMMERCE

Almost every local chamber of commerce publishes a *Manufacturer's Directory* or guide to industry for their area of responsibility. Every type of manufacturer, fabricator, contractor, etc., in the area is listed either alphabetically by company name or alphabetically in product categories and oftentimes both ways, Figure 24-3.

Categories include *Steel Fabricators, Precast Concrete Manufacturers, Contractors,* and many others of interest to the structural drafter seeking a job. Each company listing gives the complete name, address, telephone number, product, and contact person (president, personnel manager) for the company.

SUMMARY

... The primary employers of structural drafters are structural steel fabrication companies, precast concrete manufacturing companies, and structural consulting engineering firms.
... Secondary employers of structural drafters are: architects, large construction companies, and the federal civil service.
... Every city has built-in job-location aids for the aspiring structural drafter. Some of the most productive are the yellow pages, the Sunday want-ads, the state employment office, and the local chamber of commerce.

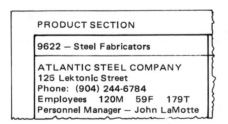

Fig. 24-3 Sample excerpt from Chamber of Commerce Manufacturer's Directory

REVIEW QUESTIONS

1. Name the primary employer of structural steel drafters.
2. Name the primary employer of precast concrete drafters.
3. List a primary employer of structural drafters that employs people who have a broad background covering all areas of structural drafting.
4. List three secondary employers of structural drafters.
5. List the four most productive job-finding aids of the structural drafter.

REINFORCEMENT ACTIVITIES

1. Examine the yellow pages of your local telephone directory and make a list of all potential employers of structural drafters that you are able to locate.

2. Examine the Sunday want-ads of your local newspaper for several weeks and cut out any ads asking for structural drafters. If yours is a small town, attempt to secure the newspapers of the nearest large town to your area.

3. Visit the local office of the state employment office in your town and identify the structural drafting jobs that are available.

4. Visit the local chamber of commerce in your town and examine the *Manufacturer's Directory*. Make a list of the potential employers of structural drafters you are able to locate.

Unit 25
Getting and Keeping a Job in Structural Drafting

OBJECTIVES

Upon completion of this unit, the student will be able to

- write a proper letter of introduction for a job.
- prepare a well-written resume.
- develop a portfolio of sample drawings.
- conduct a positive interview.
- follow up an inconclusive interview.

LETTER OF INTRODUCTION

Many times graduates of vocational training programs find it advantageous to seek employment out of town. The structural drafter seeking a job in a location other than his or her hometown will need to know how to write a proper letter of introduction.

The *letter of introduction* is simply a short statement of your interest in working for a particular company. It is used as a cover letter for a resume and should convey the following information:

- You are a structural drafter with specialized training.
- You intend to relocate to the employer's area as soon as you are able to secure a responsible position in your field.
- You would like to meet the employer and discuss present or future drafting openings with his or her company.
- A resume of your qualifications is enclosed and that you look forward to a reply.

A letter of introduction is a selling tool. It should be brief, assertive, and positively stated, Figure 25-1.

RESUME

A well-written resume is one of the structural drafter's best job-seeking aids. It is simply a categorized account of your qualifications relative to structural drafting. A properly prepared resume sells the job seeker. It amplifies those qualifications possessed by the job seeker that are most important to the job in question.

The resume should begin with personal data that includes complete name, mailing address, and telephone number. Such things as marital status, number of children (if any), height, weight, etc., need not be included.

Next you should give your occupational objective. This is a brief statement of your career goal relative to structural drafting. It should be broad enough to encompass a wide range of structural

> Mr. John Q. Employer
> Jones Structural Steel Company
> P.O. Box 229
> Houston, Texas 21609
>
> Dear Mr. Employer:
>
> I am a structural drafter with specialized training in all phases of structural steel drafting. I intend to relocate to Houston as soon as I am able to secure a responsible drafting position. I have examined the Houston job market and find that I would like to work for your company.
>
> I have enclosed a resume of my qualifications and will furnish references and samples of my work upon request. I look forward to hearing from you. Thank you very much.
>
> Sincerely,
>
> Gary Graduate
> Structural Drafter
>
>
> GG
> Enc: As stated

Fig. 25-1 Sample letter of introduction

drafting job possibilities, but specific enough to show that you have set definite plans for yourself. An example of a well-written occupational objective would be:

"To begin work at a productive level in a structural drafting position that will allow me to apply my training and to advance at a rate commensurate with my performance on the job."

The next component in a resume should be a chronological record of your education with the highest level of your drafting training listed first. The remaining entrees are presented in reverse order. There is no need to go back any further than high school.

The next component lists your experience. If you have any experience that relates to structural drafting, it should be listed first. If not, list the record of your employment in reverse order from the most recent to the earliest. Showing that you have worked, even if it has not been drafting, is a plus in your favor. If you have held a long list of jobs, select the most important only.

The two final categories are awards and hobbies. These are included to appear to the human side of the employer and by doing so, get your foot in the door for an interview. If one of your awards or hobbies interests the potential employer enough, it might just make the difference in your being granted an interview. Figure 25-2 shows an example of a resume for a student who has completed his training, but has no drafting experience. An example of a resume for a student who has actual drafting experience is shown in Figure 25-3.

Resume for

GARY GRADUATE
729 Green Street, Lot No. 1
Fort Walton Beach, Florida 32548
(904) 678-4040

Occupational Objective

To begin work at a productive level in a structural drafting position that will allow me to apply my training and advance at a rate commensurate with my performance on the job.

Education

ASSOCIATE OF SCIENCE DEGREE. Elmwood Technical School, Fort Walton Beach, Florida. My training at Elmwood Technical School covered all phases of drafting with specialized, in-depth work in structural drafting.

HIGH SCHOOL GRADUATE. Fort Walton Beach High School — June 1980.

Work Experience

LABORER. Smith Construction Company.
    Fort Walton Beach, Florida............................2 years
WAITER. Al's Seafood Restaurant. Niceville, Florida............1 year

Awards

Graduated from Elmwood Technical School on Dean's List with 3.75 G.P.A.
First Place — 10th Annual Elmwood Technical School Drafting Contest

Hobbies

Long distance running, tennis, sailing, and fishing.

Fig. 25-2 Sample resume for student with no experience

Resume for

GARY GRADUATE
729 Green Street, Lot No. 1
Fort Walton Beach, Florida 32548
(904) 678-4040

Occupational Objective

To continue working in drafting and to advance to a position of structural drafting checker at a rate commensurate with my performance and development on the job.

Education

ASSOCIATE OF SCIENCE DEGREE. Elmwood Technical School, Fort Walton Beach, Florida. My training at Elmwood Technical School covered all aspects of drafting with specialized, in-depth study in structural drafting.

HIGH SCHOOL GRADUATE. Niceville High School — June 1978.

Work Experience

STRUCTURAL STEEL DETAILER. Sanders Steel Fabrication Company, Seminole, Alabama .... 1 year

CONSTRUCTION WORKER. Sanders Steel Fabrication Company, Seminole, Alabama .... 2 years

Awards

Graduated from Elmwood Technical School on the Dean's List with a G.P.A. of 3.69

COMMUNITY SCHOLARSHIP WINNER. 1977 school year.

Hobbies

Model airplane building and antique automobile restoration.

Fig. 25-3 Sample resume for student with experience

PORTFOLIO OF DRAWINGS

Once an interview has been granted the employer will want to see samples of your work. A well-developed portfolio of structural drawings will impress a potential employer more than anything else in an interview. These drawings show the employer that you can actually do the work that needs to be done. Provided he or she is well spoken and properly dressed, the structural drafter with a good portfolio usually gets the job.

THE INTERVIEW

The interview is the most critical element in the various phases passed through in the transition from job seeker to employed drafter. The wise job seeker will rehearse over and over in his or her mind the answers to questions that might be asked during the interview. A list of sample interview questions that are commonly asked in structural drafting interviews are shown in Figure 25-4.

The job seeker should show up for the interview ten minutes early and appropriately dressed. Dress is important, but do not overdo it. When called in for the interview, be friendly but serious. Your attitude should be positive and assertive. Shake hands with your potential employer firmly, but not too tightly. Look the employer right in the eyes and introduce yourself.

While interviewing, answer all questions openly and honestly. Emphasize your positive characteristics and let the employer know you will quickly learn anything that you do not know when you start. If asked to fill out an application, remember, you are a drafter. Fill it out in your best drafting lettering.

Make sure that you conclude the interview on a positive note. If you are hired, ask the employer when you are to start. If you are turned down, try to determine exactly why. Tell the employer this information will be very helpful to you in future interviews and that you will appreciate an open, honest appraisal.

If the interview is inconclusive and you are told that you will be contacted later, shake the interviewer's hand, look the interviewer in the eyes

1. Why did you decide to become a structural drafter?
2. Why do you think you would like to work for us?
3. What do you think a beginning structural drafter will be required to do?
4. Are you willing and eager to learn?
5. Why do you think you will make a good structural drafter?
6. Do you mind starting at the bottom and working your way up?
7. What are your long-term goals in terms of structural drafting?
8. Do you have any samples of your work?
9. What would you say are your best qualities in terms of structural drafting?
10. What would you say are your weak points in terms of structural drafting?

Fig. 25-4 Sample interview questions

and say, "Thank you for your time and consideration. I would like to work for you and I think I can do the job. I hope you will give me the opportunity to prove myself."

INTERVIEW FOLLOW-UP

Interviews are often inconclusive. This does not mean that you are not going to get the job. It usually means that the employer has several people scheduled for an interview and wants to see them all before making a decision. Before leaving an inconclusive interview, ask the employer when the decision is expected to be made. Also, tell the employer that you will call back in a specified amount of time if you have not heard about the job. Be assertive but not pushy. If the employer says the decision will be made in a week, wait a week and call back. If a decision has still not been reached, you have at least reminded the employer that you want the job. Avoid becoming a pest, but on the other hand, do not be afraid to let the employer know that you want the job. Being timid rarely pays off in job interviews.

THE WISE JOB SEEKER

One of the most frustrating experiences of a person's life can be the first job search. After spending long hours training to become a structural drafter, most people are not prepared to spend additional hours trying to find a job. However, being able to do a job is only half of the process. Being able to get a job is equally important.

Hundreds of well-paying jobs are available on any given day throughout the country for well-trained structural drafters. However, most employers are looking for a person with actual drafting experience. This is the biggest problem facing the new graduate. Fortunately, it is a problem that can be overcome.

Figure 25-5 contains a list of rules for the wise job seeker. These rules can help the well-trained graduate who is lacking in experience to break into the world of work and get that all important first job. Figure 25-6 contains a list of characteristics that will insure success in a job once it has been attained.

SUMMARY

... The letter of introduction is a short, written statement introducing the job seeker to a potential employer.
... The letter of introduction is used as a cover letter for a resume.
... The letter of introduction is used as a selling tool and should be brief, positive, and assertive.
... A well-written resume is one of the job seeker's most valuable tools.
... A resume should contain personal information, an occupational objective, education information, experience, awards, and hobbies.

---

1. DO NOT BE AFRAID TO RELOCATE. For some graduates, leaving their hometown is a frightening thought. However, few things in life are more important than one's career. The bottom line in job seeking is: If you want a job, go to where the jobs are.

2. LEARN TO MARKET YOUR SKILLS. Study your job-seeking skills as you would your technical skills and develop them. It is almost as important to be a skilled job seeker as it is to be a skilled structural drafter.

3. BE POSITIVE AND ASSERTIVE IN YOUR JOB SEARCH. When interviewing, avoid appearing as a timid, unsure person willing to take any job. Go into every interview as a high-trained structural drafter and be assertive enough to say, "I would like to work for your company, I think I can do the job, and I hope you will give me the opportunity to prove myself."

4. DO NOT BE AFRAID TO START AT THE BOTTOM AND WORK YOUR WAY UP. Many companies may start you off running blueprints, doing errands, and performing minor corrections and revisions. There is nothing wrong with this. Many successful structural drafters started their careers this way.

5. DO NOT BECOME FRUSTRATED. For the unexperienced structural drafter it might take several interviews to get the first job. However, once the first job hurdle has been crossed and experience is gained, upward mobility will become the rule.

Fig. 25-5 Rules of the wise job seeker

---

*BE DEPENDABLE. Go to work on time and work while you are there. An employer needs to know you can be counted on to get the job done properly and on time.

*BE A LEARNER. Do not be afraid to tackle new and unfamiliar assignments. The broader your knowledge and skills become, the more valuable you will be to your employer.

*BE A WORKER. Avoid joining the water cooler clique, coffee pot gang, or being a clock watcher. Everyone needs a break from drawing and you will too. However, do not allow yourself to become the person the "boss" bumps into everytime he or she passes the coffee pot.

*BE SELF-SUFFICIENT. Nothing is wrong with asking questions. In fact, it is to be encouraged. However, before asking a question, try to find the answer yourself. This helps you develop into an independent-thinking problem solver.

*STRIVE TO CONSTANTLY IMPROVE. Make note of every new thing that you learn and internalize it. Also work at improving your linework, lettering, accuracy, and speed.

*BE PERSONABLE. Getting along with your fellow employees is important. Drafting tasks are completed by teams of drafters, checkers, and engineers, so developing your abilities to work with people is a must.

Fig. 25-6 Keeping a job in structural drafting

... Education and experience are listed in reverse order in one of two ways: from most recent to earliest or from most important in terms of the job in question to least important.
... Most employers want to see a portfolio of drawings during the interview.
... The interview is the most critical element in the job-seeking process.
... The job seeker should show up for an interview ten minutes early, dressed appropriately.
... When you are called in for the interview, shake hands firmly but not too tightly, look the employer right in the eyes, and introduce yourself.
... During the interview effect a positive, assertive attitude and answer all questions openly and honestly.
... During the interview emphasize your positive characteristics and let the employer know you will learn quickly.
... If asked to fill out an application, fill it out in your best drafting lettering.
... Before leaving an inconclusive interview, ask the employer when a decision is expected to be made.
... Avoid being pushy, but let the employer know you want the job.

REVIEW QUESTIONS

1. List the four items that should be conveyed by the letter of introduction.
2. Write an occupational objective for a resume of your qualifications.
3. List the personal information that would go on your resume.
4. List the various parts of a resume in order.
5. What is the portfolio of drawings?
6. When should the job seeker arrive for an interview?
7. How should a structural drafter dress for a job interview?
8. List six characteristics that will insure success on the job.

REINFORCEMENT ACTIVITIES

1. Write a letter of introduction to the following fictitious company:
   Mr. David H. Lemox
   Chief Drafter
   Arnold's Precast Concrete Company
   Ft. Smith, Arizona 96907

2. Write a complete resume of your qualifications as they will be when you complete your training.

3. Put together a portfolio of drawings from among those prepared for this course.

4. Select a fellow student or friend and conduct several practice interviews using the questions in Figure 25-4.

5. In your best drafting lettering, make two B-size charts:
   a. Five Rules of the Wise Job Seeker
   b. Keeping a Job in Drafting

Glossary

Anchor bolts – Bolts used to fasten structural members to concrete footings.

Architecture – The combined art and science of designing buildings and other structures for human use.

ASTM – American Society for Testing and Materials

Balloon framing – A type of wooden framing in which the studs extend in one piece from the sill to the double top plate. At one time popular for two-story dwellings, it is no longer widely used.

Bars – A round, square, or rectangular steel rod used for reinforcing concrete or fabricating steel members.

Beam – A large structural member of concrete, steel, or wood used to support members over openings or from column to column.

Bill of materials – A comprehensive list of all materials needed to fabricate and erect the structural members for a job.

Blueprints – Common term for copies of original drawings. The term originated with a copying method seldom used any longer that actually produced a copy with white lines on a blue background. Most modern prints have a white background and dark blue or black lines.

Channels – A general term used to describe steel C and MC shapes.

Columns – A vertical structural member usually attached to a footing and extending to the roof of the building. May be steel, concrete, or wood.

Concrete – A mixture of cement, sand, aggregate, and water. It is usually reinforced with wire mesh and steel reinforcing bars when used in heavy construction.

Contractor – The person who supplies the necessary work, materials, and coordinates subcontractors in building a structure.

Cross section – A full section cut at right angles to the longitudinal axis of a building.

Cutting plane – A hypothetical plane that cuts through a structure at designated locations to reveal inside conditions. Indicated on drawings with various types of arrows.

Details – A small part of a structure drawn separated from the structure to accentuate certain information.

Elevations – An orthographic representation of a structure or part of a structure drawn on a flat, vertical plane as if the viewer's line of sight is perpendicular to the plane.

Engineering drawings – Basic layout drawing of a structure used for design and engineering purposes.

Erection plans – Drawings prepared especially for use on the jobsite in erecting a building. Used primarily in steel and precast concrete construction to show how the building fits together and in what order each piece is to be erected.

Fabrication plans – Detailed drawings of individual structural members describing exactly how they are to be fabricated.

Footing – An enlargement at the base of a column or bottom of a wall to distribute the load over a greater portion of ground and thereby prevent settling. Most footings are made of poured concrete.

Foundation – The bottom most portion of a wall or that part of the wall that rests on the footing and upon which the rest of the wall is built.

Framing plan – A plan view drawn to scale providing a bird's eye view of the structural components of a building. Columns, beams and girders, roof members, floor members, and wall members all require separate framing plans.

Girder – A large, horizontal support member similar to a beam. The terms are differentiated in some schools of thought as follows: beams span from column to column, while girders span from beam to beam.

Joists – Horizontal structural members which support the floors and/or roof of a building.

MOD 24 framing – A wooden framing system that places joists, studs, rafters, and trusses on 24-inch modules.

Pipe – Hollow, cylindrical structural steel shape.

Plate – In structural steel, a flat steel piece rectangular in cross section. In wood construction, a term applied to 2 x 4 nailers placed on the sill and on top of the stud wall.

Post tensioning – A method of prestressing concrete by stressing the steel strands after the concrete has been poured and allowed to harden.

Precast concrete — Concrete members that are poured in forms at a plant or factory and allowed to harden. There are two types of precast products: prestressed products and reinforced products.

Prestressed concrete — Concrete products that are stressed before being erected in a job. This is accomplished by passing high-strength steel strands through the form and applying stress to the strands either before or after the concrete is poured.

Pretensioning — Stressing the steel strands in a prestressed member before the concrete is poured into the form.

Rebar — Short term used for steel reinforcing bars used to reinforce concrete.

Retaining wall — Structural wall used to hold back earth or other materials. There are two types: gravity and cantilever.

Section — A type of drawing used to clarify details of construction.

Shear walls — A wall designed to resist lateral loading from winds, underground disturbances, or blasts.

Shop drawings — Drawings prepared to guide shop personnel in the fabrication of structural members for a job. Usually includes fabrication details and a bill of materials.

Slab — A flat concrete area usually reinforced with wire mesh and/or rebar(s).

Slope — The degree of incline of a roof expressed as a ratio of the vertical rise to the horizontal run.

Specifications — Written instructions accompanying the drawings containing information about materials, workmanship, style, and other pertinent information.

Stress — Forces acting on structural members due to various types of loads. These forces are torsion, tension, compression, or shear.

Studs — The primary vertical member of a wooden wall.

Suspended floor — A concrete floor system built above and off the ground.

Tilt-up walls — Poured-in-place concrete walls that are poured in forms on the ground and then tilted up into place by cranes or hoists.

Truss — A framed structure consisting of straight members joined to form a pattern of interconnecting triangles, usually made of wood or metal.

Working drawings — A set of drawings containing all information to complete a job from start to finish.

Appendix A

DIMENSIONS

W Shapes
M Shapes
S Shapes
HP Shapes
American Standard Channels (C)
Miscellaneous Channels (MC)
Angles (L)

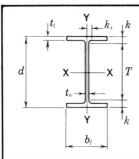

W SHAPES
Dimensions

Designation	Area A	Depth d	Web Thickness t_w		$\dfrac{t_w}{2}$	Flange Width b_f		Flange Thickness t_f		Distance T	k	k_1
	In.²	In.	In.		In.	In.		In.		In.	In.	In.
W 36×300	88.3	36.74	36¾	0.945	15/16	½	16.655	16⅝	1.680	1¹¹/₁₆	31⅛	2¹³/₁₆	1½
×280	82.4	36.52	36½	0.885	⅞	7/16	16.595	16⅝	1.570	1⁹/₁₆	31⅛	2¹¹/₁₆	1½
×260	76.5	36.26	36¼	0.840	13/16	7/16	16.550	16½	1.440	1⁷/₁₆	31⅛	2⁹/₁₆	1½
×245	72.1	36.08	36⅛	0.800	13/16	7/16	16.510	16½	1.350	1⅜	31⅛	2½	1⁷/₁₆
×230	67.6	35.90	35⅞	0.760	¾	⅜	16.470	16½	1.260	1¼	31⅛	2⅜	1⁷/₁₆
W 36×210	61.8	36.69	36¾	0.830	13/16	7/16	12.180	12⅛	1.360	1⅜	32⅛	2⁵/₁₆	1¼
×194	57.0	36.49	36½	0.765	¾	⅜	12.115	12⅛	1.260	1¼	32⅛	2³/₁₆	1³/₁₆
×182	53.6	36.33	36⅜	0.725	¾	⅜	12.075	12⅛	1.180	1³/₁₆	32⅛	2⅛	1³/₁₆
×170	50.0	36.17	36⅛	0.680	11/16	⅜	12.030	12	1.100	1⅛	32⅛	2	1³/₁₆
×160	47.0	36.01	36	0.650	⅝	5/16	12.000	12	1.020	1	32⅛	1¹⁵/₁₆	1⅛
×150	44.2	35.85	35⅞	0.625	⅝	5/16	11.975	12	0.940	15/16	32⅛	1⅞	1⅛
×135	39.7	35.55	35½	0.600	⅝	5/16	11.950	12	0.790	13/16	32⅛	1¹¹/₁₆	1⅛
W 33×241	70.9	34.18	34⅛	0.830	13/16	7/16	15.860	15⅞	1.400	1⅜	29¾	2³/₁₆	1³/₁₆
×221	65.0	33.93	33⅞	0.775	¾	⅜	15.805	15¾	1.275	1¼	29¾	2¹/₁₆	1³/₁₆
×201	59.1	33.68	33⅝	0.715	11/16	⅜	15.745	15¾	1.150	1⅛	29¾	1¹⁵/₁₆	1⅛
W 33×152	44.7	33.49	33½	0.635	⅝	5/16	11.565	11⅝	1.055	1¹/₁₆	29¾	1⅞	1⅛
×141	41.6	33.30	33¼	0.605	⅝	5/16	11.535	11½	0.960	15/16	29¾	1¾	1¹/₁₆
×130	38.3	33.09	33⅛	0.580	9/16	5/16	11.510	11½	0.855	⅞	29¾	1¹¹/₁₆	1¹/₁₆
×118	34.7	32.86	32⅞	0.550	9/16	5/16	11.480	11½	0.740	¾	29¾	1⁹/₁₆	1¹/₁₆
W 30×211	62.0	30.94	31	0.775	¾	⅜	15.105	15⅛	1.315	1⁵/₁₆	26¾	2⅛	1⅛
×191	56.1	30.68	30⅝	0.710	11/16	⅜	15.040	15	1.185	1³/₁₆	26¾	1¹⁵/₁₆	1¹/₁₆
×173	50.8	30.44	30½	0.655	⅝	5/16	14.985	15	1.065	1¹/₁₆	26¾	1⅞	1¹/₁₆
W 30×132	38.9	30.31	30¼	0.615	⅝	5/16	10.545	10½	1.000	1	26¾	1¾	1¹/₁₆
×124	36.5	30.17	30⅛	0.585	9/16	5/16	10.515	10½	0.930	15/16	26¾	1¹¹/₁₆	1
×116	34.2	30.01	30	0.565	9/16	5/16	10.495	10½	0.850	⅞	26¾	1⅝	1
×108	31.7	29.83	29⅞	0.545	9/16	5/16	10.475	10½	0.760	¾	26¾	1⁹/₁₆	1
× 99	29.1	29.65	29⅝	0.520	½	¼	10.450	10½	0.670	11/16	26¾	1⁷/₁₆	1

AMERICAN INSTITUTE OF STEEL CONSTRUCTION

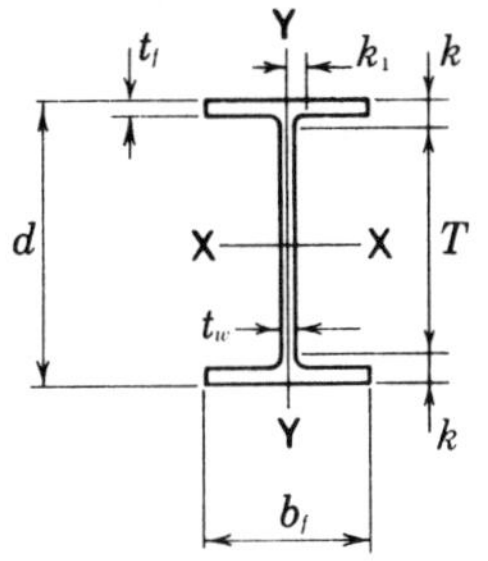

W SHAPES
Dimensions

Designation	Area A	Depth d		Web Thickness t_w		$\dfrac{t_w}{2}$	Flange Width b_f		Thickness t_f		Distance T	k	k_1
	In.²	In.		In.		In.	In.		In.		In.	In.	In.
W 27×178	52.3	27.81	27¾	0.725	¾	⅜	14.085	14⅛	1.190	1³⁄₁₆	24	1⅞	1¹⁄₁₆
×161	47.4	27.59	27⅝	0.660	¹¹⁄₁₆	⅜	14.020	14	1.080	1¹⁄₁₆	24	1¹³⁄₁₆	1
×146	42.9	27.38	27⅜	0.605	⅝	⁵⁄₁₆	13.965	14	0.975	1	24	1¹¹⁄₁₆	1
W 27×114	33.5	27.29	27¼	0.570	⁹⁄₁₆	⁵⁄₁₆	10.070	10⅛	0.930	¹⁵⁄₁₆	24	1⅝	¹⁵⁄₁₆
×102	30.0	27.09	27⅛	0.515	½	¼	10.015	10	0.830	¹³⁄₁₆	24	1⁹⁄₁₆	¹⁵⁄₁₆
× 94	27.7	26.92	26⅞	0.490	½	¼	9.990	10	0.745	¾	24	1⁷⁄₁₆	¹⁵⁄₁₆
× 84	24.8	26.71	26¾	0.460	⁷⁄₁₆	¼	9.960	10	0.640	⅝	24	1⅜	¹⁵⁄₁₆
W 24×162	47.7	25.00	25	0.705	¹¹⁄₁₆	⅜	12.955	13	1.220	1¼	21	2	1¹⁄₁₆
×146	43.0	24.74	24¾	0.650	⅝	⁵⁄₁₆	12.900	12⅞	1.090	1¹⁄₁₆	21	1⅞	1¹⁄₁₆
×131	38.5	24.48	24½	0.605	⅝	⁵⁄₁₆	12.855	12⅞	0.960	¹⁵⁄₁₆	21	1¾	1¹⁄₁₆
×117	34.4	24.26	24¼	0.550	⁹⁄₁₆	⁵⁄₁₆	12.800	12¾	0.850	⅞	21	1⅝	1
×104	30.6	24.06	24	0.500	½	¼	12.750	12¾	0.750	¾	21	1½	1
W 24× 94	27.7	24.31	24¼	0.515	½	¼	9.065	9⅛	0.875	⅞	21	1⅝	1
× 84	24.7	24.10	24⅛	0.470	½	¼	9.020	9	0.770	¾	21	1⁹⁄₁₆	¹⁵⁄₁₆
× 76	22.4	23.92	23⅞	0.440	⁷⁄₁₆	¼	8.990	9	0.680	¹¹⁄₁₆	21	1⁷⁄₁₆	¹⁵⁄₁₆
× 68	20.1	23.73	23¾	0.415	⁷⁄₁₆	¼	8.965	9	0.585	⁹⁄₁₆	21	1⅜	¹⁵⁄₁₆
W 24× 62	18.2	23.74	23¾	0.430	⁷⁄₁₆	¼	7.040	7	0.590	⁹⁄₁₆	21	1⅜	¹⁵⁄₁₆
× 55	16.2	23.57	23⅝	0.395	⅜	³⁄₁₆	7.005	7	0.505	½	21	1⁵⁄₁₆	¹⁵⁄₁₆
W 21×147	43.2	22.06	22	0.720	¾	⅜	12.510	12½	1.150	1⅛	18¼	1⅞	1¹⁄₁₆
×132	38.8	21.83	21⅞	0.650	⅝	⁵⁄₁₆	12.440	12½	1.035	1¹⁄₁₆	18¼	1¹³⁄₁₆	1
×122	35.9	21.68	21⅝	0.600	⅝	⁵⁄₁₆	12.390	12⅜	0.960	¹⁵⁄₁₆	18¼	1¹¹⁄₁₆	1
×111	32.7	21.51	21½	0.550	⁹⁄₁₆	⁵⁄₁₆	12.340	12⅜	0.875	⅞	18¼	1⅝	¹⁵⁄₁₆
×101	29.8	21.36	21⅜	0.500	½	¼	12.290	12¼	0.800	¹³⁄₁₆	18¼	1⁹⁄₁₆	¹⁵⁄₁₆
W 21× 93	27.3	21.62	21⅝	0.580	⁹⁄₁₆	⁵⁄₁₆	8.420	8⅜	0.930	¹⁵⁄₁₆	18¼	1¹¹⁄₁₆	1
× 83	24.3	21.43	21⅜	0.515	½	¼	8.355	8⅜	0.835	¹³⁄₁₆	18¼	1⁹⁄₁₆	¹⁵⁄₁₆
× 73	21.5	21.24	21¼	0.455	⁷⁄₁₆	¼	8.295	8¼	0.740	¾	18¼	1½	¹⁵⁄₁₆
× 68	20.0	21.13	21⅛	0.430	⁷⁄₁₆	¼	8.270	8¼	0.685	¹¹⁄₁₆	18¼	1⁷⁄₁₆	⅞
× 62	18.3	20.99	21	0.400	⅜	³⁄₁₆	8.240	8¼	0.615	⅝	18¼	1⅜	⅞
W 21× 57	16.7	21.06	21	0.405	⅜	³⁄₁₆	6.555	6½	0.650	⅝	18¼	1⅜	⅞
× 50	14.7	20.83	20⅞	0.380	⅜	³⁄₁₆	6.530	6½	0.535	⁹⁄₁₆	18¼	1⁵⁄₁₆	⅞
× 44	13.0	20.66	20⅝	0.350	⅜	³⁄₁₆	6.500	6½	0.450	⁷⁄₁₆	18¼	1³⁄₁₆	⅞

AMERICAN INSTITUTE OF STEEL CONSTRUCTION

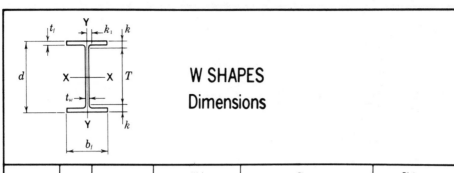

W SHAPES
Dimensions

Designation	Area A	Depth d	Web Thickness t_w		$\dfrac{t_w}{2}$	Flange Width b_f		Flange Thickness t_f		Distance T	Distance k	Distance k_1
	In.²	In.	In.		In.	In.		In.		In.	In.	In.
W 18×119	35.1	18.97	19	0.655	5/8	5/16	11.265	11¼	1.060	1 1/16	15½	1¾	15/16
×106	31.1	18.73	18¾	0.590	9/16	5/16	11.200	11¼	0.940	15/16	15½	1 5/8	15/16
× 97	28.5	18.59	18⅝	0.535	9/16	5/16	11.145	11⅛	0.870	7/8	15½	1 9/16| 7/8
× 86	25.3	18.39	18⅜	0.480	½	¼	11.090	11⅛	0.770	¾	15½	1 7/16| 7/8
× 76	22.3	18.21	18¼	0.425	7/16	¼	11.035	11	0.680	11/16	15½	1⅜	13/16
W 18× 71	20.8	18.47	18½	0.495	½	¼	7.635	7⅝	0.810	13/16	15½	1½	7/8
× 65	19.1	18.35	18⅜	0.450	7/16	¼	7.590	7⅝	0.750	¾	15½	1 7/16| 7/8
× 60	17.6	18.24	18¼	0.415	7/16	¼	7.555	7½	0.695	11/16	15½	1⅜	13/16
× 55	16.2	18.11	18⅛	0.390	3/8	3/16	7.530	7½	0.630	5/8	15½	1 5/16| 13/16
× 50	14.7	17.99	18	0.355	3/8	3/16	7.495	7½	0.570	9/16	15½	1¼	13/16
W 18× 46	13.5	18.06	18	0.360	3/8	3/16	6.060	6	0.605	5/8	15½	1¼	13/16
× 40	11.8	17.90	17⅞	0.315	5/16	3/16	6.015	6	0.525	½	15½	1 3/16| 13/16
× 35	10.3	17.70	17¾	0.300	5/16	3/16	6.000	6	0.425	7/16	15½	1⅛	¾
W 16×100	29.4	16.97	17	0.585	9/16	5/16	10.425	10⅜	0.985	1	13⅝	1 11/16| 15/16
× 89	26.2	16.75	16¾	0.525	½	¼	10.365	10⅜	0.875	7/8	13⅝	1 9/16| 7/8
× 77	22.6	16.52	16½	0.455	7/16	¼	10.295	10¼	0.760	¾	13⅝	1 7/16| 7/8
× 67	19.7	16.33	16⅜	0.395	3/8	3/16	10.235	10¼	0.665	11/16	13⅝	1⅜	13/16
W 16× 57	16.8	16.43	16⅜	0.430	7/16	¼	7.120	7⅛	0.715	11/16	13⅝	1⅜	7/8
× 50	14.7	16.26	16¼	0.380	3/8	3/16	7.070	7⅛	0.630	5/8	13⅝	1 5/16| 13/16
× 45	13.3	16.13	16⅛	0.345	3/8	3/16	7.035	7	0.565	9/16	13⅝	1¼	13/16
× 40	11.8	16.01	16	0.305	5/16	3/16	6.995	7	0.505	½	13⅝	1 3/16| 13/16
× 36	10.6	15.86	15⅞	0.295	5/16	3/16	6.985	7	0.430	7/16	13⅝	1⅛	¾
W 16× 31	9.12	15.88	15⅞	0.275	¼	⅛	5.525	5½	0.440	7/16	13⅝	1⅛	¾
× 26	7.68	15.69	15¾	0.250	¼	⅛	5.500	5½	0.345	3/8	13⅝	1 1/16| ¾

AMERICAN INSTITUTE OF STEEL CONSTRUCTION

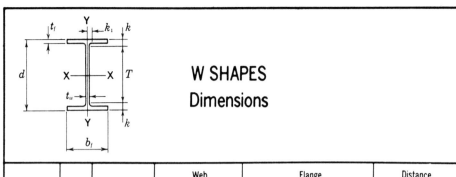

W SHAPES
Dimensions

Designation	Area A	Depth d		Web		Flange				Distance		
				Thickness t_w	$\frac{t_w}{2}$	Width b_f		Thickness t_f		T	k	k_1
	In.²	In.		In.	In.	In.		In.		In.	In.	In.
W 14×730	215.0	22.42	22³⁄₈	3.070	3¹⁄₁₆	1⁹⁄₁₆	17.890	17⁷⁄₈	4.910	4¹⁵⁄₁₆	11¹⁄₄	5⁹⁄₁₆	2³⁄₁₆
×665	196.0	21.64	21⁵⁄₈	2.830	2¹³⁄₁₆	1⁷⁄₁₆	17.650	17⁵⁄₈	4.520	4¹⁄₂	11¹⁄₄	5³⁄₁₆	2¹⁄₁₆
×605	178.0	20.92	20⁷⁄₈	2.595	2⁵⁄₈	1⁵⁄₁₆	17.415	17³⁄₈	4.160	4³⁄₁₆	11¹⁄₄	4¹³⁄₁₆	1¹⁵⁄₁₆
×550	162.0	20.24	20¹⁄₄	2.380	2³⁄₈	1³⁄₁₆	17.200	17¹⁄₄	3.820	3¹³⁄₁₆	11¹⁄₄	4¹⁄₂	1¹³⁄₁₆
×500	147.0	19.60	19⁵⁄₈	2.190	2³⁄₁₆	1¹⁄₈	17.010	17	3.500	3¹⁄₂	11¹⁄₄	4³⁄₁₆	1³⁄₄
×455	134.0	19.02	19	2.015	2	1	16.835	16⁷⁄₈	3.210	3³⁄₁₆	11¹⁄₄	3⁷⁄₈	1⁵⁄₈
W 14×426	125.0	18.67	18⁵⁄₈	1.875	1⁷⁄₈	¹⁵⁄₁₆	16.695	16³⁄₄	3.035	3¹⁄₁₆	11¹⁄₄	3¹¹⁄₁₆	1⁹⁄₁₆
×398	117.0	18.29	18¹⁄₄	1.770	1³⁄₄	⁷⁄₈	16.590	16⁵⁄₈	2.845	2⁷⁄₈	11¹⁄₄	3¹⁄₂	1¹⁄₂
×370	109.0	17.92	17⁷⁄₈	1.655	1⁵⁄₈	¹³⁄₁₆	16.475	16¹⁄₂	2.660	2¹¹⁄₁₆	11¹⁄₄	3⁵⁄₁₆	1⁷⁄₁₆
×342	101.0	17.54	17¹⁄₂	1.540	1⁹⁄₁₆	¹³⁄₁₆	16.360	16³⁄₈	2.470	2¹⁄₂	11¹⁄₄	3¹⁄₈	1³⁄₈
×311	91.4	17.12	17¹⁄₈	1.410	1⁷⁄₁₆	³⁄₄	16.230	16¹⁄₄	2.260	2¹⁄₄	11¹⁄₄	2¹⁵⁄₁₆	1⁵⁄₁₆
×283	83.3	16.74	16³⁄₄	1.290	1⁵⁄₁₆	¹¹⁄₁₆	16.110	16¹⁄₈	2.070	2¹⁄₁₆	11¹⁄₄	2³⁄₄	1¹⁄₄
×257	75.6	16.38	16³⁄₈	1.175	1³⁄₁₆	⁵⁄₈	15.995	16	1.890	1⁷⁄₈	11¹⁄₄	2⁹⁄₁₆	1³⁄₁₆
×233	68.5	16.04	16	1.070	1¹⁄₁₆	⁹⁄₁₆	15.890	15⁷⁄₈	1.720	1³⁄₄	11¹⁄₄	2³⁄₈	1³⁄₁₆
×211	62.0	15.72	15³⁄₄	0.980	1	¹⁄₂	15.800	15³⁄₄	1.560	1⁹⁄₁₆	11¹⁄₄	2¹⁄₄	1¹⁄₈
×193	56.8	15.48	15¹⁄₂	0.890	⁷⁄₈	⁷⁄₁₆	15.710	15³⁄₄	1.440	1⁷⁄₁₆	11¹⁄₄	2¹⁄₈	1¹⁄₁₆
×176	51.8	15.22	15¹⁄₄	0.830	¹³⁄₁₆	⁷⁄₁₆	15.650	15⁵⁄₈	1.310	1⁵⁄₁₆	11¹⁄₄	2	1¹⁄₁₆
×159	46.7	14.98	15	0.745	³⁄₄	³⁄₈	15.565	15⁵⁄₈	1.190	1³⁄₁₆	11¹⁄₄	1⁷⁄₈	1
×145	42.7	14.78	14³⁄₄	0.680	¹¹⁄₁₆	³⁄₈	15.500	15¹⁄₂	1.090	1¹⁄₁₆	11¹⁄₄	1³⁄₄	1

AMERICAN INSTITUTE OF STEEL CONSTRUCTION

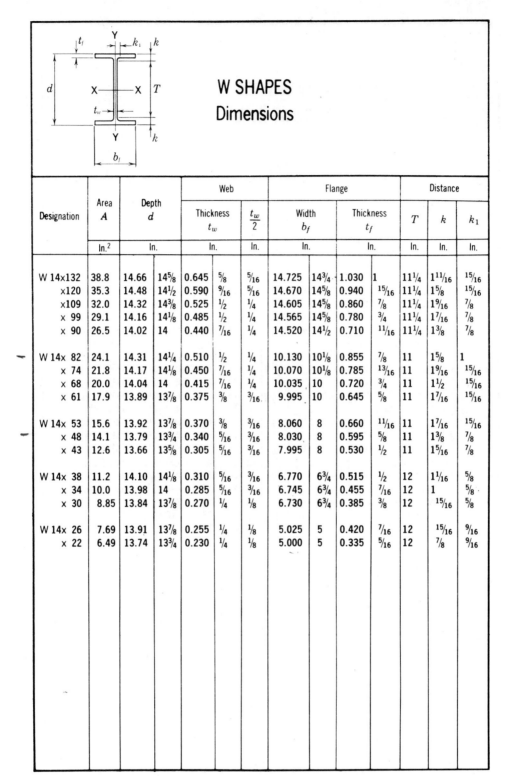

W SHAPES
Dimensions

Designation	Area A	Depth d	Web Thickness t_w	$\dfrac{t_w}{2}$	Flange Width b_f	Flange Thickness t_f	Distance T	k	k_1
	In.²	In.	In.	In.	In.	In.	In.	In.	In.
W 14×132	38.8	14.66	14⅝	0.645	⅝	5/16	14.725	14¾	1.030	1	11¼	1 11/16	15/16
×120	35.3	14.48	14½	0.590	9/16	5/16	14.670	14⅝	0.940	15/16	11¼	1⅝	15/16
×109	32.0	14.32	14⅜	0.525	½	¼	14.605	14⅝	0.860	⅞	11¼	1 9/16	⅞
× 99	29.1	14.16	14⅛	0.485	½	¼	14.565	14⅝	0.780	¾	11¼	1 7/16	⅞
× 90	26.5	14.02	14	0.440	7/16	¼	14.520	14½	0.710	11/16	11¼	1⅜	⅞
W 14× 82	24.1	14.31	14¼	0.510	½	¼	10.130	10⅛	0.855	⅞	11	1⅝	1
× 74	21.8	14.17	14⅛	0.450	7/16	¼	10.070	10⅛	0.785	13/16	11	1 9/16	15/16
× 68	20.0	14.04	14	0.415	7/16	¼	10.035	10	0.720	¾	11	1½	15/16
× 61	17.9	13.89	13⅞	0.375	⅜	3/16	9.995	10	0.645	⅝	11	1 7/16	15/16
W 14× 53	15.6	13.92	13⅞	0.370	⅜	3/16	8.060	8	0.660	11/16	11	1 7/16	15/16
× 48	14.1	13.79	13¾	0.340	5/16	3/16	8.030	8	0.595	⅝	11	1⅜	⅞
× 43	12.6	13.66	13⅝	0.305	5/16	3/16	7.995	8	0.530	½	11	1 5/16	⅞
W 14× 38	11.2	14.10	14⅛	0.310	5/16	3/16	6.770	6¾	0.515	½	12	1 1/16	⅝
× 34	10.0	13.98	14	0.285	5/16	3/16	6.745	6¾	0.455	7/16	12	1	⅝
× 30	8.85	13.84	13⅞	0.270	¼	⅛	6.730	6¾	0.385	⅜	12	15/16	⅝
W 14× 26	7.69	13.91	13⅞	0.255	¼	⅛	5.025	5	0.420	7/16	12	15/16	9/16
× 22	6.49	13.74	13¾	0.230	¼	⅛	5.000	5	0.335	5/16	12	⅞	9/16

AMERICAN INSTITUTE OF STEEL CONSTRUCTION

Appendix A

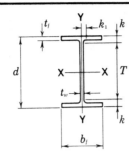

W SHAPES
Dimensions

Designation	Area A	Depth d		Web Thickness t_w		$\frac{t_w}{2}$	Flange Width b_f		Flange Thickness t_f		Distance T	k	k_1
	In.²	In.		In.		In.	In.		In.		In.	In.	In.
W 12×336	98.8	16.82	16⅞	1.775	1¾	⅞	13.385	13⅜	2.955	2¹⁵⁄₁₆	9½	3¹¹⁄₁₆	1½
×305	89.6	16.32	16⅜	1.625	1⅝	¹³⁄₁₆	13.235	13¼	2.705	2¹¹⁄₁₆	9½	3⁷⁄₁₆	1⁷⁄₁₆
×279	81.9	15.85	15⅞	1.530	1½	¾	13.140	13⅛	2.470	2½	9½	3³⁄₁₆	1⅜
×252	74.1	15.41	15⅜	1.395	1⅜	¹¹⁄₁₆	13.005	13	2.250	2¼	9½	2¹⁵⁄₁₆	1⁵⁄₁₆
×230	67.7	15.05	15	1.285	1⁵⁄₁₆	¹¹⁄₁₆	12.895	12⅞	2.070	2¹⁄₁₆	9½	2¾	1¼
×210	61.8	14.71	14¾	1.180	1³⁄₁₆	⅝	12.790	12¾	1.900	1⅞	9½	2⅝	1¼
×190	55.8	14.38	14⅜	1.060	1¹⁄₁₆	⁹⁄₁₆	12.670	12⅝	1.735	1¾	9½	2⁷⁄₁₆	1³⁄₁₆
×170	50.0	14.03	14	0.960	¹⁵⁄₁₆	½	12.570	12⅝	1.560	1⁹⁄₁₆	9½	2¼	1⅛
×152	44.7	13.71	13¾	0.870	⅞	⁷⁄₁₆	12.480	12½	1.400	1⅜	9½	2⅛	1¹⁄₁₆
×136	39.9	13.41	13⅜	0.790	¹³⁄₁₆	⁷⁄₁₆	12.400	12⅜	1.250	1¼	9½	1¹⁵⁄₁₆	1
×120	35.3	13.12	13⅛	0.710	¹¹⁄₁₆	⅜	12.320	12⅜	1.105	1⅛	9½	1¹³⁄₁₆	1
×106	31.2	12.89	12⅞	0.610	⅝	⁵⁄₁₆	12.220	12¼	0.990	1	9½	1¹¹⁄₁₆	¹⁵⁄₁₆
× 96	28.2	12.71	12¾	0.550	⁹⁄₁₆	⁵⁄₁₆	12.160	12⅛	0.900	⅞	9½	1⅝	⅞
× 87	25.6	12.53	12½	0.515	½	¼	12.125	12⅛	0.810	¹³⁄₁₆	9½	1½	⅞
× 79	23.2	12.38	12⅜	0.470	½	¼	12.080	12⅛	0.735	¾	9½	1⁷⁄₁₆	⅞
× 72	21.1	12.25	12¼	0.430	⁷⁄₁₆	¼	12.040	12	0.670	¹¹⁄₁₆	9½	1⅜	⅞
× 65	19.1	12.12	12⅛	0.390	⅜	³⁄₁₆	12.000	12	0.605	⅝	9½	1⁵⁄₁₆	¹³⁄₁₆
W 12× 58	17.0	12.19	12¼	0.360	⅜	³⁄₁₆	10.010	10	0.640	⅝	9½	1⅜	¹³⁄₁₆
× 53	15.6	12.06	12	0.345	⅜	³⁄₁₆	9.995	10	0.575	⁹⁄₁₆	9½	1¼	¹³⁄₁₆
W 12× 50	14.7	12.19	12¼	0.370	⅜	³⁄₁₆	8.080	8⅛	0.640	⅝	9½	1⅜	¹³⁄₁₆
× 45	13.2	12.06	12	0.335	⁵⁄₁₆	³⁄₁₆	8.045	8	0.575	⁹⁄₁₆	9½	1¼	¹³⁄₁₆
× 40	11.8	11.94	12	0.295	⁵⁄₁₆	³⁄₁₆	8.005	8	0.515	½	9½	1¼	¾
W 12× 35	10.3	12.50	12½	0.300	⁵⁄₁₆	³⁄₁₆	6.560	6½	0.520	½	10½	1	⁹⁄₁₆
× 30	8.79	12.34	12⅜	0.260	¼	⅛	6.520	6½	0.440	⁷⁄₁₆	10½	¹⁵⁄₁₆	½
× 26	7.65	12.22	12¼	0.230	¼	⅛	6.490	6½	0.380	⅜	10½	⅞	½
W 12× 22	6.48	12.31	12¼	0.260	¼	⅛	4.030	4	0.425	⁷⁄₁₆	10½	⅞	½
× 19	5.57	12.16	12⅛	0.235	¼	⅛	4.005	4	0.350	⅜	10½	¹³⁄₁₆	½
× 16	4.71	11.99	12	0.220	¼	⅛	3.990	4	0.265	¼	10½	¾	½
× 14	4.16	11.91	11⅞	0.200	³⁄₁₆	⅛	3.970	4	0.225	¼	10½	¹¹⁄₁₆	½

AMERICAN INSTITUTE OF STEEL CONSTRUCTION

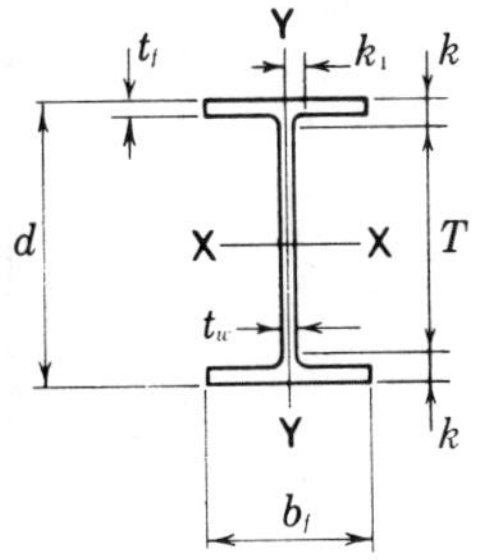

W SHAPES
Dimensions

Designation	Area A	Depth d		Web Thickness t_w		Web $\frac{t_w}{2}$	Flange Width b_f		Flange Thickness t_f		Distance T	Distance k	Distance k_1
	In.^{2}	In.		In.		In.	In.		In.		In.	In.	In.
W 10x112	32.9	11.36	11 3/8	0.755	3/4	3/8	10.415	10 3/8	1.250	1 1/4	7 5/8	1 7/8	15/16
x100	29.4	11.10	11 1/8	0.680	11/16	3/8	10.340	10 3/8	1.120	1 1/8	7 5/8	1 3/4	7/8
x 88	25.9	10.84	10 7/8	0.605	5/8	5/16	10.265	10 1/4	0.990	1	7 5/8	1 5/8	13/16
x 77	22.6	10.60	10 5/8	0.530	1/2	1/4	10.190	10 1/4	0.870	7/8	7 5/8	1 1/2	13/16
x 68	20.0	10.40	10 3/8	0.470	1/2	1/4	10.130	10 1/8	0.770	3/4	7 5/8	1 3/8	3/4
x 60	17.6	10.22	10 1/4	0.420	7/16	1/4	10.080	10 1/8	0.680	11/16	7 5/8	1 5/16	3/4
x 54	15.8	10.09	10 1/8	0.370	3/8	3/16	10.030	10	0.615	5/8	7 5/8	1 1/4	11/16
x 49	14.4	9.98	10	0.340	5/16	3/16	10.000	10	0.560	9/16	7 5/8	1 3/16	11/16
W 10x 45	13.3	10.10	10 1/8	0.350	3/8	3/16	8.020	8	0.620	5/8	7 5/8	1 1/4	11/16
x 39	11.5	9.92	9 7/8	0.315	5/16	3/16	7.985	8	0.530	1/2	7 5/8	1 1/8	11/16
x 33	9.71	9.73	9 3/4	0.290	5/16	3/16	7.960	8	0.435	7/16	7 5/8	1 1/16	11/16
W 10x 30	8.84	10.47	10 1/2	0.300	5/16	3/16	5.810	5 3/4	0.510	1/2	8 5/8	15/16	1/2
x 26	7.61	10.33	10 3/8	0.260	1/4	1/8	5.770	5 3/4	0.440	7/16	8 5/8	7/8	1/2
x 22	6.49	10.17	10 1/8	0.240	1/4	1/8	5.750	5 3/4	0.360	3/8	8 5/8	3/4	1/2
W 10x 19	5.62	10.24	10 1/4	0.250	1/4	1/8	4.020	4	0.395	3/8	8 5/8	13/16	1/2
x 17	4.99	10.11	10 1/8	0.240	1/4	1/8	4.010	4	0.330	5/16	8 5/8	3/4	1/2
x 15	4.41	9.99	10	0.230	1/4	1/8	4.000	4	0.270	1/4	8 5/8	11/16	7/16
x 12	3.54	9.87	9 7/8	0.190	3/16	1/8	3.960	4	0.210	3/16	8 5/8	5/8	7/16

AMERICAN INSTITUTE OF STEEL CONSTRUCTION

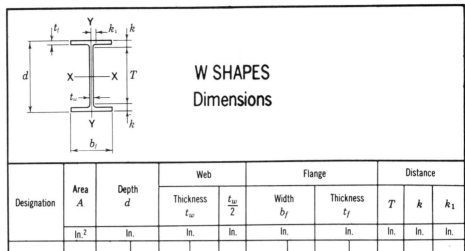

W SHAPES
Dimensions

Designation	Area A	Depth d		Web		Flange				Distance		
				Thickness t_w		$\frac{t_w}{2}$	Width b_f		Thickness t_f		T	k	k_1
	In.²	In.		In.		In.	In.		In.		In.	In.	In.
W 8×67	19.7	9.00	9	0.570	9/16	5/16	8.280	8¼	0.935	15/16	6⅛	1 7/16	11/16
×58	17.1	8.75	8¾	0.510	½	¼	8.220	8¼	0.810	13/16	6⅛	1 5/16	11/16
×48	14.1	8.50	8½	0.400	⅜	3/16	8.110	8⅛	0.685	11/16	6⅛	1 3/16	⅝
×40	11.7	8.25	8¼	0.360	⅜	3/16	8.070	8⅛	0.560	9/16	6⅛	1 1/16	⅝
×35	10.3	8.12	8⅛	0.310	5/16	3/16	8.020	8	0.495	½	6⅛	1	9/16
×31	9.13	8.00	8	0.285	5/16	3/16	7.995	8	0.435	7/16	6⅛	15/16	9/16
W 8×28	8.25	8.06	8	0.285	5/16	3/16	6.535	6½	0.465	7/16	6⅛	15/16	9/16
×24	7.08	7.93	7⅞	0.245	¼	⅛	6.495	6½	0.400	⅜	6⅛	⅞	9/16
W 8×21	6.16	8.28	8¼	0.250	¼	⅛	5.270	5¼	0.400	⅜	6⅝	13/16	½
×18	5.26	8.14	8⅛	0.230	¼	⅛	5.250	5¼	0.330	5/16	6⅝	¾	7/16
W 8×15	4.44	8.11	8⅛	0.245	¼	⅛	4.015	4	0.315	5/16	6⅝	¾	½
×13	3.84	7.99	8	0.230	¼	⅛	4.000	4	0.255	¼	6⅝	11/16	7/16
×10	2.96	7.89	7⅞	0.170	3/16	⅛	3.940	4	0.205	3/16	6⅝	⅝	7/16
W 6×25	7.34	6.38	6⅜	0.320	5/16	3/16	6.080	6⅛	0.455	7/16	4¾	13/16	7/16
×20	5.87	6.20	6¼	0.260	¼	⅛	6.020	6	0.365	⅜	4¾	¾	7/16
×15	4.43	5.99	6	0.230	¼	⅛	5.990	6	0.260	¼	4¾	⅝	⅜
W 6×16	4.74	6.28	6¼	0.260	¼	⅛	4.030	4	0.405	⅜	4¾	¾	7/16
×12	3.55	6.03	6	0.230	¼	⅛	4.000	4	0.280	¼	4¾	⅝	⅜
× 9	2.68	5.90	5⅞	0.170	3/16	⅛	3.940	4	0.215	3/16	4¾	9/16	⅜
W 5×19	5.54	5.15	5⅛	0.270	¼	⅛	5.030	5	0.430	7/16	3½	13/16	7/16
×16	4.68	5.01	5	0.240	¼	⅛	5.000	5	0.360	⅜	3½	¾	7/16
W 4×13	3.83	4.16	4⅛	0.280	¼	⅛	4.060	4	0.345	⅜	2¾	11/16	7/16

AMERICAN INSTITUTE OF STEEL CONSTRUCTION

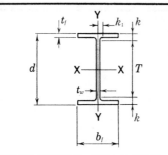

M SHAPES
Dimensions

Designation	Area A	Depth d		Web		Flange				Distance		Grip	Max. Flge. Fastener
				Thickness t_w		$\dfrac{t_w}{2}$	Width b_f		Thickness t_f		T	k		
	In.²	In.		In.		In.	In.		In.		In.	In.	In.	In.
M 14x18	5.10	14.00	14	0.215	3/16	1/8	4.000	4	0.270	1/4	12 3/4	5/8	1/4	3/4
M 12x11.8	3.47	12.00	12	0.177	3/16	1/8	3.065	3 1/8	0.225	1/4	10 7/8	9/16	1/4	—
M 10x9	2.65	10.00	10	0.157	3/16	1/8	2.690	2 3/4	0.206	3/16	8 7/8	9/16	3/16	—
M 8x6.5	1.92	8.00	8	0.135	1/8	1/16	2.281	2 1/4	0.189	3/16	7	1/2	3/16	—
M 6x20	5.89	6.00	6	0.250	1/4	1/8	5.938	6	0.379	3/8	4 1/4	7/8	3/8	7/8
M 6x4.4	1.29	6.00	6	0.114	1/8	1/16	1.844	1 7/8	0.171	3/16	5 1/8	7/16	3/16	—
M 5x18.9	5.55	5.00	5	0.316	5/16	3/16	5.003	5	0.416	7/16	3 1/4	7/8	7/16	7/8
M 4x13	3.81	4.00	4	0.254	1/4	1/8	3.940	4	0.371	3/8	2 3/8	13/16	3/8	3/4

AMERICAN INSTITUTE OF STEEL CONSTRUCTION

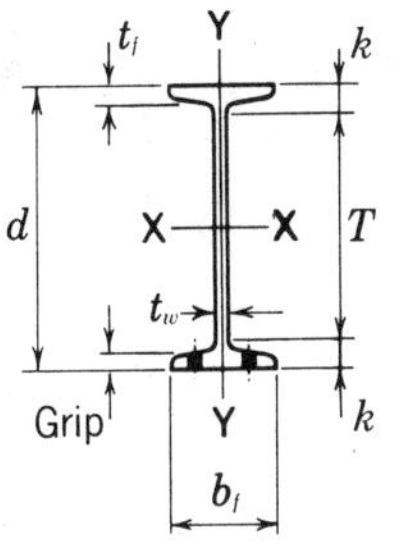

S SHAPES
Dimensions

Designation	Area A	Depth d		Web Thickness t_w		Web $\frac{t_w}{2}$	Flange Width b_f		Flange Thickness t_f		Distance T	Distance k	Grip	Max. Flge. Fastener
	In.^{2}	In.		In.		In.	In.		In.		In.	In.	In.	In.
S 24x121	35.6	24.50	24 1/2	0.800	13/16	7/16	8.050	8	1.090	1 1/16	20 1/2	2	1 1/8	1
x106	31.2	24.50	24 1/2	0.620	5/8	5/16	7.870	7 7/8	1.090	1 1/16	20 1/2	2	1 1/8	1
S 24x100	29.3	24.00	24	0.745	3/4	3/8	7.245	7 1/4	0.870	7/8	20 1/2	1 3/4	7/8	1
x90	26.5	24.00	24	0.625	5/8	5/16	7.125	7 1/8	0.870	7/8	20 1/2	1 3/4	7/8	1
x80	23.5	24.00	24	0.500	1/2	1/4	7.000	7	0.870	7/8	20 1/2	1 3/4	7/8	1
S 20x96	28.2	20.30	20 1/4	0.800	13/16	7/16	7.200	7 1/4	0.920	15/16	16 3/4	1 3/4	15/16	1
x86	25.3	20.30	20 1/4	0.660	11/16	3/8	7.060	7	0.920	15/16	16 3/4	1 3/4	15/16	1
S 20x75	22.0	20.00	20	0.635	5/8	5/16	6.385	6 3/8	0.795	13/16	16 3/4	1 5/8	13/16	7/8
x66	19.4	20.00	20	0.505	1/2	1/4	6.255	6 1/4	0.795	13/16	16 3/4	1 5/8	13/16	7/8
S 18x70	20.6	18.00	18	0.711	11/16	3/8	6.251	6 1/4	0.691	11/16	15	1 1/2	11/16	7/8
x54.7	16.1	18.00	18	0.461	7/16	1/4	6.001	6	0.691	11/16	15	1 1/2	11/16	7/8
S 15x50	14.7	15.00	15	0.550	9/16	5/16	5.640	5 5/8	0.622	5/8	12 1/4	1 3/8	9/16	3/4
x42.9	12.6	15.00	15	0.411	7/16	1/4	5.501	5 1/2	0.622	5/8	12 1/4	1 3/8	9/16	3/4
S 12x50	14.7	12.00	12	0.687	11/16	3/8	5.477	5 1/2	0.659	11/16	9 1/8	1 7/16	11/16	3/4
x40.8	12.0	12.00	12	0.462	7/16	1/4	5.252	5 1/4	0.659	11/16	9 1/8	1 7/16	5/8	3/4
S 12x35	10.3	12.00	12	0.428	7/16	1/4	5.078	5 1/8	0.544	9/16	9 5/8	1 3/16	1/2	3/4
x31.8	9.35	12.00	12	0.350	3/8	3/16	5.000	5	0.544	9/16	9 5/8	1 3/16	1/2	3/4
S 10x35	10.3	10.00	10	0.594	5/8	5/16	4.944	5	0.491	1/2	7 3/4	1 1/8	1/2	3/4
x25.4	7.46	10.00	10	0.311	5/16	3/16	4.661	4 5/8	0.491	1/2	7 3/4	1 1/8	1/2	3/4
S 8x23	6.77	8.00	8	0.441	7/16	1/4	4.171	4 1/8	0.426	7/16	6	1	7/16	3/4
x18.4	5.41	8.00	8	0.271	1/4	1/8	4.001	4	0.426	7/16	6	1	7/16	3/4
S 7x20	5.88	7.00	7	0.450	7/16	1/4	3.860	3 7/8	0.392	3/8	5 1/8	15/16	3/8	5/8
x15.3	4.50	7.00	7	0.252	1/4	1/8	3.662	3 5/8	0.392	3/8	5 1/8	15/16	3/8	5/8
S 6x17.25	5.07	6.00	6	0.465	7/16	1/4	3.565	3 5/8	0.359	3/8	4 1/4	7/8	3/8	5/8
x12.5	3.67	6.00	6	0.232	1/4	1/8	3.332	3 3/8	0.359	3/8	4 1/4	7/8	3/8	—
S 5x14.75	4.34	5.00	5	0.494	1/2	1/4	3.284	3 1/4	0.326	5/16	3 3/8	13/16	5/16	—
x10	2.94	5.00	5	0.214	3/16	1/8	3.004	3	0.326	5/16	3 3/8	13/16	5/16	—
S 4x9.5	2.79	4.00	4	0.326	5/16	3/16	2.796	2 3/4	0.293	5/16	2 1/2	3/4	5/16	—
x7.7	2.26	4.00	4	0.193	3/16	1/8	2.663	2 5/8	0.293	5/16	2 1/2	3/4	5/16	—
S 3x7.5	2.21	3.00	3	0.349	3/8	3/16	2.509	2 1/2	0.260	1/4	1 5/8	11/16	1/4	—
x5.7	1.67	3.00	3	0.170	3/16	1/8	2.330	2 3/8	0.260	1/4	1 5/8	11/16	1/4	—

AMERICAN INSTITUTE OF STEEL CONSTRUCTION

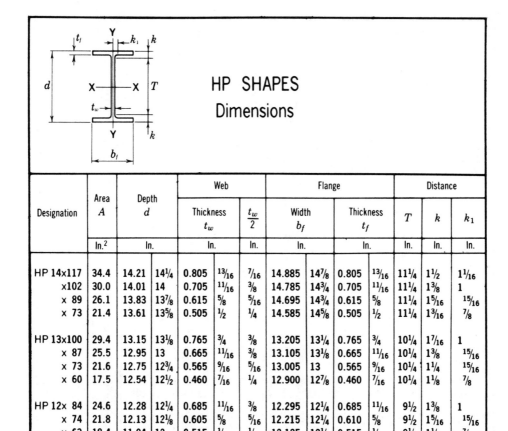

HP SHAPES
Dimensions

Designation	Area A	Depth d	Web Thickness t_w		$\frac{t_w}{2}$	Flange Width b_f		Flange Thickness t_f		Distance T	k	k_1
	In.²	In.	In.		In.	In.		In.		In.	In.	In.
HP 14×117	34.4	14.21	14¼	0.805	13/16	7/16	14.885	14⅞	0.805	13/16	11¼	1½	1 1/16
×102	30.0	14.01	14	0.705	11/16	⅜	14.785	14¾	0.705	11/16	11¼	1⅜	1
× 89	26.1	13.83	13⅞	0.615	⅝	5/16	14.695	14¾	0.615	⅝	11¼	1 5/16	15/16
× 73	21.4	13.61	13⅝	0.505	½	¼	14.585	14⅝	0.505	½	11¼	1 3/16	⅞
HP 13×100	29.4	13.15	13⅛	0.765	¾	⅜	13.205	13¼	0.765	¾	10¼	1 7/16	1
× 87	25.5	12.95	13	0.665	11/16	⅜	13.105	13⅛	0.665	11/16	10¼	1⅜	15/16
× 73	21.6	12.75	12¾	0.565	9/16	5/16	13.005	13	0.565	9/16	10¼	1¼	15/16
× 60	17.5	12.54	12½	0.460	7/16	¼	12.900	12⅞	0.460	7/16	10¼	1⅛	⅞
HP 12× 84	24.6	12.28	12¼	0.685	11/16	⅜	12.295	12¼	0.685	11/16	9½	1⅜	1
× 74	21.8	12.13	12⅛	0.605	⅝	5/16	12.215	12¼	0.610	⅝	9½	1 5/16	15/16
× 63	18.4	11.94	12	0.515	½	¼	12.125	12⅛	0.515	½	9½	1¼	⅞
× 53	15.5	11.78	11¾	0.435	7/16	¼	12.045	12	0.435	7/16	9½	1⅛	⅞
HP 10× 57	16.8	9.99	10	0.565	9/16	5/16	10.225	10¼	0.565	9/16	7⅝	1 3/16	13/16
× 42	12.4	9.70	9¾	0.415	7/16	¼	10.075	10⅛	0.420	7/16	7⅝	1 1/16	¾
HP 8× 36	10.6	8.02	8	0.445	7/16	¼	8.155	8⅛	0.445	7/16	6⅛	15/16	⅝

AMERICAN INSTITUTE OF STEEL CONSTRUCTION

CHANNELS AMERICAN STANDARD Dimensions

Designation	Area A	Depth d	Web Thickness t_w		$\dfrac{t_w}{2}$	Flange Width b_f		Average thickness t_f		Distance T	k	Grip	Max. Flge. Fastener
	In.²	In.	In.		In.	In.		In.		In.	In.	In.	In.
C 15×50	14.7	15.00	0.716	11/16	3/8	3.716	3 3/4	0.650	5/8	12 1/8	1 7/16	5/8	1
×40	11.8	15.00	0.520	1/2	1/4	3.520	3 1/2	0.650	5/8	12 1/8	1 7/16	5/8	1
×33.9	9.96	15.00	0.400	3/8	3/16	3.400	3 3/8	0.650	5/8	12 1/8	1 7/16	5/8	1
C 12×30	8.82	12.00	0.510	1/2	1/4	3.170	3 1/8	0.501	1/2	9 3/4	1 1/8	1/2	7/8
×25	7.35	12.00	0.387	3/8	3/16	3.047	3	0.501	1/2	9 3/4	1 1/8	1/2	7/8
×20.7	6.09	12.00	0.282	5/16	1/8	2.942	3	0.501	1/2	9 3/4	1 1/8	1/2	7/8
C 10×30	8.82	10.00	0.673	11/16	5/16	3.033	3	0.436	7/16	8	1	7/16	3/4
×25	7.35	10.00	0.526	1/2	1/4	2.886	2 7/8	0.436	7/16	8	1	7/16	3/4
×20	5.88	10.00	0.379	3/8	3/16	2.739	2 3/4	0.436	7/16	8	1	7/16	3/4
×15.3	4.49	10.00	0.240	1/4	1/8	2.600	2 5/8	0.436	7/16	8	1	7/16	3/4
C 9×20	5.88	9.00	0.448	7/16	1/4	2.648	2 5/8	0.413	7/16	7 1/8	15/16	7/16	3/4
×15	4.41	9.00	0.285	5/16	1/8	2.485	2 1/2	0.413	7/16	7 1/8	15/16	7/16	3/4
×13.4	3.94	9.00	0.233	1/4	1/8	2.433	2 3/8	0.413	7/16	7 1/8	15/16	7/16	3/4
C 8×18.75	5.51	8.00	0.487	1/2	1/4	2.527	2 1/2	0.390	3/8	6 1/8	15/16	3/8	3/4
×13.75	4.04	8.00	0.303	5/16	1/8	2.343	2 3/8	0.390	3/8	6 1/8	15/16	3/8	3/4
×11.5	3.38	8.00	0.220	1/4	1/8	2.260	2 1/4	0.390	3/8	6 1/8	15/16	3/8	3/4
C 7×14.75	4.33	7.00	0.419	7/16	3/16	2.299	2 1/4	0.366	3/8	5 1/4	7/8	3/8	5/8
×12.25	3.60	7.00	0.314	5/16	3/16	2.194	2 1/4	0.366	3/8	5 1/4	7/8	3/8	5/8
× 9.8	2.87	7.00	0.210	3/16	1/8	2.090	2 1/8	0.366	3/8	5 1/4	7/8	3/8	5/8
C 6×13	3.83	6.00	0.437	7/16	3/16	2.157	2 1/8	0.343	5/16	4 3/8	13/16	5/16	5/8
×10.5	3.09	6.00	0.314	5/16	3/16	2.034	2	0.343	5/16	4 3/8	13/16	3/8	5/8
× 8.2	2.40	6.00	0.200	3/16	1/8	1.920	1 7/8	0.343	5/16	4 3/8	13/16	5/16	5/8
C 5× 9	2.64	5.00	0.325	5/16	3/16	1.885	1 7/8	0.320	5/16	3 1/2	3/4	5/16	5/8
× 6.7	1.97	5.00	0.190	3/16	1/8	1.750	1 3/4	0.320	5/16	3 1/2	3/4	—	—
C 4× 7.25	2.13	4.00	0.321	5/16	3/16	1.721	1 3/4	0.296	5/16	2 5/8	11/16	5/16	5/8
× 5.4	1.59	4.00	0.184	3/16	1/16	1.584	1 5/8	0.296	5/16	2 5/8	11/16	—	—
C 3× 6	1.76	3.00	0.356	3/8	3/16	1.596	1 5/8	0.273	1/4	1 5/8	11/16	—	—
× 5	1.47	3.00	0.258	1/4	1/8	1.498	1 1/2	0.273	1/4	1 5/8	11/16	—	—
× 4.1	1.21	3.00	0.170	3/16	1/16	1.410	1 3/8	0.273	1/4	1 5/8	11/16	—	—

AMERICAN INSTITUTE OF STEEL CONSTRUCTION

Appendix A

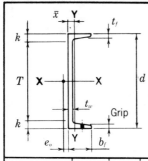

CHANNELS
MISCELLANEOUS
Dimensions

Designation	Area A	Depth d	Web Thickness t_w	$\frac{t_w}{2}$	Flange Width b_f	Flange Average thickness t_f	Distance T	Distance k	Grip	Max. Flge. Fastener
	In.²	In.	In.	In.	In.	In.	In.	In.	In.	In.
MC 18×58	17.1	18.00	0.700 11/16	3/8	4.200 4¼	0.625 5/8	15¼	1 3/8	5/8	1
×51.9	15.3	18.00	0.600 5/8	5/16	4.100 4⅛	0.625 5/8	15¼	1 3/8	5/8	1
×45.8	13.5	18.00	0.500 ½	¼	4.000 4	0.625 5/8	15¼	1 3/8	5/8	1
×42.7	12.6	18.00	0.450 7/16	¼	3.950 4	0.625 5/8	15¼	1 3/8	5/8	1
MC 13×50	14.7	13.00	0.787 13/16	3/8	4.412 4 3/8	0.610 5/8	10¼	1 3/8	5/8	1
×40	11.8	13.00	0.560 9/16	¼	4.185 4⅛	0.610 5/8	10¼	1 3/8	9/16	1
×35	10.3	13.00	0.447 7/16	¼	4.072 4⅛	0.610 5/8	10¼	1 3/8	9/16	1
×31.8	9.35	13.00	0.375 3/8	3/16	4.000 4	0.610 5/8	10¼	1 3/8	9/16	1
MC 12×50	14.7	12.00	0.835 13/16	7/16	4.135 4⅛	0.700 11/16	9 3/8	1 5/16	11/16	1
×45	13.2	12.00	0.712 11/16	3/8	4.012 4	0.700 11/16	9 3/8	1 5/16	11/16	1
×40	11.8	12.00	0.590 9/16	5/16	3.890 3 7/8	0.700 11/16	9 3/8	1 5/16	11/16	1
×35	10.3	12.00	0.467 7/16	¼	3.767 3¾	0.700 11/16	9 3/8	1 5/16	11/16	1
MC 12×37	10.9	12.00	0.600 5/8	5/16	3.600 3 5/8	0.600 5/8	9 3/8	1 5/16	5/8	7/8
×32.9	9.67	12.00	0.500 ½	¼	3.500 3½	0.600 5/8	9 3/8	1 5/16	9/16	7/8
×30.9	9.07	12.00	0.450 7/16	¼	3.450 3½	0.600 5/8	9 3/8	1 5/16	9/16	7/8
MC 12×10.6	3.10	12.00	0.190 3/16	⅛	1.500 1½	0.309 5/16	10 5/8	11/16	—	—
MC 10×41.1	12.1	10.00	0.796 13/16	3/8	4.321 4 3/8	0.575 9/16	7½	1¼	9/16	7/8
×33.6	9.87	10.00	0.575 9/16	5/16	4.100 4⅛	0.575 9/16	7½	1¼	9/16	7/8
×28.5	8.37	10.00	0.425 7/16	3/16	3.950 4	0.575 9/16	7½	1¼	9/16	7/8
MC 10×28.3	8.32	10.00	0.477 ½	¼	3.502 3½	0.575 9/16	7½	1¼	9/16	7/8
×25.3	7.43	10.00	0.425 7/16	3/16	3.550 3½	0.500 ½	7¾	1⅛	½	7/8
×24.9	7.32	10.00	0.377 3/8	3/16	3.402 3 3/8	0.575 9/16	7½	1¼	9/16	7/8
×21.9	6.43	10.00	0.325 5/16	3/16	3.450 3½	0.500 ½	7¾	1⅛	½	7/8
MC 10× 8.4	2.46	10.00	0.170 3/16	1/16	1.500 1½	0.280 ¼	8 5/8	11/16	—	—
MC 10× 6.5	1.91	10.00	0.152 ⅛	1/16	1.127 1⅛	0.202 3/16	9⅛	7/16	—	—

American Institute of Steel Construction

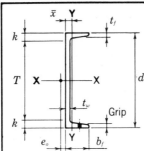

CHANNELS
MISCELLANEOUS
Dimensions

Designation	Area A	Depth d	Web Thickness t_w		$\frac{t_w}{2}$	Flange Width b_f		Average thickness t_f		Distance T	k	Grip	Max. Flge. Fastener
	In.²	In.	In.		In.	In.		In.		In.	In.	In.	In.
MC 9x25.4	7.47	9.00	0.450	7/16	1/4	3.500	3½	0.550	9/16	6⅝	1 3/16	9/16	7/8
x23.9	7.02	9.00	0.400	3/8	3/16	3.450	3½	0.550	9/16	6⅝	1 3/16	9/16	7/8
MC 8x22.8	6.70	8.00	0.427	7/16	3/16	3.502	3½	0.525	1/2	5⅝	1 3/16	1/2	7/8
x21.4	6.28	8.00	0.375	3/8	3/16	3.450	3½	0.525	1/2	5⅝	1 3/16	1/2	7/8
MC 8x20	5.88	8.00	0.400	3/8	3/16	3.025	3	0.500	1/2	5¾	1⅛	1/2	7/8
x18.7	5.50	8.00	0.353	3/8	3/16	2.978	3	0.500	1/2	5¾	1⅛	1/2	7/8
MC 8x 8.5	2.50	8.00	0.179	3/16	1/16	1.874	1⅞	0.311	5/16	6½	3/4	5/16	5/8
MC 7x22.7	6.67	7.00	0.503	1/2	1/4	3.603	3⅝	0.500	1/2	4¾	1⅛	1/2	7/8
x19.1	5.61	7.00	0.352	3/8	3/16	3.452	3½	0.500	1/2	4¾	1⅛	1/2	7/8
MC 7x17.6	5.17	7.00	0.375	3/8	3/16	3.000	3	0.475	1/2	4⅞	1 1/16	1/2	3/4
MC 6x18	5.29	6.00	0.379	3/8	3/16	3.504	3½	0.475	1/2	3⅞	1 1/16	1/2	7/8
x15.3	4.50	6.00	0.340	5/16	3/16	3.500	3½	0.385	3/8	4¼	7/8	3/8	7/8
MC 6x16.3	4.79	6.00	0.375	3/8	3/16	3.000	3	0.475	1/2	3⅞	1 1/16	1/2	3/4
x15.1	4.44	6.00	0.316	5/16	3/16	2.941	3	0.475	1/2	3⅞	1 1/16	1/2	3/4
MC 6x12	3.53	6.00	0.310	5/16	1/8	2.497	2½	0.375	3/8	4⅜	13/16	3/8	5/8

AMERICAN INSTITUTE OF STEEL CONSTRUCTION

WELDED JOINTS
Standard symbols

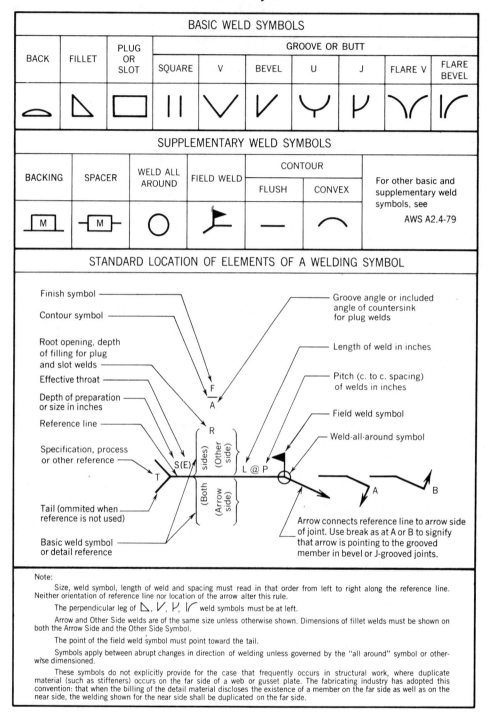

AMERICAN INSTITUTE OF STEEL CONSTRUCTION

Appendix B

USE OF THE SPAN TABLES

Spans for floor and ceiling joists are calculated on the basis of the modulus of elasticity (E) with the required fiber bending stress (F_b) listed below each span. Spans for rafters are calculated on the basis of fiber bending stress (F_b) with the required modulus of elasticity (E) listed below each span. Use of the tables is illustrated in the examples which follow.

Example 1. Floor Joists. Assume a required span of 12'-9", a live load of 40 psf and joists spaced 16 inches on centers. Table J-1 shows that a grade of 2 x 8 having an E value of 1,600,000 psi and an F_b value of 1250 psi would have a span of 12'-10", which satisfies the condition.

Example 2. Rafters. Assume a horizontal projection span of 13'-0", a live load of 20 psf, dead load of 7 psf, no attached ceiling and rafters spaced 16 inches on centers. Table R-13 shows that a 2 x 6 having an F_b value of 1200 psi and an E value of 940,000 psi would have a span of 13'-0" of horizontal projection. Conversion of horizontal to sloping distance is shown in the chart, Conversion Diagram for Rafters.

Since many combinations of size, spacing, E and F_b values are possible, it is recommended that the user examine the tables to determine which combination fits his particular case most effectively.

The spans for nominal 2 x 5 joists or rafters are 82 percent of the spans tabulated for the same spacing of nominal 2 x 6 joists or rafters. For each joist or rafter spacing, the required values of F_b or E for 2 x 5's are the same as the tabulated values of 2 x 6's.

Acknowledgment:

The material contained in this Appendix is taken from: SPAN TABLES FOR JOISTS AND RAFTERS (American Softwood Lumber, Standard Sizes PS 20-70) National Forest Products Association, 1619 Massachusetts Avenue, N.W., Washington, DC 20036

Appendix B

**TABLE J-1
FLOOR JOISTS**

40 Lbs. Per Sq. Ft. Live Load
(All rooms except those used for sleeping areas and attic floors.)

DESIGN CRITERIA:
Deflection - For 40 lbs. per sq. ft. live load.
Limited to span in inches divided by 360.
Strength - Live Load of 40 lbs. per sq. ft. plus dead load of 10 lbs. per sq. ft. determines the required fiber stress value.

Modulus of Elasticity, "E", in 1,000,000 psi

JOIST SIZE	SPACING (IN)	0.4	0.5	0.6	0.7	0.8	0.9	1.0	1.1	1.2	1.3	1.4	1.5	1.6	1.7	1.8	1.9	2.0	2.2	2.4
2×6	12.0	6-9 / 450	7-3 / 520	7-9 / 590	8-2 / 660	8-6 / 720	8-10 / 780	9-2 / 830	9-6 / 890	9-9 / 940	10-0 / 990	10-3 / 1040	10-6 / 1090	10-9 / 1140	10-11 / 1190	11-2 / 1230	11-4 / 1280	11-7 / 1320	11-11 / 1410	12-3 / 1490
	13.7	6-6 / 470	7-0 / 550	7-5 / 620	7-9 / 690	8-2 / 750	8-6 / 810	8-9 / 870	9-1 / 930	9-4 / 980	9-7 / 1040	9-10 / 1090	10-0 / 1140	10-3 / 1190	10-6 / 1240	10-8 / 1290	10-10 / 1340	11-1 / 1380	11-5 / 1470	11-9 / 1560
	16.0	6-2 / 500	6-7 / 580	7-0 / 650	7-5 / 720	7-9 / 790	8-0 / 860	8-4 / 920	8-7 / 980	8-10 / 1040	9-1 / 1090	9-4 / 1150	9-6 / 1200	9-9 / 1250	9-11 / 1310	10-2 / 1360	10-4 / 1410	10-6 / 1460	10-10 / 1550	11-2 / 1640
	19.2	5-9 / 530	6-3 / 610	6-7 / 690	7-0 / 770	7-3 / 840	7-7 / 910	7-10 / 970	8-1 / 1040	8-4 / 1100	8-7 / 1160	8-9 / 1220	9-0 / 1280	9-2 / 1330	9-4 / 1390	9-6 / 1440	9-8 / 1500	9-10 / 1550	10-2 / 1650	10-6 / 1750
	24.0	5-4 / 570	5-9 / 660	6-2 / 750	6-6 / 830	6-9 / 900	7-0 / 980	7-3 / 1050	7-6 / 1120	7-9 / 1190	7-11 / 1250	8-2 / 1310	8-4 / 1380	8-6 / 1440	8-8 / 1500	8-10 / 1550	9-0 / 1610	9-2 / 1670	9-6 / 1780	9-9 / 1880
	32.0					6-2 / 1010	6-5 / 1090	6-7 / 1150	6-10 / 1230	7-0 / 1300	7-3 / 1390	7-5 / 1450	7-7 / 1520	7-9 / 1590	7-11 / 1660	8-0 / 1690	8-2 / 1760	8-4 / 1840	8-7 / 1950	8-10 / 2060
2×8	12.0	8-11 / 450	9-7 / 520	10-2 / 590	10-9 / 660	11-3 / 720	11-8 / 780	12-1 / 830	12-6 / 890	12-10 / 940	13-2 / 990	13-6 / 1040	13-10 / 1090	14-2 / 1140	14-5 / 1190	14-8 / 1230	15-0 / 1280	15-3 / 1320	15-9 / 1410	16-2 / 1490
	13.7	8-6 / 470	8-2 / 550	9-9 / 620	10-3 / 690	10-9 / 750	11-2 / 810	11-7 / 870	11-11 / 930	12-3 / 980	12-7 / 1040	12-11 / 1090	13-3 / 1140	13-6 / 1190	13-10 / 1240	14-1 / 1290	14-4 / 1340	14-7 / 1380	15-0 / 1470	15-6 / 1560
	16.0	8-1 / 500	8-9 / 580	9-3 / 650	9-9 / 720	10-2 / 790	10-7 / 850	11-0 / 920	11-4 / 980	11-8 / 1040	12-0 / 1090	12-3 / 1150	12-7 / 1200	12-10 / 1250	13-1 / 1310	13-4 / 1360	13-7 / 1410	13-10 / 1460	14-3 / 1550	14-8 / 1640
	19.2	7-7 / 530	8-2 / 610	8-9 / 690	9-2 / 770	9-7 / 840	10-0 / 910	10-4 / 970	10-8 / 1040	11-0 / 1100	11-3 / 1160	11-7 / 1220	11-10 / 1280	12-1 / 1330	12-4 / 1390	12-7 / 1440	12-10 / 1500	13-0 / 1550	13-5 / 1650	13-10 / 1750
	24.0	7-1 / 570	7-7 / 660	8-1 / 750	8-6 / 830	8-11 / 900	9-3 / 980	9-7 / 1050	9-11 / 1120	10-2 / 1190	10-6 / 1250	10-9 / 1310	11-0 / 1380	11-3 / 1440	11-5 / 1500	11-8 / 1550	11-11 / 1610	12-1 / 1670	12-6 / 1780	12-10 / 1880
	32.0					8-1 / 990	8-5 / 1080	8-9 / 1170	9-0 / 1230	9-3 / 1300	9-6 / 1370	9-9 / 1450	10-0 / 1520	10-2 / 1570	10-5 / 1650	10-7 / 1700	10-10 / 1790	11-0 / 1840	11-4 / 1950	11-8 / 2070
2×10	12.0	11-4 / 450	12-3 / 520	13-0 / 590	13-8 / 660	14-4 / 720	14-11 / 780	15-5 / 830	15-11 / 890	16-5 / 940	16-10 / 990	17-3 / 1040	17-8 / 1090	18-0 / 1140	18-5 / 1190	18-9 / 1230	19-1 / 1280	19-5 / 1320	20-1 / 1410	20-8 / 1490
	13.7	10-10 / 470	11-8 / 550	12-5 / 620	13-1 / 690	13-8 / 750	14-3 / 810	14-9 / 870	15-3 / 930	15-8 / 980	16-1 / 1040	16-6 / 1090	16-11 / 1140	17-3 / 1190	17-7 / 1240	17-11 / 1290	18-3 / 1340	18-7 / 1380	19-2 / 1470	19-9 / 1560
	16.0	10-4 / 500	11-1 / 580	11-10 / 650	12-5 / 720	13-0 / 790	13-6 / 850	14-0 / 920	14-6 / 980	14-11 / 1040	15-3 / 1090	15-8 / 1150	16-0 / 1200	16-5 / 1250	16-9 / 1310	17-0 / 1360	17-4 / 1410	17-8 / 1460	18-3 / 1550	18-9 / 1640
	19.2	9-9 / 530	10-6 / 610	11-1 / 690	11-8 / 770	12-3 / 840	12-9 / 910	13-2 / 970	13-7 / 1040	14-0 / 1100	14-5 / 1160	14-9 / 1220	15-1 / 1280	15-5 / 1330	15-9 / 1390	16-0 / 1440	16-4 / 1500	16-7 / 1550	17-2 / 1650	17-8 / 1750
	24.0	9-0 / 570	9-9 / 660	10-4 / 750	10-10 / 830	11-4 / 900	11-10 / 980	12-3 / 1050	12-8 / 1120	13-0 / 1190	13-4 / 1250	13-8 / 1310	14-0 / 1380	14-4 / 1440	14-7 / 1500	14-11 / 1550	15-2 / 1610	15-5 / 1670	15-11 / 1780	16-5 / 1880
	32.0					10-4 / 1000	10-9 / 1080	11-1 / 1150	11-6 / 1240	11-10 / 1310	12-2 / 1380	12-5 / 1440	12-9 / 1520	13-0 / 1580	13-3 / 1640	13-6 / 1700	13-9 / 1770	14-0 / 1830	14-6 / 1970	14-11 / 2080
2×12	12.0	13-10 / 450	14-11 / 520	15-10 / 590	16-8 / 660	17-5 / 720	18-1 / 780	18-9 / 830	19-4 / 890	19-11 / 940	20-6 / 990	21-0 / 1040	21-6 / 1090	21-11 / 1140	22-5 / 1190	22-10 / 1230	23-3 / 1280	23-7 / 1320	24-5 / 1410	25-1 / 1490
	13.7	13-3 / 470	14-3 / 550	15-2 / 620	15-11 / 690	16-8 / 750	17-4 / 810	17-11 / 870	18-6 / 930	19-1 / 980	19-7 / 1040	20-1 / 1090	20-6 / 1140	21-0 / 1190	21-5 / 1240	21-10 / 1290	22-3 / 1340	22-7 / 1380	23-4 / 1470	24-0 / 1560
	16.0	12-7 / 500	13-6 / 580	14-4 / 650	15-2 / 720	15-10 / 790	16-5 / 860	17-0 / 920	17-7 / 980	18-1 / 1040	18-7 / 1090	19-1 / 1150	19-6 / 1200	19-11 / 1250	20-4 / 1310	20-9 / 1360	21-1 / 1410	21-6 / 1460	22-2 / 1550	22-10 / 1640
	19.2	11-10 / 530	12-9 / 610	13-6 / 690	14-3 / 770	14-11 / 840	15-6 / 910	16-0 / 970	16-7 / 1040	17-0 / 1100	17-6 / 1160	17-11 / 1220	18-4 / 1280	18-9 / 1330	19-2 / 1390	19-6 / 1440	19-10 / 1500	20-2 / 1550	20-10 / 1650	21-6 / 1750
	24.0	11-0 / 570	11-10 / 660	12-7 / 750	13-3 / 830	13-10 / 900	14-4 / 980	14-11 / 1050	15-4 / 1120	15-10 / 1190	16-3 / 1250	16-8 / 1310	17-0 / 1380	17-5 / 1440	17-9 / 1500	18-1 / 1550	18-5 / 1610	18-9 / 1670	19-4 / 1780	19-11 / 1880
	32.0					12-7 / 1000	13-1 / 1080	13-6 / 1150	13-11 / 1220	14-4 / 1300	14-9 / 1380	15-2 / 1450	15-6 / 1520	15-10 / 1580	16-2 / 1650	16-5 / 1700	16-9 / 1770	17-0 / 1830	17-7 / 1950	18-1 / 2070

Note: The required extreme fiber stress in bending, "F_b," in pounds per square inch is shown below each span.

TABLE R-13
MEDIUM OR HIGH SLOPE RAFTERS
No Ceiling Load
Slope over 3 in 12
Live Load - 20 lb. per. sq. ft.
(Light roof covering)

DESIGN CRITERIA:
Strength - 7 lbs. per sq. ft. dead load plus 20 lbs. per sq. ft. live load determines required fiber stress.
Deflection - For 20 lbs. per sq. ft. live load. Limited to span in inches divided by 180.

RAFTER SIZE (IN)	SPACING (IN)	Extreme Fiber Stress in Bending, "F_b" (psi).											
		200	300	400	500	600	700	800	900	1000	1100	1200	1300
2x4	12.0	3-11 / 0.07	4-9 / 0.14	5-6 / 0.21	6-2 / 0.29	6-9 / 0.38	7-3 / 0.49	7-9 / 0.59	8-3 / 0.71	8-8 / 0.83	9-1 / 0.96	9-6 / 1.09	9-11 / 1.23
2x4	13.7	3-8 / 0.07	4-5 / 0.13	5-2 / 0.20	5-9 / 0.27	6-4 / 0.36	6-10 / 0.45	7-3 / 0.55	7-9 / 0.66	8-2 / 0.77	8-6 / 0.89	8-11 / 1.02	9-3 / 1.15
2x4	16.0	3-4 / 0.06	4-1 / 0.12	4-9 / 0.18	5-4 / 0.25	5-10 / 0.33	6-4 / 0.42	6-9 / 0.51	7-2 / 0.61	7-6 / 0.72	7-11 / 0.83	8-3 / 0.94	8-7 / 1.06
2x4	19.2	3-1 / 0.06	3-9 / 0.11	4-4 / 0.17	4-10 / 0.23	5-4 / 0.30	5-9 / 0.38	6-2 / 0.47	6-6 / 0.56	6-10 / 0.65	7-3 / 0.76	7-6 / 0.86	7-10 / 0.97
2x4	24.0	2-9 / 0.05	3-4 / 0.10	3-11 / 0.15	4-4 / 0.21	4-9 / 0.27	5-2 / 0.34	5-6 / 0.42	5-10 / 0.50	6-2 / 0.59	6-5 / 0.68	6-9 / 0.77	7-0 / 0.87
2x6	12.0	6-1 / 0.07	7-6 / 0.14	8-8 / 0.21	9-8 / 0.29	10-7 / 0.38	11-5 / 0.49	12-3 / 0.59	13-0 / 0.71	13-8 / 0.83	14-4 / 0.96	15-0 / 1.09	15-7 / 1.23
2x6	13.7	5-9 / 0.07	7-0 / 0.13	8-1 / 0.20	9-0 / 0.27	9-11 / 0.36	10-8 / 0.45	11-5 / 0.55	12-2 / 0.66	12-9 / 0.77	13-5 / 0.89	14-0 / 1.02	14-7 / 1.15
2x6	16.0	5-4 / 0.06	6-6 / 0.12	7-6 / 0.18	8-4 / 0.25	9-2 / 0.33	9-11 / 0.42	10-7 / 0.51	11-3 / 0.61	11-10 / 0.72	12-5 / 0.83	13-0 / 0.94	13-6 / 1.06
2x6	19.2	4-10 / 0.06	5-11 / 0.11	6-10 / 0.17	7-8 / 0.23	8-4 / 0.30	9-0 / 0.38	9-8 / 0.47	10-3 / 0.56	10-10 / 0.65	11-4 / 0.76	11-10 / 0.86	12-4 / 0.97
2x6	24.0	4-4 / 0.05	5-4 / 0.10	6-1 / 0.15	6-10 / 0.21	7-6 / 0.27	8-1 / 0.34	8-8 / 0.42	9-2 / 0.50	9-8 / 0.59	10-2 / 0.68	10-7 / 0.77	11-0 / 0.87
2x8	12.0	8-1 / 0.07	9-10 / 0.14	11-5 / 0.21	12-9 / 0.29	13-11 / 0.38	15-1 / 0.49	16-1 / 0.59	17-1 / 0.71	18-0 / 0.83	18-11 / 0.96	19-9 / 1.09	20-6 / 1.23
2x8	13.7	7-6 / 0.07	9-3 / 0.13	10-8 / 0.20	11-11 / 0.27	13-1 / 0.36	14-1 / 0.45	15-1 / 0.55	16-0 / 0.66	16-10 / 0.77	17-8 / 0.89	18-5 / 1.02	19-3 / 1.15
2x8	16.0	7-0 / 0.06	8-7 / 0.12	9-10 / 0.18	11-0 / 0.25	12-1 / 0.33	13-1 / 0.42	13-11 / 0.51	14-10 / 0.61	15-7 / 0.72	16-4 / 0.83	17-1 / 0.94	17-9 / 1.06
2x8	19.2	6-4 / 0.06	7-10 / 0.11	9-0 / 0.17	10-1 / 0.23	11-0 / 0.30	11-11 / 0.38	12-9 / 0.47	13-6 / 0.56	14-3 / 0.65	14-11 / 0.76	15-7 / 0.86	16-3 / 0.97
2x8	24.0	5-8 / 0.05	7-0 / 0.10	8-1 / 0.15	9-0 / 0.21	9-10 / 0.27	10-8 / 0.34	11-5 / 0.42	12-1 / 0.50	12-9 / 0.59	13-4 / 0.68	13-11 / 0.77	14-6 / 0.87
2x10	12.0	10-3 / 0.07	12-7 / 0.14	14-6 / 0.21	16-3 / 0.29	17-10 / 0.38	19-3 / 0.49	20-7 / 0.59	21-10 / 0.71	23-0 / 0.83	24-1 / 0.96	25-2 / 1.09	26-2 / 1.23
2x10	13.7	9-7 / 0.07	11-9 / 0.13	13-7 / 0.20	15-2 / 0.27	16-8 / 0.36	18-0 / 0.45	19-3 / 0.55	20-5 / 0.66	21-6 / 0.77	22-7 / 0.89	23-7 / 1.02	24-6 / 1.15
2x10	16.0	8-11 / 0.06	10-11 / 0.12	12-7 / 0.18	14-1 / 0.25	15-5 / 0.33	16-8 / 0.42	17-10 / 0.51	18-11 / 0.61	19-11 / 0.72	20-10 / 0.83	21-10 / 0.94	22-8 / 1.06
2x10	19.2	8-2 / 0.06	9-11 / 0.11	11-6 / 0.17	12-10 / 0.23	14-1 / 0.30	15-2 / 0.38	16-3 / 0.47	17-3 / 0.56	18-2 / 0.65	19-1 / 0.76	19-11 / 0.86	20-9 / 0.97
2x10	24.0	7-3 / 0.05	8-11 / 0.10	10-3 / 0.15	11-6 / 0.21	12-7 / 0.27	13-7 / 0.34	14-6 / 0.42	15-5 / 0.50	16-3 / 0.59	17-1 / 0.68	17-10 / 0.77	18-6 / 0.87

Note: The required modulus of elasticity, "E", in 1,000,000 pounds per square inch is shown below each span.

TABLE R-13 (cont.)

RAFTERS: Spans are measured along the horizontal projection and loads are considered as applied on the horizontal projection.

Extreme Fiber Stress in Bending, "F_b" (psi).											RAFTER SPACING (IN)	SIZE (IN)
1400	1500	1600	1700	1800	1900	2000	2100	2200	2400	2700		
10-3 1.37	10-8 1.52	11-0 1.68	11-4 1.84	11-8 2.00	12-0 2.17	12-4 2.34	12-7 2.52				12.0	
9-7 1.28	10-0 1.42	10-3 1.57	10-7 1.72	10-11 1.87	11-3 2.03	11-6 2.19	11-9 2.36	12-1 2.53			13.7	
8-11 1.19	9-3 1.32	9-6 1.45	9-10 1.59	10-1 1.73	10-5 1.88	10-8 2.03	10-11 2.18	11-2 2.34			16.0	2x4
8-2 1.08	8-5 1.20	8-8 1.33	9-0 1.45	9-3 1.58	9-6 1.71	9-9 1.85	10-0 1.99	10-2 2.14	10-8 2.43		19.2	
7-3 0.97	7-6 1.08	7-9 1.19	8-0 1.30	8-3 1.41	8-6 1.53	8-8 1.66	8-11 1.78	9-1 1.91	9-6 2.18	10-1 2.60	24.0	
16-2 1.37	16-9 1.52	17-3 1.68	17-10 1.84	18-4 2.00	18-10 2.17	19-4 2.34	19-10 2.52				12.0	
15-1 1.28	15-8 1.42	16-2 1.57	16-8 1.72	17-2 1.87	17-7 2.03	18-1 2.19	18-6 2.36	19-0 2.53			13.7	
14-0 1.19	14-6 1.32	15-0 1.45	15-5 1.59	15-11 1.73	16-4 1.88	16-9 2.03	17-2 2.18	17-7 2.34			16.0	2x6
12-9 1.08	13-3 1.20	13-8 1.33	14-1 1.45	14-6 1.58	14-11 1.71	15-3 1.85	15-8 1.99	16-0 2.14	16-9 2.43		19.2	
11-5 0.97	11-10 1.08	12-3 1.19	12-7 1.30	13-0 1.41	13-4 1.53	13-8 1.66	14-0 1.78	14-4 1.91	15-0 2.18	15-11 2.60	24.0	
21-4 1.37	22-1 1.52	22-9 1.68	23-6 1.84	24-2 2.00	24-10 2.17	25-6 2.34	26-1 2.52				12.0	
19-11 1.28	20-8 1.42	21-4 1.57	22-0 1.72	22-7 1.87	23-3 2.03	23-10 2.19	24-5 2.36	25-0 2.53			13.7	
18-5 1.19	19-1 1.32	19-9 1.45	20-4 1.59	20-11 1.73	21-6 1.88	22-1 2.03	22-7 2.18	23-2 2.34			16.0	2x8
16-10 1.08	17-5 1.20	18-0 1.33	18-7 1.45	19-1 1.58	19-8 1.71	20-2 1.85	20-8 1.99	21-1 2.14	22-1 2.43		19.2	
15-1 0.97	15-7 1.08	16-1 1.19	16-7 1.30	17-1 1.41	17-7 1.53	18-0 1.66	18-5 1.78	18-11 1.91	19-9 2.18	20-11 2.60	24.0	
27-2 1.37	28-2 1.52	29-1 1.68	30-0 1.84	30-10 2.00	31-8 2.17	32-6 2.34	33-4 2.52				12.0	
25-5 1.28	26-4 1.42	27-2 1.57	28-0 1.72	28-10 1.87	29-8 2.03	30-5 2.19	31-2 2.36	31-11 2.53			13.7	
23-7 1.19	24-5 1.32	25-2 1.45	25-11 1.59	26-8 1.73	27-5 1.88	28-2 2.03	28-10 2.18	29-6 2.34			16.0	2x10
21-6 1.08	22-3 1.20	23-0 1.33	23-8 1.45	24-5 1.58	25-1 1.71	25-8 1.85	26-4 1.99	26-11 2.14	28-2 2.43		19.2	
19-3 0.97	19-11 1.08	20-7 1.19	21-2 1.30	21-10 1.41	22-5 1.53	23-0 1.66	23-7 1.78	24-1 1.91	25-2 2.18	26-8 2.60	24.0	

Note: The required modulus of elasticity, "E", in 1,000,000 pounds per square inch is shown below each span.

CONVERSION DIAGRAM FOR RAFTERS

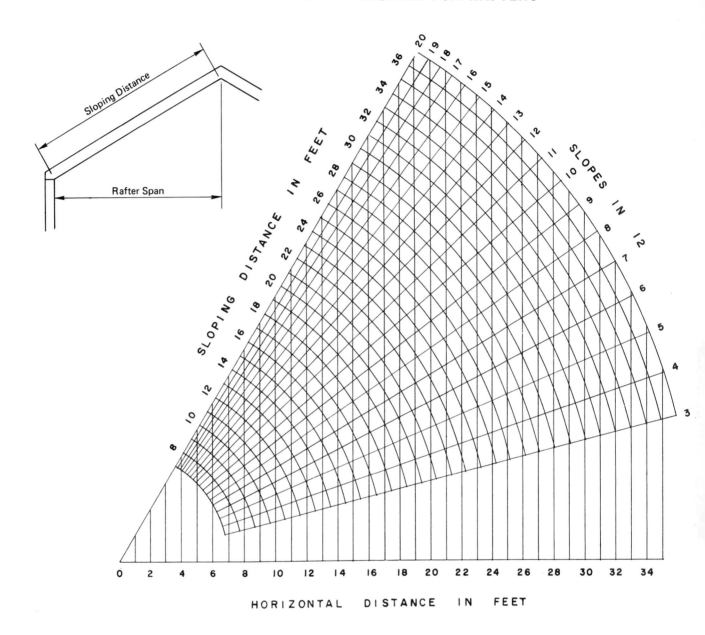

TABLE J-3
CEILING JOISTS
20 Lbs. Per Sq. Ft. Live Load
(Limited attic storage where development of future rooms is not possible)
(Plaster Ceiling)

DESIGN CRITERIA:
Deflection - For 20 lbs. per sq. ft. live load.
Limited to span in inches divided by 360.
Strength - Live load of 20 lbs. per sq. ft. plus dead load of 10 lbs. per sq. ft. determines required fiber stress value.

Note: The required extreme fiber stress in bending, "F_b", in pounds per square inch is shown below each span.

JOIST SIZE (IN)	SPACING (IN)	\multicolumn{21}{c}{Modulus of Elasticity, "E", in 1,000,000 psi}
		0.4	0.5	0.6	0.7	0.8	0.9	1.0	1.1	1.2	1.3	1.4	1.5	1.6	1.7	1.8	1.9	2.0	2.2	2.4
2x4	12.0	5-5 / 430	5-10 / 500	6-2 / 560	6-6 / 630	6-10 / 680	7-1 / 740	7-4 / 790	7-7 / 850	7-10 / 900	8-0 / 950	8-3 / 990	8-5 / 1040	8-7 / 1090	8-9 / 1130	8-11 / 1170	9-1 / 1220	9-3 / 1260	9-7 / 1340	9-10 / 1420
	13.7	5-2 / 450	5-7 / 520	5-11 / 590	6-3 / 650	6-6 / 720	6-9 / 770	7-0 / 830	7-3 / 880	7-6 / 940	7-8 / 990	7-10 / 1040	8-1 / 1090	8-3 / 1140	8-5 / 1180	8-7 / 1230	8-8 / 1270	8-10 / 1320	9-2 / 1400	9-5 / 1490
	16.0	4-11 / 470	5-4 / 550	5-8 / 620	5-11 / 690	6-2 / 750	6-5 / 810	6-8 / 870	6-11 / 930	7-1 / 990	7-3 / 1040	7-6 / 1090	7-8 / 1140	7-10 / 1200	8-0 / 1240	8-1 / 1290	8-3 / 1340	8-5 / 1390	8-8 / 1480	8-11 / 1570
	19.2	4-8 / 500	5-0 / 580	5-4 / 660	5-7 / 730	5-10 / 800	6-1 / 870	6-3 / 930	6-6 / 990	6-8 / 1050	6-10 / 1110	7-0 / 1160	7-2 / 1220	7-4 / 1270	7-6 / 1320	7-8 / 1370	7-9 / 1420	7-11 / 1470	8-2 / 1570	8-5 / 1660
	24.0	4-4 / 540	4-8 / 630	4-11 / 710	5-2 / 790	5-5 / 860	5-8 / 930	5-10 / 1000	6-0 / 1070	6-2 / 1130	6-4 / 1190	6-6 / 1250	6-8 / 1310	6-10 / 1370	7-0 / 1420	7-1 / 1480	7-3 / 1530	7-4 / 1590	7-7 / 1690	7-10 / 1790
2x6	12.0	8-6 / 430	9-2 / 500	9-9 / 560	10-3 / 630	10-9 / 680	11-2 / 740	11-7 / 790	11-11 / 850	12-3 / 900	12-7 / 950	12-11 / 990	13-3 / 1040	13-6 / 1090	13-9 / 1130	14-1 / 1170	14-4 / 1220	14-7 / 1260	15-0 / 1340	15-6 / 1420
	13.7	8-2 / 450	8-9 / 520	9-4 / 590	9-10 / 650	10-3 / 720	10-8 / 770	11-1 / 830	11-5 / 880	11-9 / 940	12-1 / 990	12-4 / 1040	12-8 / 1090	12-11 / 1140	13-2 / 1180	13-5 / 1230	13-8 / 1270	13-11 / 1320	14-4 / 1400	14-9 / 1490
	16.0	7-9 / 470	8-4 / 550	8-10 / 620	9-4 / 690	9-9 / 750	10-2 / 810	10-6 / 870	10-10 / 930	11-2 / 990	11-5 / 1040	11-9 / 1090	12-0 / 1140	12-3 / 1200	12-6 / 1240	12-9 / 1290	13-0 / 1340	13-3 / 1390	13-8 / 1480	14-1 / 1570
	19.2	7-3 / 500	7-10 / 580	8-4 / 660	8-9 / 730	9-2 / 800	9-6 / 870	9-10 / 930	10-2 / 990	10-6 / 1050	10-9 / 1110	11-1 / 1160	11-4 / 1220	11-7 / 1270	11-9 / 1320	12-0 / 1370	12-3 / 1420	12-5 / 1470	12-10 / 1570	13-3 / 1660
	24.0	6-9 / 540	7-3 / 630	7-9 / 710	8-2 / 790	8-6 / 860	8-10 / 930	9-2 / 1000	9-6 / 1070	9-9 / 1130	10-0 / 1190	10-3 / 1250	10-6 / 1310	10-9 / 1370	10-11 / 1420	11-2 / 1480	11-4 / 1530	11-7 / 1590	11-11 / 1690	12-3 / 1790
2x8	12.0	11-3 / 430	12-1 / 500	12-10 / 560	13-6 / 630	14-2 / 680	14-8 / 740	15-3 / 790	15-9 / 850	16-2 / 900	16-7 / 950	17-0 / 990	17-5 / 1040	17-10 / 1090	18-2 / 1130	18-6 / 1170	18-10 / 1220	19-2 / 1260	19-10 / 1340	20-5 / 1420
	13.7	10-9 / 450	11-7 / 520	12-3 / 590	12-11 / 650	13-6 / 720	14-1 / 770	14-7 / 830	15-0 / 880	15-6 / 940	15-11 / 990	16-3 / 1040	16-8 / 1090	17-0 / 1140	17-5 / 1180	17-9 / 1230	18-0 / 1270	18-4 / 1320	18-11 / 1400	19-6 / 1490
	16.0	10-2 / 470	11-0 / 550	11-8 / 620	12-3 / 690	12-10 / 750	13-4 / 810	13-10 / 870	14-3 / 930	14-8 / 990	15-1 / 1040	15-6 / 1090	15-10 / 1140	16-2 / 1200	16-6 / 1240	16-10 / 1290	17-2 / 1340	17-5 / 1390	18-0 / 1480	18-6 / 1570
	19.2	9-7 / 500	10-4 / 580	11-0 / 660	11-7 / 730	12-1 / 800	12-7 / 870	13-0 / 930	13-5 / 990	13-10 / 1050	14-2 / 1110	14-7 / 1160	14-11 / 1220	15-3 / 1270	15-6 / 1320	15-10 / 1370	16-1 / 1420	16-5 / 1470	16-11 / 1570	17-5 / 1660
	24.0	8-11 / 540	9-7 / 630	10-2 / 710	10-9 / 790	11-3 / 860	11-8 / 930	12-1 / 1000	12-6 / 1070	12-10 / 1130	13-2 / 1190	13-6 / 1250	13-10 / 1310	14-2 / 1370	14-5 / 1420	14-8 / 1480	15-0 / 1530	15-3 / 1590	15-9 / 1690	16-2 / 1790
2x10	12.0	14-4 / 430	15-5 / 500	16-5 / 560	17-3 / 630	18-0 / 680	18-9 / 740	19-5 / 790	20-1 / 850	20-8 / 900	21-2 / 950	21-9 / 990	22-3 / 1040	22-9 / 1090	23-2 / 1130	23-8 / 1170	24-1 / 1220	24-6 / 1260	25-3 / 1340	26-0 / 1420
	13.7	13-8 / 450	14-9 / 520	15-8 / 590	16-6 / 650	17-3 / 720	17-11 / 770	18-7 / 830	19-2 / 880	19-9 / 940	20-3 / 990	20-9 / 1040	21-3 / 1090	21-9 / 1140	22-2 / 1180	22-7 / 1230	23-0 / 1270	23-5 / 1320	24-2 / 1400	24-10 / 1490
	16.0	13-0 / 470	14-0 / 550	14-11 / 620	15-8 / 690	16-5 / 750	17-0 / 810	17-8 / 870	18-3 / 930	18-9 / 990	19-3 / 1040	19-9 / 1090	20-2 / 1140	20-8 / 1200	21-1 / 1240	21-6 / 1290	21-10 / 1340	22-3 / 1390	22-11 / 1480	23-8 / 1570
	19.2	12-3 / 500	13-2 / 580	14-0 / 660	14-9 / 730	15-5 / 800	16-0 / 870	16-7 / 930	17-2 / 990	17-8 / 1050	18-1 / 1110	18-7 / 1160	19-0 / 1220	19-5 / 1270	19-10 / 1320	20-2 / 1370	20-7 / 1420	20-11 / 1470	21-7 / 1570	22-3 / 1660
	24.0	11-4 / 540	12-3 / 630	13-0 / 710	13-8 / 790	14-4 / 860	14-11 / 930	15-5 / 1000	15-11 / 1070	16-5 / 1130	16-10 / 1190	17-3 / 1250	17-8 / 1310	18-0 / 1370	18-5 / 1420	18-9 / 1480	19-1 / 1530	19-5 / 1590	20-1 / 1690	20-8 / 1790

TABLE J-4
CEILING JOISTS
20 Lbs. Per Sq. Ft. Live Load
(Limited attic storage where development of future rooms is not possible)
(Drywall Ceiling)

DESIGN CRITERIA:
Deflection - For 20 lbs. per sq. ft. live load.
Limited to span in inches divided by 240.
Strength - live load of 20 lbs. per sq. ft. plus dead load of 10 lbs. per sq. ft. determines required fiber stress value.

JOIST SIZE (IN)	SPACING (IN)	\multicolumn{21}{c}{Modulus of Elasticity, "E", in 1,000,000 psi}
		0.4	0.5	0.6	0.7	0.8	0.9	1.0	1.1	1.2	1.3	1.4	1.5	1.6	1.7	1.8	1.9	2.0	2.2	2.4
2x4	12.0	6-2 / 560	6-8 / 660	7-1 / 740	7-6 / 820	7-10 / 900	8-1 / 970	8-5 / 1040	8-8 / 1110	8-11 / 1170	9-2 / 1240	9-5 / 1300	9-8 / 1360	9-10 / 1420	10-0 / 1480	10-3 / 1540	10-5 / 1600	10-7 / 1650	10-11 / 1760	11-3 / 1860
	13.7	5-11 / 590	6-5 / 690	6-9 / 770	7-2 / 860	7-6 / 940	7-9 / 1010	8-1 / 1090	8-4 / 1160	8-7 / 1230	8-9 / 1300	9-0 / 1360	9-3 / 1420	9-5 / 1490	9-7 / 1550	9-9 / 1610	10-0 / 1670	10-2 / 1730	10-6 / 1840	10-9 / 1950
	16.0	5-8 / 620	6-1 / 720	6-5 / 810	6-9 / 900	7-1 / 990	7-5 / 1070	7-8 / 1140	7-11 / 1220	8-1 / 1290	8-4 / 1360	8-7 / 1430	8-9 / 1500	8-11 / 1570	9-1 / 1630	9-4 / 1690	9-6 / 1760	9-8 / 1820	9-11 / 1940	10-3 / 2050
	19.2	5-4 / 660	5-9 / 770	6-1 / 870	6-5 / 960	6-8 / 1050	6-11 / 1130	7-2 / 1220	7-5 / 1300	7-8 / 1370	7-10 / 1450	8-1 / 1520	8-3 / 1590	8-5 / 1660	8-7 / 1730	8-9 / 1800	8-11 / 1870	9-1 / 1930	9-4 / 2060	9-8 / 2180
	24.0	4-11 / 710	5-4 / 830	5-8 / 930	5-11 / 1030	6-2 / 1130	6-5 / 1220	6-8 / 1310	6-11 / 1400	7-1 / 1480	7-3 / 1560	7-6 / 1640	7-8 / 1720	7-10 / 1790	8-0 / 1870	8-1 / 1940	8-3 / 2010	8-5 / 2080	8-8 / 2220	8-11 / 2350
2x6	12.0	9-9 / 560	10-6 / 660	11-2 / 740	11-9 / 820	12-3 / 900	12-9 / 970	13-3 / 1040	13-8 / 1110	14-1 / 1170	14-5 / 1240	14-9 / 1300	15-2 / 1360	15-6 / 1420	15-9 / 1480	16-1 / 1540	16-4 / 1600	16-8 / 1650	17-2 / 1760	17-8 / 1860
	13.7	9-4 / 590	10-0 / 690	10-8 / 770	11-3 / 860	11-9 / 940	12-3 / 1010	12-8 / 1090	13-1 / 1160	13-5 / 1230	13-10 / 1300	14-2 / 1360	14-6 / 1420	14-9 / 1490	15-1 / 1550	15-5 / 1610	15-8 / 1670	15-11 / 1730	16-5 / 1840	16-11 / 1950
	16.0	8-10 / 620	9-6 / 720	10-2 / 810	10-8 / 900	11-2 / 990	11-7 / 1070	12-0 / 1140	12-5 / 1220	12-9 / 1290	13-1 / 1360	13-5 / 1430	13-9 / 1500	14-1 / 1570	14-4 / 1630	14-7 / 1690	14-11 / 1760	15-2 / 1820	15-7 / 1940	16-1 / 2050
	19.2	8-4 / 660	9-0 / 770	9-6 / 870	10-0 / 960	10-6 / 1050	10-11 / 1130	11-4 / 1220	11-8 / 1300	12-0 / 1370	12-4 / 1450	12-8 / 1520	12-11 / 1590	13-3 / 1660	13-6 / 1730	13-9 / 1800	14-0 / 1870	14-3 / 1930	14-8 / 2060	15-2 / 2180
	24.0	7-9 / 710	8-4 / 830	8-10 / 930	9-4 / 1030	9-9 / 1130	10-2 / 1220	10-6 / 1310	10-10 / 1400	11-2 / 1480	11-5 / 1560	11-9 / 1640	12-0 / 1720	12-3 / 1790	12-6 / 1870	12-9 / 1940	13-0 / 2010	13-3 / 2080	13-8 / 2220	14-1 / 2350
2x8	12.0	12-10 / 560	13-10 / 660	14-8 / 740	15-6 / 820	16-2 / 900	16-10 / 970	17-5 / 1040	18-0 / 1110	18-6 / 1170	19-0 / 1240	19-6 / 1300	19-11 / 1360	20-5 / 1420	20-10 / 1480	21-2 / 1540	21-7 / 1600	21-11 / 1650	22-8 / 1760	23-4 / 1860
	13.7	12-3 / 590	13-3 / 690	14-1 / 770	14-10 / 860	15-6 / 940	16-1 / 1010	16-8 / 1090	17-2 / 1160	17-9 / 1230	18-2 / 1300	18-8 / 1360	19-1 / 1420	19-6 / 1490	19-11 / 1550	20-3 / 1610	20-8 / 1670	21-0 / 1730	21-8 / 1840	22-4 / 1950
	16.0	11-8 / 620	12-7 / 720	13-4 / 810	14-1 / 900	14-8 / 990	15-3 / 1070	15-10 / 1140	16-4 / 1220	16-10 / 1290	17-3 / 1360	17-9 / 1430	18-2 / 1500	18-6 / 1570	18-11 / 1630	19-3 / 1690	19-7 / 1760	19-11 / 1820	20-7 / 1940	21-2 / 2050
	19.2	11-0 / 660	11-10 / 770	12-7 / 870	13-3 / 960	13-10 / 1050	14-5 / 1130	14-11 / 1220	15-5 / 1300	15-10 / 1370	16-3 / 1450	16-8 / 1520	17-1 / 1590	17-5 / 1660	17-9 / 1730	18-2 / 1800	18-5 / 1870	18-9 / 1930	19-5 / 2060	19-11 / 2180
	24.0	10-2 / 710	11-0 / 830	11-8 / 930	12-3 / 1030	12-10 / 1130	13-4 / 1220	13-10 / 1310	14-3 / 1400	14-8 / 1480	15-1 / 1560	15-6 / 1640	15-10 / 1720	16-2 / 1790	16-6 / 1870	16-10 / 1940	17-2 / 2010	17-5 / 2080	18-0 / 2220	18-6 / 2350
2x10	12.0	16-5 / 560	17-8 / 660	18-9 / 740	19-9 / 820	20-8 / 900	21-6 / 970	22-3 / 1040	22-11 / 1110	23-8 / 1170	24-3 / 1240	24-10 / 1300	25-5 / 1360	26-0 / 1420	26-6 / 1480	27-1 / 1540	27-6 / 1600	28-0 / 1650	28-11 / 1760	29-9 / 1860
	13.7	15-8 / 590	16-11 / 690	17-11 / 770	18-11 / 860	19-9 / 940	20-6 / 1010	21-3 / 1090	21-11 / 1160	22-7 / 1230	23-3 / 1300	23-9 / 1360	24-4 / 1420	24-10 / 1490	25-5 / 1550	25-10 / 1610	26-4 / 1670	26-10 / 1730	27-8 / 1840	28-6 / 1950
	16.0	14-11 / 620	16-0 / 720	17-0 / 810	17-11 / 900	18-9 / 990	19-6 / 1070	20-2 / 1140	20-10 / 1220	21-6 / 1290	22-1 / 1360	22-7 / 1430	23-2 / 1500	23-8 / 1570	24-1 / 1630	24-7 / 1690	25-0 / 1760	25-5 / 1820	26-3 / 1940	27-1 / 2050
	19.2	14-0 / 660	15-1 / 770	16-0 / 870	16-11 / 960	17-8 / 1050	18-4 / 1130	19-0 / 1220	19-7 / 1300	20-2 / 1370	20-9 / 1450	21-3 / 1520	21-9 / 1590	22-3 / 1660	22-8 / 1730	23-2 / 1800	23-7 / 1870	23-11 / 1930	24-9 / 2060	25-5 / 2180
	24.0	13-0 / 710	14-0 / 830	14-11 / 930	15-8 / 1030	16-5 / 1130	17-0 / 1220	17-8 / 1310	18-3 / 1400	18-9 / 1480	19-3 / 1560	19-9 / 1640	20-2 / 1720	20-8 / 1790	21-1 / 1870	21-6 / 1940	21-10 / 2010	22-3 / 2080	22-11 / 2220	23-8 / 2350

Note: The required extreme fiber stress in bending, "F_b", in pounds per square inch is shown below each span.

TABLE J-5
CEILING JOISTS
10 Lbs. Per Sq. Ft. Live Load
(No attic storage and roof slope not steeper than 3 in 12)
(Plaster Ceiling)

DESIGN CRITERIA:
Deflection - For 10 lbs. per sq. ft. live load.
Limited to span in inches divided by 360.
Strength - live load of 10 lbs. per sq. ft. plus dead load of 5 lbs. per sq. ft. determines required fiber stress value.

JOIST SIZE (IN)	SPACING (IN)	0.4	0.5	0.6	0.7	0.8	0.9	1.0	1.1	1.2	1.3	1.4	1.5	1.6	1.7	1.8	1.9	2.0	2.2	2.4
2x4	12.0	6-10 / 340	7-4 / 400	7-10 / 450	8-3 / 500	8-7 / 540	8-11 / 590	9-3 / 630	9-7 / 670	9-10 / 710	10-1 / 750	10-4 / 790	10-7 / 830	10-10 / 860	11-1 / 900	11-3 / 930	11-6 / 970	11-8 / 1000	12-1 / 1070	12-5 / 1130
2x4	13.7	6-6 / 360	7-0 / 410	7-6 / 470	7-10 / 520	8-3 / 570	8-7 / 610	8-10 / 660	9-2 / 700	9-5 / 740	9-8 / 780	9-11 / 820	10-2 / 860	10-4 / 900	10-7 / 940	10-9 / 970	11-0 / 1010	11-2 / 1050	11-6 / 1110	11-10 / 1180
2x4	16.0	6-2 / 380	6-8 / 440	7-1 / 490	7-6 / 550	7-10 / 600	8-1 / 650	8-5 / 690	8-8 / 740	8-11 / 780	9-2 / 830	9-5 / 870	9-8 / 910	9-10 / 950	10-0 / 990	10-3 / 1030	10-5 / 1060	10-7 / 1100	10-11 / 1170	11-3 / 1240
2x4	19.2	5-10 / 400	6-3 / 460	6-8 / 520	7-0 / 580	7-4 / 630	7-8 / 690	7-11 / 740	8-2 / 790	8-5 / 830	8-8 / 880	8-10 / 920	9-1 / 970	9-3 / 1010	9-5 / 1050	9-8 / 1090	9-10 / 1130	10-0 / 1170	10-4 / 1250	10-7 / 1320
2x4	24.0	5-5 / 430	5-10 / 500	6-2 / 560	6-6 / 630	6-10 / 680	7-1 / 740	7-4 / 790	7-7 / 850	7-10 / 900	8-0 / 950	8-3 / 990	8-5 / 1040	8-7 / 1090	8-9 / 1130	8-11 / 1170	9-1 / 1220	9-3 / 1260	9-7 / 1340	9-10 / 1420
2x6	12.0	10-9 / 340	11-7 / 400	12-3 / 450	12-11 / 500	13-6 / 540	14-1 / 590	14-7 / 630	15-0 / 670	15-6 / 710	15-11 / 750	16-3 / 790	16-8 / 830	17-0 / 860	17-4 / 900	17-8 / 930	18-0 / 970	18-4 / 1000	18-11 / 1070	19-6 / 1130
2x6	13.7	10-3 / 360	11-1 / 410	11-9 / 470	12-4 / 520	12-11 / 570	13-5 / 610	13-11 / 660	14-4 / 700	14-9 / 740	15-2 / 780	15-7 / 820	15-11 / 860	16-3 / 900	16-7 / 940	16-11 / 970	17-3 / 1010	17-6 / 1050	18-1 / 1110	18-8 / 1180
2x6	16.0	9-9 / 380	10-6 / 440	11-2 / 490	11-9 / 550	12-3 / 600	12-9 / 650	13-3 / 690	13-8 / 740	14-1 / 780	14-5 / 830	14-9 / 870	15-2 / 910	15-6 / 950	15-9 / 990	16-1 / 1030	16-4 / 1060	16-8 / 1100	17-2 / 1170	17-8 / 1240
2x6	19.2	9-2 / 400	9-10 / 460	10-6 / 520	11-1 / 580	11-7 / 630	12-0 / 690	12-5 / 740	12-10 / 790	13-3 / 830	13-7 / 880	13-11 / 920	14-3 / 970	14-7 / 1010	14-10 / 1050	15-2 / 1090	15-5 / 1130	15-8 / 1170	16-2 / 1250	16-8 / 1320
2x6	24.0	8-6 / 430	9-2 / 500	9-9 / 560	10-3 / 630	10-9 / 680	11-2 / 740	11-7 / 790	11-11 / 850	12-3 / 900	12-7 / 950	12-11 / 990	13-3 / 1040	13-6 / 1090	13-9 / 1130	14-1 / 1170	14-4 / 1220	14-7 / 1260	15-0 / 1340	15-6 / 1420
2x8	12.0	14-2 / 340	15-3 / 400	16-2 / 450	17-0 / 500	17-10 / 540	18-6 / 590	19-2 / 630	19-10 / 670	20-5 / 710	20-11 / 750	21-5 / 790	21-11 / 830	22-5 / 860	22-11 / 900	23-4 / 930	23-9 / 970	24-2 / 1000	24-11 / 1070	25-8 / 1130
2x8	13.7	13-6 / 360	14-7 / 410	15-6 / 470	16-3 / 520	17-0 / 570	17-9 / 610	18-4 / 660	18-11 / 700	19-6 / 740	20-0 / 780	20-6 / 820	21-0 / 860	21-5 / 900	21-11 / 940	22-4 / 970	22-9 / 1010	23-1 / 1050	23-10 / 1110	24-7 / 1180
2x8	16.0	12-10 / 380	13-10 / 440	14-8 / 490	15-6 / 550	16-2 / 600	16-10 / 650	17-5 / 690	18-0 / 740	18-6 / 780	19-0 / 830	19-6 / 870	19-11 / 910	20-5 / 950	20-10 / 990	21-2 / 1030	21-7 / 1060	21-11 / 1100	22-8 / 1170	23-4 / 1240
2x8	19.2	12-1 / 400	13-0 / 460	13-10 / 520	14-7 / 580	15-3 / 630	15-10 / 690	16-5 / 740	16-11 / 790	17-5 / 830	17-11 / 880	18-4 / 920	18-9 / 970	19-2 / 1010	19-7 / 1050	19-11 / 1090	20-4 / 1130	20-8 / 1170	21-4 / 1250	21-11 / 1320
2x8	24.0	11-3 / 430	12-1 / 500	12-10 / 560	13-6 / 630	14-2 / 680	14-8 / 740	15-3 / 790	15-9 / 850	16-2 / 900	16-7 / 950	17-0 / 990	17-5 / 1040	17-10 / 1090	18-2 / 1130	18-6 / 1170	18-10 / 1220	19-2 / 1260	19-10 / 1340	20-5 / 1420
2x10	12.0	18-0 / 340	19-5 / 400	20-8 / 450	21-9 / 500	22-9 / 540	23-8 / 590	24-6 / 630	25-3 / 670	26-0 / 710	26-9 / 750	27-5 / 790	28-0 / 830	28-7 / 860	29-2 / 900	29-9 / 930	30-4 / 970	30-10 / 1000	31-10 / 1070	32-9 / 1130
2x10	13.7	17-3 / 360	18-7 / 410	19-9 / 470	20-9 / 520	21-9 / 570	22-7 / 610	23-5 / 660	24-2 / 700	24-10 / 740	25-7 / 780	26-2 / 820	26-10 / 860	27-5 / 900	27-11 / 940	28-6 / 970	29-0 / 1010	29-6 / 1050	30-5 / 1110	31-4 / 1180
2x10	16.0	16-5 / 380	17-8 / 440	18-9 / 490	19-9 / 550	20-8 / 600	21-6 / 650	22-3 / 690	22-11 / 740	23-8 / 780	24-3 / 830	24-10 / 870	25-5 / 910	26-0 / 950	26-6 / 990	27-1 / 1030	27-6 / 1060	28-0 / 1100	28-11 / 1170	29-9 / 1240
2x10	19.2	15-5 / 400	16-7 / 460	17-8 / 520	18-7 / 580	19-5 / 630	20-2 / 690	20-11 / 740	21-7 / 790	22-3 / 830	24-3 / 830	23-5 / 920	23-11 / 970	24-6 / 1010	25-0 / 1050	25-5 / 1090	25-11 / 1130	26-4 / 1170	27-3 / 1250	28-0 / 1320
2x10	24.0	14-4 / 430	15-5 / 500	16-5 / 560	17-3 / 630	18-0 / 680	18-9 / 740	19-5 / 790	20-1 / 850	20-8 / 900	21-2 / 950	21-9 / 990	22-3 / 1040	22-9 / 1090	23-2 / 1130	23-8 / 1170	24-1 / 1220	24-6 / 1260	25-3 / 1340	26-0 / 1420

Modulus of Elasticity, "E", in 1,000,000 psi

Note: The required extreme fiber stress in bending, "F_b" in pounds per square inch is shown below each span.

TABLE J-6
CEILING JOISTS
10 Lbs. Per Sq. Ft. Live Load
(No attic storage and roof slope not steeper than 3 in 12)
(Drywall Ceiling)

DESIGN CRITERIA:
Deflection - For 10 lbs. per sq. ft. live load.
  Limited to span in inches divided by 240.
Strength - live load of 10 lbs. per sq. ft. plus dead load of 5 lbs. per sq. ft. determines required fiber stress value.

Modulus of Elasticity, "E", in 1,000,000 psi

JOIST SIZE (IN)	SPACING (IN)	0.4	0.5	0.6	0.7	0.8	0.9	1.0	1.1	1.2	1.3	1.4	1.5	1.6	1.7	1.8	1.9	2.0	2.2	2.4
2x4	12.0	7-10 / 450	8-5 / 520	8-11 / 590	9-5 / 650	9-10 / 710	10-3 / 770	10-7 / 830	10-11 / 880	11-3 / 930	11-7 / 980	11-10 / 1030	12-2 / 1080	12-5 / 1130	12-8 / 1180	12-11 / 1220	13-2 / 1270	13-4 / 1310	13-9 / 1400	14-2 / 1480
2x4	13.7	7-6 / 470	8-1 / 540	8-7 / 610	9-0 / 680	9-5 / 740	9-9 / 800	10-2 / 860	10-6 / 920	10-9 / 970	11-1 / 1030	11-4 / 1080	11-7 / 1130	11-10 / 1180	12-1 / 1230	12-4 / 1280	12-7 / 1320	12-9 / 1370	13-2 / 1460	13-7 / 1550
2x4	16.0	7-1 / 490	7-8 / 570	8-1 / 650	8-7 / 720	8-11 / 780	9-4 / 850	9-8 / 910	9-11 / 970	10-3 / 1030	10-6 / 1080	10-9 / 1140	11-0 / 1190	11-3 / 1240	11-6 / 1290	11-9 / 1340	11-11 / 1390	12-2 / 1440	12-6 / 1540	12-11 / 1630
2x4	19.2	6-8 / 520	7-2 / 610	7-8 / 690	8-1 / 760	8-5 / 830	8-9 / 900	9-1 / 970	9-4 / 1030	9-8 / 1090	9-11 / 1150	10-2 / 1210	10-4 / 1270	10-7 / 1320	10-10 / 1380	11-0 / 1430	11-3 / 1480	11-5 / 1530	11-9 / 1630	12-2 / 1730
2x4	24.0	6-2 / 560	6-8 / 660	7-1 / 740	7-6 / 820	7-10 / 900	8-1 / 970	8-5 / 1040	8-8 / 1110	8-11 / 1170	9-2 / 1240	9-5 / 1300	9-8 / 1360	9-10 / 1420	10-0 / 1480	10-3 / 1540	10-5 / 1600	10-7 / 1650	10-11 / 1760	11-3 / 1860
2x6	12.0	12-3 / 450	13-3 / 520	14-1 / 590	14-9 / 650	15-6 / 710	16-1 / 770	16-8 / 830	17-2 / 880	17-8 / 930	18-2 / 980	18-8 / 1030	19-1 / 1080	19-6 / 1130	19-11 / 1180	20-3 / 1220	20-8 / 1270	21-0 / 1310	21-8 / 1400	22-4 / 1480
2x6	13.7	11-9 / 470	12-8 / 540	13-5 / 610	14-2 / 680	14-9 / 740	15-5 / 800	15-11 / 860	16-5 / 920	16-11 / 970	17-5 / 1030	17-10 / 1080	18-3 / 1130	18-8 / 1180	19-0 / 1230	19-5 / 1280	19-9 / 1320	20-1 / 1370	20-9 / 1460	21-4 / 1550
2x6	16.0	11-2 / 490	12-0 / 570	12-9 / 650	13-5 / 720	14-1 / 780	14-7 / 850	15-2 / 910	15-7 / 970	16-1 / 1030	16-6 / 1080	16-11 / 1140	17-4 / 1190	17-8 / 1240	18-1 / 1290	18-5 / 1340	18-9 / 1390	19-1 / 1440	19-8 / 1540	20-3 / 1630
2x6	19.2	10-6 / 520	11-4 / 610	12-0 / 690	12-8 / 760	13-3 / 830	13-9 / 900	14-3 / 970	14-8 / 1030	15-2 / 1090	15-7 / 1150	15-11 / 1210	16-4 / 1270	16-8 / 1320	17-0 / 1380	17-4 / 1430	17-8 / 1480	17-11 / 1530	18-6 / 1630	19-1 / 1730
2x6	24.0	9-9 / 560	10-6 / 660	11-2 / 740	11-9 / 820	12-3 / 900	12-9 / 970	13-3 / 1040	13-8 / 1110	14-1 / 1170	14-5 / 1240	14-9 / 1300	15-2 / 1360	15-6 / 1420	15-9 / 1480	16-1 / 1540	16-4 / 1600	16-8 / 1650	17-2 / 1760	17-8 / 1860
2x8	12.0	16-2 / 450	17-5 / 520	18-6 / 590	19-6 / 650	20-5 / 710	21-2 / 770	21-11 / 830	22-8 / 880	23-4 / 930	24-0 / 980	24-7 / 1030	25-2 / 1080	25-8 / 1130	26-2 / 1180	26-9 / 1220	27-2 / 1270	27-8 / 1310	28-7 / 1400	29-5 / 1480
2x8	13.7	15-6 / 470	16-8 / 540	17-9 / 610	18-8 / 680	19-6 / 740	20-3 / 800	21-0 / 860	21-8 / 920	22-4 / 970	22-11 / 1030	23-6 / 1080	24-0 / 1130	24-7 / 1180	25-1 / 1230	25-7 / 1280	26-0 / 1320	26-6 / 1370	27-4 / 1460	28-1 / 1550
2x8	16.0	14-8 / 490	15-10 / 570	16-10 / 650	17-9 / 720	18-6 / 780	19-3 / 850	19-11 / 910	20-7 / 970	21-2 / 1030	21-9 / 1080	22-4 / 1140	22-10 / 1190	23-4 / 1240	23-10 / 1290	24-3 / 1340	24-8 / 1390	25-2 / 1440	25-11 / 1540	26-9 / 1630
2x8	19.2	13-10 / 520	14-11 / 610	15-10 / 690	16-8 / 760	17-5 / 830	18-2 / 900	18-9 / 970	19-5 / 1030	19-11 / 1090	20-6 / 1150	21-0 / 1210	21-6 / 1270	21-11 / 1320	22-5 / 1380	22-10 / 1430	23-3 / 1480	23-8 / 1530	24-5 / 1630	25-2 / 1730
2x8	24.0	12-10 / 560	13-10 / 660	14-8 / 740	15-6 / 820	16-2 / 900	16-10 / 970	17-5 / 1040	18-0 / 1110	18-6 / 1170	19-0 / 1240	19-6 / 1300	19-11 / 1360	20-5 / 1420	20-10 / 1480	21-2 / 1540	21-7 / 1600	21-11 / 1650	22-8 / 1760	23-4 / 1860
2x10	12.0	20-8 / 450	22-3 / 520	23-8 / 590	24-10 / 650	26-0 / 710	27-1 / 770	28-0 / 830	28-11 / 880	29-9 / 930	30-7 / 980	31-4 / 1030	32-1 / 1080	32-9 / 1130	33-5 / 1180	34-1 / 1220	34-8 / 1270	35-4 / 1310	36-5 / 1400	37-6 / 1480
2x10	13.7	19-9 / 470	21-3 / 540	22-7 / 610	23-9 / 680	24-10 / 740	25-10 / 800	26-10 / 860	27-8 / 920	28-6 / 970	29-3 / 1030	30-0 / 1080	30-8 / 1130	31-4 / 1180	32-0 / 1230	32-7 / 1280	33-2 / 1320	33-9 / 1370	34-10 / 1460	35-10 / 1550
2x10	16.0	18-9 / 490	20-2 / 570	21-6 / 650	22-7 / 720	23-8 / 780	24-7 / 850	25-5 / 910	26-3 / 970	27-1 / 1030	27-9 / 1080	28-6 / 1140	29-2 / 1190	29-9 / 1240	30-5 / 1290	31-0 / 1340	31-6 / 1390	32-1 / 1440	33-1 / 1540	34-1 / 1630
2x10	19.2	17-8 / 520	19-0 / 610	20-2 / 690	21-3 / 760	22-3 / 830	23-2 / 900	23-11 / 970	24-9 / 1030	25-5 / 1090	26-2 / 1150	26-10 / 1210	27-5 / 1270	28-0 / 1320	28-7 / 1380	29-2 / 1430	29-8 / 1480	30-2 / 1530	31-2 / 1630	32-1 / 1730
2x10	24.0	16-5 / 560	17-8 / 660	18-9 / 740	19-9 / 820	20-8 / 900	21-6 / 970	22-3 / 1040	22-11 / 1110	23-8 / 1170	24-3 / 1240	24-10 / 1300	25-5 / 1360	26-0 / 1420	26-6 / 1480	27-1 / 1540	27-6 / 1600	28-0 / 1650	28-11 / 1760	29-9 / 1860

Note: The required extreme fiber stress in bending, "F_b", in pounds per square inch is shown below each span.

Index

A

A-frame arch, 304
Abbreviations
   commonly used, 240
   in poured-in-concrete drawings, 239
   in precast concrete drafting, 151
Advance bill(s), 128-130
   preparing, 133
American Concrete Institute (ACI), 237-238, 243
American Institute of Steel Construction (AISC), 42, 75
American Society for Testing Materials (ASTM), 62
American Welding Society (AWS), 62
Anchor bolts, 181, 325
Angles, 76
Apprentice drafters, 22
Approval(s), of job by originator, 35
Arc welding, 62
Arch(es)
   A-frame, 304
   laminated, 304
   structural drawings for, 306
   three-hinged, 304
   tudor, 304
Architect(s), 28, 314
Architectural plans
   for commercial or industrial buildings, 314
   floor plan for small bank, 29
Architectural scale, 15
Army Corps of Engineers, 314

B

Back weld, 63
Balloon framing, 285-286
Bars, 73, 325
Baseplate connection details
   columns, 181-182
   precast concrete, drawing, 182
   structural steel, 103-106, 107
Beam(s)
   laminated, 304-305
   for post-and-beam construction, 303
   structural drawings for, 306
Beam fabrication detail, 119-120
Beam framing plans, 85
Beam-to-beam connection, 106. *See also* Framed connection
Beam-to-column connection, 106. *See also* Framed connection
   basic, 184
   bolted, drawing, 184
   in precast concrete construction, 183

Beginning drafter, 22
Bend press, 45
Bending steel members, 45
Bills of materials
   advance bill, 128-130
   defined, 219
   precast concrete, 219-225
   preparing, 133, 219
   purpose of, 219
   shop bills, 130, 132
   structural steel, 128-134
Blockout, 150
Bolt(s)
   anchor, 181, 325
   common, 60
   high-strength, 45, 60, 65
Bolted connections, 60-62
   uses for, 103
Bolting, of structural steel, 45
Brace(s)
   let-in, 288
   plywood, 289
Bracing wood walls, 288-289
Break lines, 10
Bridging, 280
   types of, 280
Built-up girder, 78

C

Cantilever walls, 250-251
Centerlines, 10
Chamber of commerce, local, 316
Channels, 76, 325
   American Standard, dimensions, 339
   miscellaneous, dimensions, 340, 341
Checker, 24
   primary responsibilities of, 24
Checking process, 31-34
Checkprint(s), 31-34
Chief drafter, 24
   primary responsibilities of, 24
Civil service, federal, 314
Column(s)
   defined, 325
   spiral, 249, 251-252
   tied, 249, 251-252
Column fabrication detail, 116-119
Column framing plan, 82, 83
   schedule, 84
Connection detail(s)
   defined, 103, 181
   drawing, 111
   precast concrete, 181
   structural steel, 61, 103-111

Connections, in heavy construction, 60-67
   methods of, 60
   riveted, 65
   seated, 106
Consulting engineering firms, 313-314
Conventional framing, defined, 296
Cornice, defined, 295
Correcting process, 35
Cutting plane lines, 10, 325

D

Designations for structural steel products, 75
Designing structural connections, 111
Detail punch, 45
Detailers, 22
Dimension lines, 10
Drafter, 22
   apprentice, 22
   beginning, 22
   chief, 24
   junior, 22
   primary responsibilities of, 22
   senior, 23
Drafting clerk, 20-21
   primary responsibilities of, 21
   work order for, 21
Drafting department, 20-27
   checker, 24
   chief drafter, 24
   drafting clerk, 20-21
   junior drafter, 22
   senior drafter, 23
Drafting symbols, common, 240
Draftsman, 22
Drawing process, 28-29
Drawings
   engineering, 3, 237, 243, 253, 265
   placing, 237, 241, 243, 257-259, 265
   shop, 3, 52, 115
   working, 3
Drills, used in steel fabrication shops, 45

E

Eave, defined, 295
Electrodes, welding, 62
Employer(s) of structural drafters, 3, 6
   potential, locating, 314
   primary, 313
   secondary, 313, 314
Employment in structural drafting, 313
Engineering/drafting secretary, 21
Engineering drawings, 3
   poured-in-place concrete, 237
   floor systems, 265

Engineering drawings (Continued)
   precast concrete, 5
   preparing, for foundations, 243
   prestressed concrete, 30
   schedules, 241
   shop drawings and, 73
   structural steel, 4
   wall and column, 253-256
Engineer's scale, 16

F

Fabrication
   glued-laminated wood, 53
   poured-in-place concrete, 52
   precast concrete, 46-51
   steps during, 42
   structural steel, 42-46
Fabrication details, 115-127
   beam, 119-120
   column, 116-119
   constructing, 115-116
   defined, 115, 195
   precast concrete, 194-211
   structural steel, 6
Fastening processes, structural steel, 45
Fillet weld, 63
Finishing, of structural steel during fabrication, 46
Flame cutting torch, 44
Flange weld plates, 186
Floor system(s)
   floor truss, 281
   joist and girder, 277-278
   plywood, 279, 280
   poured-in-place concrete, 262-274
   wood, 277-283
Floor truss system, 281
   advantages of, 281
Framed connections, 106
Framing methods, for structural wood walls, 285-288
Framing plan(s), 325
   beam, 85, 143, 155
   column, 142, 152
   column-and-beam, 81
   double-tee wall panel, 145, 146, 147
   drawing, 82
   floor, 157
   precast concrete, 141
   roof, 144, 157
   of structural steel products, 73, 74, 79-82
   wall panel, 159
Framing products, 73
   heavy load/long-span, 78

G

Gag press, 45
Girder(s), 277, 325
   built-up, 78
   laminated, 304-305
   structural drawings for, 306
   types of, 278
Glossary, 325-326
Glue, high-strength, waterproof, 53
Glue-laminated wood fabrication, 53
Gravity walls, 250
Groove weld, 63

H

Handling and cutting, of steel products, 44
Haunch, defined, 188
Haunch connection(s), 188-189
   beam-to-column, 189
   bolted, drawing, 191
Hidden lines, 10
Hot cutting saw, 44
HP shapes, dimensions, 338

I

Impact wrench, 45
Interview(s), 322
   follow-up, 322

J

Job(s)
   getting and keeping, 318-324
   interview, 322
Job-finding, 313-324
   aids, 314
Job seeker, rules for the wise, 322-323
Joist(s), 277, 278, 325
   selection of, in wooden floor systems, 282
Joist and girder floor systems, 277-278
Junior drafter, 22
   primary responsibilities of, 22

L

Laminated, defined, 304
Legend, 150
Let-in brace, 288
Letter of introduction, 318
Lettering, 13
Line types, 7-12
   commonly used, 8
Linework, 7

M

M shapes, dimensions, 336
Manual of Steel Construction, 42, 79, 85, 99
Mark numbers
   on poured-in-concrete drawings, 239-241
   in precast concrete, 148
   rules for assigning, 148
Material quantities, counting, 223-224
MOD 24 framing, 287-288
Multiple punch, 45

N

Notes, general, 150

O

Object line(s), 7
   samples, 9
One-way ribbed (or joist slab) floor system, 262-264
One-way solid slab and beam floor system, 262
Organizational chart, typical structural drafting department, 20
Original drawing, 28-31

P

Paper sizes, 16
Phantom lines, 10
Pipe, 78, 325
Pitch, of roof, 293
Placing drawings, 237
   poured-in-place concrete floor systems, 265
   preparing, for foundations, 243
   schedule, 241
   wall and column, 257-259
Plates, 73, 325
Platform framing, 285
Plug (or slot) weld, 63
Plywood, tongue-and-groove, 279
Plywood floor system, 279, 280
Portfolio of drawings, 322
Post-and-beam construction, 303-304
   metal fasteners for, 304
Posts, structural drawings for, 306
Poured-in-place concrete
   construction, 237
   drawings, 237
   fabrication, 52
   floor systems, 262-274
      drawings, 265-267
      one-way ribbed or joist slab, 262-264
      one-way solid slab and beam, 262
      two-way flat-plate, 264
      two-way solid slab and beam, 264
      waffle slab, 265
   foundations, 236-246
      drawings, 243
   mark numbering system, 239-241
   sheet layout and scales, 237-238
   stairs and ramps, 270-283
   symbols and abbreviations, 239
   walls and columns, 249-260
Precast concrete, 326
   abbreviations in, 151
   bills of materials, 219-225
      constructing, 222
   connection details, 181-193
      defined, 181
   fabrication, 46-51
   fabrication detail, 7, 195-211
      beam, 198, 199
      column, 196, 197
      constructing, 196
      floor/roof member, 200
      metal connectors for, 204, 205
      wall panel, 202
   framing plans, 141-161
      basic types, 141
      drawing, 152
      general notes, 150
      legend, 150
      mark numbers, 148
      product schedule, 148, 149
   sections, 167-176
      conventions, 172
      drawing, 173
      full sections, 169
      offset, 170
      partial, 170
      structural, 169
Precast concrete companies, 314
Prestressed concrete
   engineering drawing, 30
   products, 50

Printing clerk, 21
Product
    fabrication, 42–53
    schedules, 148, 149
    shipping, 54
Punching and drilling, structural steel, 44–45

R

Rafters, conversion diagram for, 347
Ramps, 272–273
    defined, 272
Rebars, 50, 223, 326
Reinforced concrete
    precast and poured-in-place, difference between, 52
    products, 50
Reinforcing bar(s), nonstressed, 50
Reinforcing bar schedule, 204, 208
    completed, 209
Reproduction clerk, 20
Resume(s), 318–321
Retaining walls, 250–251, 326
Revisions, 35–37
Ridge detail, defined, 295
Riveted connections, 45, 65
Riveting. *See* Riveted connections
Roof(s)
    common classification, 293
    eave and ridge details, 295
    flat, 293
    slope and pitch, 293
    structural wood, 293–301
    trusses, 295
Roof truss(es)
    common, 297
    defined, 296
    flat, 298
    mono pitch, 298
    scissors, 297
    span table, 299
    types of, 295

S

S shapes, dimensions, 337
Scale(s), 15–16
Schedules
    importance of, 241
    samples of, 242
Seated connections, 106
Section(s)
    architectural, 96
    conventions, 99
    cross, 92, 325
    defined, 167, 326
    drawing, 99
    full, 92–96
    longitudinal, 92
    offset, 98

partial, 96–97
    precast concrete, 167–176
    structural, 96
    structural steel drafting, 91–100
Sectional drawings, preparation sources, 99
Security walls, 249
    common applications of, 249
Senior drafter, 23
    primary responsibilities of, 23
Shapes, 73
    C, 77
    L, 76
    structural pipe, 78
    structural tees, 77
    structural tubing, 77–78
    W, S, and M, 75–76
Shear plates, 62
Shear walls, 249, 326
Shipping, to jobsite, 54
Shop bill, 130, 132
    preparing, 133
Shop drawings, 3, 52, 115, 326
    defined, 115, 195
    and engineering drawings, 73
    for heavy construction jobs, 6
Slope, of roof, 293, 326
Span tables
    ceiling joists, 348, 349, 350, 351
    floor joists, 344
    medium or high slope rafters, 345
    rafters, 346
    use of, 343
Split ring connectors, 62
Stair(s)
    defined, 270
    design of, 270–271
    design computations, 272
    types of, 270
State employment office, 315–316
Steel, cutting and handling, 44
Steel fabrication, 42
Steel shapes, straightening, 45
Steel strap plate, single, 187
Stem plate-weld strip connection, 187–188
Stick framing, defined, 296
Structural drafting
    defined, 1
    overview of, 1–69
    techniques, 7
Structural drawings, types of, 3–6
Structural Steel Detailing, 78
Structural steel fabrication companies, 314
Structural steel sections, 91–100
Structure, defined, 1
Student, in structural drafting, 6
Symbols
    poured-in-concrete drawings, 239
    precast concrete drafting, 151
    weld, 63–65

T

Tees, 77
Threaded rod, 60
Three-hinged arch, 304
Tilt-up walls, 249–250, 326
Title blocks and borders, 18
Trainees, 22
Trucks, use of in shipping, 54
Truss(es), 326. *See also* Roof truss(es)
Tubing, 77–78
Tudor arch, 304
Two-way flat-plate floor systems, 264–265
Two-way solid slab-and-beam floor system, 264

W

W shapes, dimensions, 328–335
Waffle-slab floor system, 265
Wall and column
    engineering drawings, 253–256
    placing drawings, 257–259
Walls
    cantilever, 250
    gravity, 250
    load-bearing, 285
    non-load bearing, 285
    retaining, 250–251
    security, 249
    shear, 249
    tilt-up, 249–250
    wood, 285–291
Want-ads, in Sunday papers, 315
Weld plate-strap plate connection, 187
Weld plate-weld angle connections, 187
Weld symbols, 63–65
    annotated, 66
    basic, 65
    explanation of, 64
    sample applications of, 67
Weld types, 63
Welded connections, 62
    drawing, in precast concrete construction, 188
    in precast concrete, 186
    uses for, 103
Welded joints, standard symbols, 342
Welding, 45
Wood floor systems, 277–283
    joist and girder, 277–278
Wood roofs, 293–301
    configuration diagrams, 300
    framing, 296
Wood walls, 285–291
    balloon framing, 285–287
    bracing, 288–289
    MOD 24 framing, 287–288
    platform framing, 285
Working drawings, 3, 326

Y

Yellow pages, the, 315

DELMAR PUBLISHERS INC.

ISBN 0-8273-1930-4

DELMAR PUBLISHERS INC.
50 Wolf Road, Albany, New York 12205

ISBN 0-8273-1930-4